D0936903

Developing an Effective Safety Culture:
A Leadership Approach

Developing an Effective Safety Culture:
A Leadership Approach

James E. Roughton
James J. Mercurio

Boston   Oxford   Auckland   Johannesburg   Melbourne   New Delhi

Recognizing the importance of preserving what has been written, Butterworth–Heinemann prints its books on acid-free paper whenever possible.

Library of Congress Cataloging-in-Publication Data
Roughton, James E.
 Developing an effective safety culture : a leadership approach / James E. Roughton,
 James J. Mercurio.
  p. cm.
 Includes bibliographical references and index.
 ISBN 0-7506-7411-3 (alk. paper)
 1. Industrial safety—Management. I. Mercurio, James, 1959– II. Title.

 T55 .R69 2002
 658.3′82—dc21   2001043911

British Library Cataloguing-in-Publication Data
A catalogue record for this book is available from the British Library.

The publisher offers special discounts on bulk orders of this book.
For information, please contact:

Manager of Special Sales
Butterworth–Heinemann
225 Wildwood Avenue
Woburn, MA 01801–2041
Tel: 781-904-2500
Fax: 781-904-2620

For information on all Butterworth–Heinemann publications available, contact our World Wide Web home page at: http://www.bh.com

10 9 8 7 6 5 4 3 2 1

Printed in the United States of America

Dedication

To my loving wife, my friend, my lifelong partner, who has always been patient with me in my endeavors to enhance the safety profession. She has always given me the freedom to pursue my dreams.

Contents

Introduction, 3
Cost Impact to Business, 3
The Direct and Indirect Costs of Accidents, 8
Incident Costs, 9
Case Histories, 11
Summary, 12
References, 14

Introduction, 16
Culture and Safety, 16
What Sets the Culture? 17
Why Do Cultures Fail? 18
What Are Values? 20
Changes in Behavior, 25
Recognizing Success, 25
What Can We Learn? 27
A New Management Safety System Begins to Emerge, 28
Safety Climate (Culture) Defined, 29
Audits, 31
Summary, 31
References, 32

13. Developing and Administering a Medical Surveillance Program 258

14. Defining Safety and Health Training Needs 266

Acronyms

ABC	Above and beyond compliance
ABSS	Activity-based safety system
ADA	Americans with Disabilities Act
BBS	Behavioral-based safety
BMP	Best management practices
CFR	Code of Federal Regulations
CPR	Cardiopulmonary resuscitation
CTD	Cumulative trauma disorder
EMR	Experience modification rate
EMTs	Emergency medical technicians
FAA	Federal Aviation Administration
ILCI	International Loss Control Institute, Inc.
JHA	Job Hazard Analysis
JSA	Job Safety Analysis
LPN	Licensed practical nurse
LVN	Licensed vocational nurse
LWDIR	Lost Workday Incident Rate
MSD	Musculoskeletal disorder
MSDS	Material Safety Data Sheet
NIOSH	National Institute of Occupational Safety and Health
OIR	OSHA Incident Rate
OSHA	Occupational Safety and Health Administration
PEP	Performance Evaluation Program
PPE	Personal protective equipment
SOPs	Standard Operating Practices
SWAMP	Safety without a Management Process
VPP	OSHA's Voluntary Protection Program
WCCR	Workers' Compensation Cost per Hour Worked

About the Author

James E. Roughton has a Master of Science from Indiana University of Pennsylvania (IUP) in safety science and is a Certified Safety Professional (CSP), a Canadian Register Safety Science Professional (CRSP), and a Certified Hazardous Materials Manager (CHMM). His experience includes 4 years in the military and 35 years' experience in industry, with the past 27 years in the safety area developing and implementing safety management systems.

Mr. Roughton has worked for various corporations where he has served in the following capacities: providing consulting services from medium to large manufacturing facilities; developing and implementing safety programs and management systems; providing consulting services for hazardous waste remedial investigation and site cleanup regarding developing and implementing site-specific safety plans; conducting site health and safety assessments; and providing internal support for multiple office locations.

In addition, he is an accomplished author in various areas of safety, environment, quality, security, computers, etc. He also provides mentoring to other professionals who want to get published and is a frequent co-author with those professionals. He is a frequent speaker at conferences and professional meetings.

He is also a member of the American Society of Safety Engineers (ASSE), where he is the past president of the Georgia Chapter. He also has received several management and professional awards for safety related activities, including the Safety Professional of the Year (SPY).

James J. Mercurio graduated in 1981 from Indiana University of Pennsylvania (IUP) as a Safety Sciences major. He has been a Certified Safety Professional (CSP) since 1993 and has served as a Special Government Employee (SGE) since 1994 as an OSHA Voluntary Protection Program volunteer. His work experience includes the surface coal mining industry, manufacturing industry, industrial minerals industry, as well as the forest products industry. He received the MSHA Sentinels of Safety Award for the safest mining facility, as well as achieving the OSHA Voluntary Protection Program's (VPP) Star approval for seven forest products facilities.

Foreword

When you begin a new project such as writing this book, you know that it will be a lot of work and you will be spending a lot of time researching, reading, writing, editing, etc. You begin to look for professionals who can contribute to the quality of the book and can help to convey the message that the authors are trying to communicate. This is how I got my co-author. He has a wealth of knowledge from various backgrounds. I am lucky to find such a co-author, who complements me well. Although we both have similar backgrounds, our experience is in different industries, which helps create a different perspective. This book will provide a more realistic approach to building a safety culture and a management system because both of us have lived the experience, both good and bad.

The other thing that you look for once you have a good co-author is other quality authors who have been instrumental in developing safety cultures and have published their experiences. I was fortunate to find both who were willing to share their written text and graphics to make this book a better reference. You have the benefit of proven culture building from industry experts and the authors who have lived the experience.

A special thanks goes to all of the tremendous authors who have developed techniques with other safety professionals. I acknowledge the following authors who were kind enough to provide me permission to use their material (text and graphics) and incorporate it into this book: Dan Petersen and Scott Geller of Safety Performance Solutions, Michael Topf and Donald H. Theune of the Topf Group, Frank Bird, George Germain, Det Norske Veritas (DNV), L. L. Hansen, Steven Geigle, James C. Manjella, Tom and Sue Cox, Bob Veazie, and Anne R. French. I also acknowledge other fine authors who provided me permission to reproduce their work and are referenced in the book. In addition, there are several great Web sites where some of this material is available: the OSHA Web site, www.osha.gov; Oklahoma Department of Labor, Safety and Health Management: Safety Pays, 2000, http://www.state.ok.us/~okdol/osha/index. htm; and the Oregon Web site http://www.cbs.state.or.us/external/osha/. All Web sites are public domain.

With the knowledge and expertise of these individuals and the authors of this book, college professors, managers, safety professionals, and students of safety management will have a good road map to assist with their efforts at developing a successful safety management system.

We hope that you will enjoy this book and recommend it to your friends and to other professionals.

Good luck!!

—James E. Roughton, MS, CSP, CRSP, CHMM,
safeday@mindspring.com
—James J. Mercurio, CSP

Preface

Many managers and safety professionals consider a written policy as a solution to safety issues. Merely developing procedures does not set the stage for a safety culture. This takes many years of hard work to accomplish. In the development stage, you will see peaks and valleys (ups and downs) in your injuries. This is natural. You must be patient. It is important that you do not react hastily to each situation. If you feel that your process is working then you must learn to manage the process. To help you understand the importance of developing a safety management system, we have divided this book into four parts to help you understand the process.

PART 1: CHARACTERISTICS OF AN EFFECTIVE SAFETY CULTURE

Chapter 1: Does Management Commitment Make a Difference?

Most managers can give many good reasons for improving the way they manage their safety system, but many cannot tell you how to develop and enhance their safety management system. A well-designed management system can help to reduce incidents along with the associated hidden costs; increase efficiency; improve productivity, morale, and quality of products; and reduce the potential for regulatory citations.

This chapter will help you understand the related cost of an incident. We discuss several methods of cost reduction, showing you where some of the cost is, and the advantages of controlling cost.

Chapter 2: Defining a Value System

Some professionals in the safety field all but ignored the concept of safety culture through out the 1980s. As management attempted to improve culture through changing their styles of leadership and through employee participation, professional safety tended to change their approaches very little (possibly afraid of change). They were using the same elements in their safety "programs" that they had always used. Safety programs typically consisted of the usual items: meetings, inspections, incident investigations, and limited use of Job Hazard Analyses (JHAs). Most professionals perceived these tools as the essential elements of a safety program. Not many managers looked at the process as a management system.

According to Dan Petersen, "while OSHA and states' programs were going down the 'essential element' track to safety (as was much of the safety profession), a number of research pieces began to come in with totally different answers to the safety problem. Most of the research results were consistent in saying 'There are no essential elements': what works in one organization will not in another. Each organization must determine for itself what will work for it. There are no magic pills. The answer seems to be clear: it is the culture of the organization that determines what will work in that organization."

In this chapter we try to help you understand the difference between a priority and a value.

Chapter 3: OSHA's Voluntary Guidelines for Safety Management

As you continue through this book, you will begin to understand why top management must be committed. This commitment must be strong so that a successful management system can be developed, continually improved, and sustained. The OSHA voluntary program that we discuss will be a good place to start. In addition, the voluntary program will provide you a specific perspective on management systems that is suitable for a successful management system.

PART 2: MANAGEMENT ASPECTS OF AN EFFECTIVE SAFETY CULTURE

Chapter 4: Management's Role in Developing an Effective Safety Culture

Too often safety and health program books talk only about the technical aspects of a safety and health program. Some books may include innovative or proactive techniques of incident prevention, while others focus purely on more basic, fundamental incident prevention techniques. One of the basic tenets of developing and ultimately sustaining a safety culture argues that an effective safety culture rests on the shoulders of the management team at an organization, regardless of the structure. Management, starting from the individuals with the most authority, all the way through to front-line management (supervisors, superintendents, leaders, etc.), then all the way to the employee, must be engaged in the effort to systematically reduce and/or eliminate exposure to hazardous situations. These situations encompass both exposures to physical hazards and work practices that put the employee at a greater risk of an injury.

In this chapter, we review a list of principal management leadership traits we have seen effectively utilized throughout our careers in various industries and work environments. For practical purposes, we will focus only on a few of the more apparently critical attributes we have observed, which have helped to maintain a keen focus on the impact of leadership and the safety culture's management system.

Chapter 5: Journey to a Safety Culture: Determining the Direction of Your Management System

When you plan a trip, you and your family have an objective in mind for making the journey. You make extensive plans to make sure that you are going in the right direction by mapping out your destination. When planning and developing a safety and health program, you must also make sure that your objective for establishing such a program is crystal clear. The first step is to decide how successful you want your program to be and what you want your program to accomplish. The next step is to put your plan in writing. Once you have completed these steps, you can then map out the path toward your established goals and objectives. This is the method used to determine the direction of your program when developing your policy, goals, and objectives.

In this chapter we will help you begin your journey by detailing how to write and communicate your safety and health policy to your employees. In addition, this chapter will help you understand how to establish and evaluate your goals and objectives.

Chapter 6: Management Leadership: Demonstrating Commitment

According to Dan Petersen, safety results require support and behaviors from the entire organization, especially top management. But although commitment starts with top management, it is necessary to get employees to participate to make the management system work. It is important that this process be driven down to the employee level to be successful. Employees are one of the keys to a successful process and must not be forgotten. This is an indicator of the perception of how well the organization is working to create a safe work environment.

Protecting employees from hazards takes top management commitment. This commitment is essential and must be clearly demonstrated.

In this chapter, we will describe ways to provide visible leadership. Ideally, this means participation in a process that demonstrates concern for every aspect of the safety of all employees throughout the organization. In addition, we have included a description of a management system to make sure that contract employees are both protected from hazards and prevented from endangering employees of the owner-company.

Chapter 7: Employee Participation

The success of any business depends on all employees who work in an organization. Protecting employees from hazards on the job not only makes good business sense, but in all cases is the right thing to do. As part of management, you do not have to face this task alone. In this chapter, we outline how employee participation can strengthen your management system and safety program. We provide several case histories for your review.

In addition, we take a look at some of the reasons behind this employee participation and some of the ways you can implement a successful employee-driven program.

Chapter 8: Assigning Safety Responsibilities

Management is the function responsible for establishing the purpose of an operation, determining measurable objectives, and taking the actions necessary to accomplish those objectives.

As a member of management (no matter the environment, large or small), you must understand the importance of assuming accountability for the safety of employees. This could be new employees, transferred employees, contractors (including temporary employees), etc. To be successful you must assign this accountability to someone else in the organization. It is important to understand the difference between assigning responsibilities and delegating responsibilities. You cannot delegate your responsibilities. However, if you assign your responsibilities you will still have control. What you can do is to expect other individuals in the organization to share the responsibility for certain elements of the safety program. Many managers question why they should be assigned the responsibility for safety. In most cases, you have a working knowledge of your business issues and you should be close to your employees. However, as a business grows and the numbers of employees increase, being responsible for all of the details of an effective safety management system may become less feasible. It is important to understand how to have a management system in place for assigning some of the safety responsibilities to others.

In this chapter, we discuss ways to assign the required responsibilities to the appropriate individuals.

Chapter 9: Developing Accountability

Why is it important to develop an accountability management system? Imagine a sports organization with a coach (manager) and players (employees, contractors, temporary employees). Each player is assigned specific tasks and responsibilities that are critical to the success of the team.

In this chapter, we will help you understand why a clearly defined accountability management system is the mechanism to make sure that employees fulfill their assigned responsibilities.

PART 3: SAFETY AND PROGRAMS THAT SUPPORT THE SAFETY CULTURE

Chapter 10: Developing a Hazard Inventory

Do you know all of the potential hazards that are associated with your type of industry and your site-specific working conditions? A means of systematically identifying hazards is useful. Three major actions can be used to develop a hazard control inventory. The activities in this chapter form the basis for a good hazard recognition, prevention, and control program and describe the three major actions needed to control hazards.

Chapter 11: Developing a Hazard Prevention and Control System

Once you have conducted a survey of your workplace and you know your hazards, what are you going to do now? This is an important question that you must ask yourself.

In this chapter, we discuss the management systems that can be used to help minimize and control hazards, as well as tools that will help to close any potential gaps in the inspection system.

Chapter 12: Conducting Effective Incident Investigations

Incident investigations are an important element in any effective management system. Any incident investigation is a fact-finding management tool to help prevent future incidents. The investigation is an analysis and account of an incident based on factual information gathered by a thorough and conscientious examination of *all* of the facts. It is not a mere repetition of the employee's explanation of the incident. Effective investigations include the objective evaluation of all the facts, opinions, statements, and related information, as well as the identification of the root cause(s) and actions to be taken to prevent recurrence. Facts should be reported without regard to personalities, individual responsibilities, or actions. Blame and fault-finding should never be a part of the investigation proceedings or results.

This chapter provides a brief summary of the root cause analysis process and will help you understand and conduct successful incident investigations. Incident investigation is an important element in an effective safety management system. The basic reason for investigating and reporting the causes of occurrences is to identify action plans to prevent recurrence of incidents.

Chapter 13: Developing and Administering a Medical Surveillance Program

A medical surveillance program is a system that is put in place to make sure that the level of occupational health expertise identified in the safety and health program is sufficient. Having a medical surveillance program does not mean that you have to hire a doctor to work at your facility. There are many ways for you to find and use occupational health expertise. This chapter provides some guidance to help you decide what will work best for your operation.

Chapter 14: Defining Safety and Health Training Needs

A common knee-jerk response to employee performance deficiencies is to provide training. If it is improved performance we seek, then training must be based on specific, measurable, performance-based objectives (not what the student will know, but what the student will be able to do upon successful completion of the course). The appropriate training must be designed with specific guidelines and must supplement and enhance other educational and training objectives.

This chapter will help you design, revise, implement, and evaluate your safety and health training. It also provides information on OSHA requirements for training and tells you where to find further OSHA references and other assistance.

Chapter 15: Understanding Job Hazard Analysis

Job-related incidents occur every day in the workplace. These incidents, which include injuries and fatalities, often occur because employees are not trained in the proper job procedure(s) or the task they perform. One way to reduce these workplace incidents is to develop proper job procedures and train all employees in safer and more efficient methods. This chapter discusses why developing a JHA for each task is important.

Chapter 16: Making Sense of the Behavior-Based Safety Process

Pick up any safety literature today and you will read an article concerning behavior-based safety (BBS). You will note that every author has a different take on the process and the implementation. Some are better than others. In one author's opinion, there are several consultants who have a good understanding of the concept, whereas others have processes that are intimidating to some companies. If not implemented properly and integrated into the management system, this process stands out like a sore thumb. This is the worst thing that you can do. If this occurs, all the focus will be on one program and not the complete process.

This chapter is designed to provide an overview of the BBS process. It will highlight some of the expert opinion in the field.

PART 4: MEASURING THE SAFETY CULTURE

Chapter 17: Safety and Health Program Evaluation: Assessing the Management System

After you have developed and implemented a management system, it is time to establish your safety and health program with measurable goals that have established objectives. These measurable goals and objectives should be accompanied by procedures, activities, and resources to achieve them. This process should involve all employees, including managers and supervisors. What happens next?

According to Dan Petersen, to hold someone accountable you must know whether they are performing their job functions correctly. To understand these accomplishments, we must measure their performance. Without measurement, accountability becomes an empty, meaningless, and unenforceable concept.

Some elements of the safety program are best reviewed using one of the suggested methods. Other elements lend themselves to being reviewed by any of several selected methods.

This chapter provides some proven methods of assessing your workplace and gives you some additional information from seasoned safety professionals.

Final Words

This section provides an overview of the concepts presented in this book.

Introduction

We believe that the style of writing in this book and the information presented will encourage you to read it from cover to cover. In attempting to reach large numbers of diverse readers, many writers have a tendency to consider only part of their audience, despite the vital importance of the topic. Safety and health are no exception.

In this case, the readers and the authors engage in risk taking; we hope that it will offer the most comprehensive approach to developing an effective safety culture, and you hope to get something out of it to help build your own successful safety culture. We have tried to provide a book that can be a reference in all phases of building your safety management system (process) and ultimately developing a safety program that supports your safety culture.

A significant number of incidents are a consequence of our daily actions, habits, and lifestyles. For example, we add to the probability of having an incident every time we get in our car to go to work or run an errand, board an airplane to go on vacation or on a business trip, cross the street, lift a heavy object, etc. The list is endless.

What you will learn from this book is that people (employees) alter their behavior in response to safety measures, but everyday risk will not change, unless the management system is capable of motivating and allowing employees to alter the amount of risk they are willing to incur.

As an alternative to the enforcement, educational, and engineering approaches of the past, a systematic, motivational approach to incident prevention is presented in this book. This is an approach that offers employees a reason to reduce incidents and to adopt safer ways of life.

In *Quality Is Free* [1], Crosby outlines five points of what the term real quality means, emphasizing the absolutes of quality management. Table 0-1, Comparing Quality to Safety, lists these points. In addition, the authors have related these elements as they would fit into a safety management system.

Crosby further notes that to eliminate this waste (employees getting hurt), to improve the operation (incidents), to become more efficient, we must concentrate on preventing the defects and errors that plague us.

Table 0-1
Comparing Quality to Safety

Quality	Safety
Quality means conformance, not elegance	Development of management system
There is no such thing as a quality problem	There is no such thing as a safety issue
There is no such thing as the economics of quality; it is always cheaper to do the job right the first time	If the management system is flexible and proactive, incidents that occur are controlled and minimized
The only performance measurement is the cost of quality	Reducing incidents
The only performance standard is Zero Defects	Zero Incidents
Ref. 1, p. 131	This volume

The defect that is prevented doesn't need repair, examination, or explanation [1].

He continues his discussion by stating that the first step is to examine and adopt the attitude of defect prevention (incidents). This attitude is called, symbolically [1], Zero Defects (incidents). Zero Defects is a standard for management, a standard that management can convey to the employees to help them to decide to "do the job right the first time" (no injuries due to incidents). However, some people still think that you cannot reduce incidents [1].

Crosby in his book *Quality Is Free* makes a statement that one author believes is worth repeating to let others think about it. We believe that it puts things in perspective. "People are conditioned to believe that error is inevitable. We not only accept error, we anticipate it" [1]. "It does not bother us to make a few errors, and management plans for these errors to occur. We feel that human beings have a 'built-in' error factor" [1].

Think about this. How many companies have you heard of building service (repair) centers as they are designing a new product? We have a tendency to anticipate that the product we are building is going to fail. Do we want this in our management system? This is the same as someone developing safety policies and procedures or guarding machines *after* an employee gets hurt or planning to have X number of employees get hurt. We need to be smarter in some respects and understand how to build management systems that will sustain themselves. This is what we provide you in this book.

"However, we do not maintain the same standard when it comes to our personal life. If we did, we would resign ourselves to being shortchanged now and then as we cash our paychecks. We would expect hospital nurses to drop a certain percentage of all newborn babies. We would expect to periodically drive to the wrong house. As individuals we do not tolerate these things. Thus we have a double standard, one for ourselves, one for the company" [1]. There is a lot to be said about responsibilities and accountabilities. Let's look at an example. It is a common perception that employees who drive a forklift tend to damage walls, overhead doors, production equipment, etc. in their daily routine. Hence, we spend a lot of money putting barriers up to stop employees from hitting these walls, equipment, etc.

On a personal side employees would never think about damaging their own property, but do not give it a second thought on an employer's site. The usual comment is "It is just one of those things." Why do you think that there is a difference? In the author's opinion, people are trained when they grow up to cherish toys, cars, etc. To illustrate this point, one author worked several years in construction with one major customer who was building a new 3 million square foot distribution warehouse. This company spent $300,000 on protective devices for fire equipment, overhead doors, walls, etc. before they started operations. Why do you ask? Think about it. Management has this preconceived notion that the building and equipment would be damaged in a short time after startup. So they would rather spend the money up front—"solve the problem, rather than manage the problem." After all this is how managers perceive employees. It is a typical stereotype that we need to deal with all of the time. This is where the culture building should start with each employee on the first day of the job and continue.

The reason for this is that the family creates a higher performance standard for us than the company does [1].

Crosby's Zero Defect concept is based on the fact that two things cause mistakes: lack of knowledge and lack of attention [1]. Lack of knowledge can be measured and attacked by tried and true means. However, attention is a state of mind. It is a preconceived attitude (at-risk behavior) that must be changed by the individual (employee) [1].

When presented with the challenge of Zero Defects and the encouragement to attempt it, individuals will respond enthusiastically. Remember that Zero Defects is not a motivation method, it is a performance standard. It is not just for production employees, it is for all employees [1].

Therefore, employees receive their standards from their leaders. Top management must personally direct the Zero Defects program. To gain the benefits of Zero Defects, you must make a personal commitment to improve your management system. You must want it. The first step is to make the attitude of Zero Defects your personal standard [1], and to tell your employees. They will perform to the requirements (standards) given to them.

As we continue to our quest to develop a safety management system that will support a successful safety culture, we want you to think about what we have just discussed. There are individuals who believe in Zero Incidents (Zero Defects), while others believe it is not possible. It is now time for you to make your decision.

ZERO INCIDENT: OTHER PERSPECTIVES

The following is a summary of various quotes on zero relationship to safety. You should try to understand what is stated and then make your own mind up what you want to accomplish.

"When objectives are set unrealistically high (such as 'Having zero accidents and incidents on our 47 construction sites this year') they have little or no positive motivational value. People laugh at them, ignore them, or are demoralized by them. Motivational objectives are attainable objectives (such as 'Reduce our lost time to accident by at least 30% this fiscal period')" [2].

"Indeed a zero injury goal is fact! We see it every year here at (company name removed), when operating departments go with zero injuries and zero vehicle accidents. I mean none, not just 'recordable' cases, or 'preventable' vehicle incidents" [3]. "If you set out not to have an incident, you won't have one" [3].

"Having a 'zero' accident goal could be compared to attempting to get to heaven when you die. If your program is honest, sincere, and forthcoming, you may just get there. If not, and you're just playing a numbers game, you'll just wind up in a 'fool's paradise'" [3].

"Having a zero goal only makes the pressure worse and drives reporting underground. Most safety programs are reward programs which celebrate good safety records with very little or no idea how those records were achieved" [3].

" 'Zero' is a real goal, not a punishable offense when not reached in a particular time frame" [3].

"Fact: Every employer should want zero accidents. Right? Why would anyone want to have an injury goal that is not zero?" [3].

"We must strive for zero rates, which seems impossible. But why set other goals that are harder to meet? Set high numbers, you get a high number of injuries. Educate workers to zero rates. Let them know their injuries will not be held against them. Explain the use of injury reports to change or establish new procedures" [3].

"Zero is a noble but unattainable goal! We should set challenging injury prevention goals, but zero workplace injuries is a dream" [3].

"Zero injuries must be the goal. Accepting injuries to employees as part of business cannot be morally justified. As with any goal, it is the desired outcome, not always the attainable outcome" [3].

"Zero injuries must be your goal. Once your mind-set says that some accidents are acceptable, you've lost the battle. Then you've established the crutch excuse of the 'unavoidable' accident. How many accidents are acceptable? One but not two? Ten but not eleven?" [3].

"I believe a zero injury/illness goal is attainable for some facilities. For others more realistic goals should be selected" [3].

"With health, safety, and environmental strategies so completely thought out and sophisticated, 'zero injury' is the only real goal left. Most often philosophies in safety are defined and conquered already. To be satisfied with the status quo would result in the safety profession becoming little more than an incident rate accountant. I personally feel that a company without a 'zero injury' goal is a stagnant company with no future" [3].

"Zero injury goals are enforced with safety training and accident/near-miss investigations coupled with a lucrative injury incentive program" [3].

"A zero injury goal is a must. Anything less indicates acceptance of injuries that we have said for years are preventable. So if they are, how could we say those injuries cannot be reduced to zero?" [3].

Robert F. Mager, Experience Trainer, once said "To go from a zero to a management hero to answer these question you'll strive: Where am I going? How will I get there? and How will I know I've arrived?"

With these thoughts in mind, you must decide what direction you want your program to go. Do you want to become compliance driven or do you want to develop a system that will provide you support of your safety process?

Let's explore the zero-incident question. What is zero incident? As we have seen there are some professionals who say zero is not a credible or realistic goal, while others say that it is the only direction to go. It is the opinion of one of the authors that you have to lead people to believe that zero is possible. Then, with management support, it is possible. We can implement all of the programs in the world, create all of the activities we want, add all of the guarding to prevent employees from getting hurt, but if employees don't believe that they can work safely, then they probably will not. One of the authors is firmly a believer that if you don't have a mindset of zero, you will never achieve a workplace without injuries. This has been proven in many of the personal experiences of one author. Some of the behaviorists would like you to believe that you must implement a safety behavior process or change will not happen. This is all a matter of opinion and should be explored in more detail. One must remember that behavior-based safety is not the total answer. We will explore this in more detail in Chapter 16.

Some say, "How can you celebrate zero incidents? The fact is that there are many elements that make up zero incidents (workplace injuries, vehicle incidents, property damage, first aid cases with medical only, etc.). We still have a long way to go to understand how to change and sustain a safety culture."

What you can do is to celebrate milestones in the journey to zero incidents in your workplace. Celebrating these milestones will allow you to focus on many elements that produce losses in an organization. How many times have you heard of a company that only focuses on no lost time cases, recordable incidents, or no injuries? This is a narrow view, because to get to no recordable injuries takes a lot more work. There will have to be some good upstream measurement programs that will drive this process.

Let's take a final look at this controversial subject of zero incidents highlighted by a luncheon speech at the Workplace Safety Summit, held in late March at Georgetown University in Washington, D.C. The speech was given by U.S. Secretary of the Treasury Paul O'Neill and was printed in the *Industrial Safety and Hygiene News* (ISHN), March 2001 [4]. We believe that this will set the stage for the entire book. According to the article O'Neill began his speech by stating "Workplace safety is a subject I've spent a lot of time thinking about and studying." He backed up these words when he proudly displayed a bar graph showing Alcoa's lost workday rate steadily declining during his years at the helm, from 1.86 cases per 100 workers in 1987 to 0.14 in February 2001.

O'Neill explained his interest in safety as follows: "A truly great organization must be aligned around values that bind the organization together. This is how companies withstand competitive pressures and operate consistently on a far-flung global basis." He stated that: great organizations have three characteristics:

- Employees are treated with dignity and respect.
- They are encouraged to make contributions that give meaning to their lives.
- Those contributions are recognized.

According to O'Neill: "Safety is a tangible way to show that human beings really matter." He continued, "Leadership uses safety to make human connections across the organization. Stamping out accidents" (which at Alcoa O'Neill called "incidents") "and telling employees we can get to zero incidents is a way to show caring about people. This is leadership."

O'Neill's safety philosophy includes these points:

Leadership accepts no excuses, and does not excuse itself when safety problems arise.

Simply caring about safety is "not nearly enough, not nearly enough." O'Neill went on to say: "At the end of the day, caring alone is not enough to make sure that an incident never happens again."

What's needed is for "safety to be as automatic as breathing. . . . It has to be something unconscious almost."

This won't happen by leadership simply giving orders, according to O'Neill. "You need a process in place to get results." A process based on leadership, commitment, understanding, and no excuses, he said.

"Safety is not a priority at Alcoa, it is a precondition," O'Neill explained. If a hazard needs to be fixed, it's understood by supervisors and employees that "you do it today. You don't budget for it next year."

The challenge he poses is: How do you get an organization to believe this? "You always must be constantly thinking about ways of refreshing the organization's thinking about safety," O'Neill said.

O'Neill outlined five steps he took soon after coming to Alcoa:

- He called in the safety director to review the company's performance. O'Neill was told Alcoa's rates were below industry average. "That's good," said O'Neill. "But the goal is for no Alcoan to be hurt at work." No injuries down to first aid cases. "The only legitimate goal is zero," O'Neill said. Otherwise, who's going to volunteer to be that one annual case, or whatever? Getting to zero is a journey of discovery, O'Neill said, and at no point can you stop and say, "We've reached the point of diminishing returns and can't afford to get better."
- O'Neill met with employees and gave them his home phone number. "I told them to call me if their managers didn't fix safety problems. What I was doing was making a point to my managers."

- O'Neill had 26 business units. Vice presidents called him personally whenever their group experienced a lost-workday case. "This constantly engaged them about safety," he said. It forced them to confront themselves: "Why do I have to make this call I hate to make?"
- When Alcoa launched an internal computer network, safety information came online first, before marketing, sales, or finance, according to O'Neill. Just another way to keep safety in front of employees and managers and reinforce that it is a precondition, he explained.
- O'Neill told his financial people, "If you ever try to calculate how much money we save in safety, you're fired." Why? He didn't want employees looking at safety as a "management scheme" to save money. "Safety needs to be about a human value. Cost savings suggest something else. Safety is not about money; it's about constantly reinforcing its value as a precondition."

We will explore many options, opinions, and information from other credible safety experts. With the authors' expertise in various industries and the other safety experts who contributed to the making of this book, you will have a comprehensive resource to develop a successful safety culture.

REFERENCES

1. Crosby, Phillip B., *Quality Is Free*. McGraw-Hill Book Company, New York, 1979.

2. Germain, George L., "Beyond Behaviorism to Holistic Motivation," in *Safety and The Bottom Line*, Frank E. Bird, Jr. and Ray J. Davies, 1996, p. 215.

3. *Industrial Safety and Hygiene*, May 1997, Reader Response, p. 8.

4. *Industrial Safety and Hygiene News*, "Treasury Chief Gets Blunt about Safety," May 2001, pp. 16, 19.

Part 1
Characteristics of an Effective Safety Culture

1

Does Management Commitment Make a Difference?

INTRODUCTION

Most managers can give many good reasons why they want to improve the way they manage their safety systems. However, few can tell you how to develop and enhance their safety management system. A well-designed safety management system can help to reduce injuries and illnesses along with the associated hidden costs. One of the most compelling reasons that management must consider is to minimize injuries. In addition, a good management system can increase efficiency; improve productivity, morale, and/or quality of products; and reduce the potential for regulatory citations. Evidence of a comprehensive, well-managed safety and health program that supports the management system will help to strengthen your position during any Occupational Safety and Health Administration (OSHA) visit.

COST IMPACT TO BUSINESS

Injuries alone cost U.S. businesses more than $110 billion in 1993. This does not include occupational illnesses that cost many times more [5]. Are you surprised? Before we get into developing a safety culture you should understand the implication of cost to your business. Let's take a closer look at the impact of incidents on companies' profits and sales.

3

The following list provides an overview of what a typical company might have to do to pay for an incident costing $500:

- A soft drink bottler would have to bottle and sell more than 61,000 cans of soda
- A food packer would have to can and sell more than 235,000 cans of corn
- A bakery would have to bake and sell more than 235,000 donuts
- A contractor would have to pour and finish 3,000 square feet of concrete
- A ready-mix company would have to deliver 20 truckloads of concrete
- A paving contractor would have to lay 900 feet of two-lane asphalt road [5]

Refer to Table 1-1 for a review of the sales required to cover specific losses. This should bring the cost into perspective. In times of keen competition and low profit margins, loss control may contribute significantly more to profits. It is necessary for additional sales of $1,667,000 in products to pay the cost of $50,000 in annual losses from injury, illness, damage or theft, assuming an average profit on sales of 3 percent. The amount of sales required to pay for losses will vary with the profit margin.

Table 1-1
Sales Required to Cover Losses

Yearly Incident Costs	Profit Margin				
	1%	2%	3%	4%	5%
$1,000	$100,000	$50,000	$33,000	$25,000	$20,000
5,000	500,000	250,000	167,000	125,000	100,000
10,000	1,000,000	500,000	333,000	250,000	200,000
25,000	2,500,000	1,250,000	833,000	625,000	500,000
50,000	5,000,000	2,500,000	1,667,000	1,250,000	1,000,000
100,000	10,000,000	5,000,000	3,333,000	2,500,000	2,000,000
150,000	15,000,000	7,500,000	5,000,000	3,750,000	3,000,000
200,000	20,000,000	10,000,000	6,666,000	5,000,000	4,000,000

This table shows the dollars of sales required to pay for different amounts of costs for accident losses, i.e., if an organization's profit margin is 5%, it would have to make sales of $500,000 to pay for $25,000 worth of losses. With a 1% margin, $10,000,000 of sales would be necessary to pay for $100,000 of the costs involved with accidents.

Bird, Frank, George L. Germain, *Loss Control Management: Practical Loss Control Leadership*, Revised Edition, Det Norske Veritas, 1996, Sales Required To Cover Losses, Figure 1-9, p. 10. Reprinted with permission.

Other useful statistics that may help to sell your program are given in Table 1-2.

Are you shocked at the examples stated in Table 1-2? If the answer is yes, then you must understand that a management system will help you to control these associated costs. You cannot rely on mandated requirements to drive your safety process. The costs of incidents can harm a company's profitability.

$afety Pays is a tool developed by OSHA to assist employers in assessing the impact of injuries and illnesses on their profitability. It uses a

Table 1-2
Useful Statistics

Nearly 50 employees are injured every minute of a 40-hour workweek and almost 17 employees die each day.

Since OSHA was created 28 years ago, workplace fatalities have been cut in half. Occupational injury and illness rates have been declining for the past five years. In 1997, injuries dropped to the lowest level since the U.S. began collecting this information.

According to OSHA, their premier partnership, the Voluntary Protection Program (VPP), continues to pay big dividends. Today more than 500 workplaces, representing 180 industries, save $110 million each year because their injury rates are 50 percent below the average for their industries.

Nearly one-third of all serious injuries and illnesses stem from overexertion or repetitive motion. These are disabling, expensive injuries. They cost our economy as much as $20 billion in direct costs and billions more in indirect costs.

Only about 30 percent of businesses have established safety programs. According to OSHA about half of the 95 million workers who would be covered under an OSHA safety program standard don't have that protection today. Establishing a safety program to prevent injuries is not only the right thing to do, it's the profitable thing to do. Studies have shown a $4 to $6 return for every dollar invested in safety. Together federal and state OSHA programs have about 2,500 inspectors to cover more than 100 million workers at 6 million sites. That's one inspector for every 2,400 worksites and every 40,000 employees. As you can see, OSHA cannot depend on inspections alone to achieve the mission of protecting employees. At a rate of roughly 90,000 inspections per year, OSHA would visit each worksite once every 66 years!

OSHA Web Site: http://www.osha-slc.gov/SLTC/safetyhealth_ecat/mod1.htm#.

company's profit margin, the average costs of an injury or illness, and an indirect cost multiplier to project the amount of sales a company would need to generate to cover those costs. Since averages are used, the actual costs may be higher or lower. Costs calculated using the example do not reflect the pain and suffering of an injured employee. Several example worksheets using OSHA's $afety Pays are included in this chapter to help you to calculate your injury cost. Refer to Figure 1-1 for a sample cost calculation worksheet. These worksheets are tools that can help you to assess and understand the financial impact of an incident at your company [3].

Estimated Annual Accident Costs			
	Number	Average Cost*	Results
Annual Fatalities	0	X $910,000	0
Annual lost workdays	20	X $28,000	$560,000
Annual number of recordable cases without lost workdays	50	X $7,000	$350,000
Total estimated annual cost of occupational fatalities, injuries, and illnesses			

*Using National Safety Council average cost for 1998 includes both direct and indirect cost, excluding property damage.

Impact of Accidents on Profits and Sales (Compare accident costs to company profits)	
Comparison	Results
Sales Volume	$10,000,000
Profit Margin	5%
Sales Volume × % Profit = Annual Profits	$500,000
Accident Cost as a Percent of Profits	182%
** Amount of Sales Needed to Replace Lost Profits	
If Profit margin is 5%, then it takes $20 of sales to replace every dollar of loss.	

Figure 1-1 Sample cost calculation worksheet. Adapted from OSHA web site, Cost Calculation Worksheet, Using the OSHA $afety Pays Advisor, http://www.osha-slc.gov/SLTC/safetyhealth_ecat/mod1_estimating_costs.htm, public domain [9].

Not all indirect costs will affect a company's bottom line on a dollar-for-dollar basis. Therefore, using all indirect costs in the calculation over-estimates the amount of sales needed to replace lost profits. Likewise, only using direct costs in this comparison would underestimate the amount of sales needed. Actual costs will be somewhere in between the two figures.

Refer to Figure 1-2 for a sample worksheet to calculate your cost per incident.

HOW TO ESTIMATE THE IMPACT OF INCIDENTS ON YOUR PROFITS AND SALES

Direct Cost

To calculate the direct cost, enter the following information

- Total value of the insurance claims for an incident (injury/illness, loss producing events). This cost will consist of medical cost and indemnity payments, reserves on accounts, etc.

$_____

Indirect Cost

To calculate the indirect cost of this incident, multiply the direct cost by a cost multi-plier. The cost multiplier that you use will depend on the size of the direct cost.

If your direct cost is:	Use this cost multiplier
$0–2,999	4.5
$3,000–4,000	1.6
$5,000–9,999	1.2
$10,000 or more	1.1

Direct Cost   ×   Cost Multiplier   =   Indirect Cost
$_____ $_____ $_____

Total Cost

Direct Cost   +   Cost Multiplier   =   Indirect Cost
$_____ $_____ $_____

Adapted from OSHA, Sample Worksheets to Calculate Incident Cost, http://www.osha-slc.gov/SLTC/safetyhealth_ecat/images/safpay3.gif.

Figure 1-2 Sample worksheets to calculate incident cost. Adapted from OSHA Web site, Cost Calculation Worksheet, Using the OSHA $afety Pays Advisor, http://www.osha-slc.gov/SLTC/safetyhealth_ecat/mod1_estimating_costs.htm, public domain [9].

THE DIRECT AND INDIRECT COSTS OF ACCIDENTS

As you can see, incidents are more expensive than many managers realize. Why? Because there are many hidden costs that are not truly understood. Some costs are obvious while other costs are transparent. Your workers' compensation claims cover medical costs and indemnity payments for an injured employee. These are the direct costs of incidents.

However, what about the costs to train and compensate a replacement employee, repair damaged property or equipment, downtime of equipment, investigating the incident, and implementing corrective actions? Even less apparent are the costs related to product schedule delays, added administrative time, lower morale, increased absenteeism, pain and suffering of the employee, and impaired customer relations. These are the indirect costs and, as such, have been described by many professionals as an iceberg. You cannot see the bottom until it is too late. We will discuss these costs later in more detail [3].

According to OSHA, studies show that the ratio of indirect costs to direct costs can vary widely, from a high ratio of 20:1 to a low of 1:1. OSHA provides a conservative approach where they estimate that the lower the direct costs of an incident, the higher the ratio of indirect to direct costs. Table 1-3 demonstrates the effects of the hidden cost that is buried below the surface. In many cases, we don't understand how to capture these costs [3, 4].

The Business Roundtable's Report A-3, "Improving Construction Safety Performance," concludes that in the construction industry, an effec-

Table 1-3
Direct Cost vs Indirect Cost

Direct Cost of Claims	Ratio of Indirect to Direct Cost
$0–2,999	4.5
$3,000–4,999	1.6
$5,000–9,999	1.2
$10,000 or more	1.1

The Business Roundtable, Improving Construction Safety Performance Report A-3, New York, January 1982 and Oregon OSHA, The Cost of Accidents, Safety Training Module,http://www.cbs.state.or.us/osha/educate/training/pages/materials.html, public domain.

tively administered safety and health program will produce savings of 3.2 times the program's cost [3, 6].

A subsequent companion publication takes an in-depth look at the upward spiral of contractor workers' compensation costs and finds that:

> Implementing an effective safety program to reduce work site incidents can influence workers' compensation premium costs. Lowering the frequency and severity of construction accidents will help to lower experience modification rates (EMR) and manual rates that, in turn, lower workers' compensation insurance premiums [3, 7].

INCIDENT COSTS

In their book subtitled *Practical Loss Control Leadership* [1], Frank Bird, Jr., and George Germain use the same analogy of the iceberg theory to describe accident (incident) costs. Bird and Germain contend that the medical or insurance compensation costs are the visible tip of the iceberg. For example, for every $1 spent on an obvious incident, from $5 to $50 or more is likely to be spent on below the surface (below the iceberg). According to Bird and Germain, the below-the-surface cost can include the following:

- Uninsured costs for building repairs
- Tool or equipment damage
- Replacement of damaged products and/or materials
- Losses from production delays and interruptions of a business
- Legal expenses
- Replacement of emergency supplies and equipment
- Renting of needed interim equipment
- Paying for time spent on the investigation [1]

Refer to Figure 1-3 for the cost of accidents at work.

Even more hidden costs are the additional $1 to $3 that will be spent for uninsured miscellaneous costs. These can include the following:

- Wages paid for lost time of an injured employee and co-workers
- Cost of hiring and training replacements
- Overtime costs
- Extra supervisory time

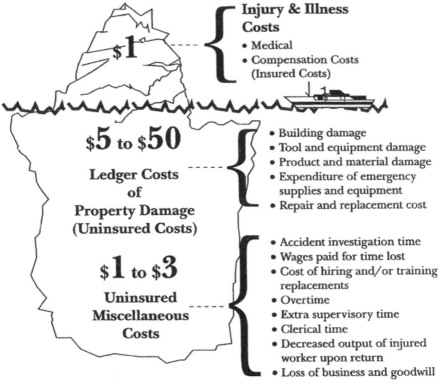

Injury & Illness Costs
- Medical
- Compensation Costs (Insured Costs)

$5 to $50

Ledger Costs of Property Damage (Uninsured Costs)

- Building damage
- Tool and equipment damage
- Product and material damage
- Expenditure of emergency supplies and equipment
- Repair and replacement cost

$1 to $3

Uninsured Miscellaneous Costs

- Accident investigation time
- Wages paid for time lost
- Cost of hiring and/or training replacements
- Overtime
- Extra supervisory time
- Clerical time
- Decreased output of injured worker upon return
- Loss of business and goodwill

This represents the statistical average of accident costs reported by clients.

ILCI: International Loss Control Institute, Inc.

Figure 1-3 The cost of incidents at work: ILCI cost studies. Bird, Frank E., Germain, George L., *Loss Control Management: Practical Loss Control Leadership*, 4th edition, p. 8, Figure 1-7. Det Norske Veritas (USA), Inc., August 1996. Reprinted with permission.

- Clerical time
- Decreased output of the injured employees upon return to work
- Loss of business and goodwill [1]

Bird and Germain do not consider the impact of reduced commitment to work when employees operate in an environment where injuries are common. Nor do they address the costs specifically associated with illnesses that may be associated with health hazards [1]. It is OSHA's belief that the financial impact of work-related illnesses may be even greater than that of work-related injuries, because certain illnesses frequently involve longer absences.

CASE HISTORIES

To help you understand the impact of cost, the following case histories represent workplaces participating in OSHA's Voluntary Protection Programs (VPP), over the years since VPP began (1982.) OSHA has verified lost workday cases at rates 60 to 80 percent lower than their industry averages. VPP requirements consist of management systems consistent with this manual. To understand how you would go through the process of becoming a VPP certified site refer to Chapter 3. We emphasize again: OSHA has verified all of these rates to be accurate [2].

Case 1. Between 1983 and 1987, a major chemical company brought all of its existing plants into the VPP process. During that period, recordable injuries for the company dropped 32 percent. Lost workday cases dropped 39 percent, and the company saw a 24 percent decrease in the severity of cases.

The chemical company cut its workers' compensation costs by 70 percent, or more than $1.6 million. This reduction occurred between 1983 and 1986, the years that the company was qualifying its plants for VPP. In the ensuing years, the company has continued to enjoy impressive savings. In addition, its sister oil company had average annual workers' compensation costs of almost $300,000 during the early 1980s. By 1991, workers' compensation payments were down to $69,000, and the following year the cost dropped to $33,000. Additional savings have come from reductions in third-party lawsuits from contractor employees.

The oil company saw an already low absenteeism rate drop an additional 50 percent between 1988 and 1992.

Case 2. In the construction industry, a major utility company brought two large power plant construction sites into VPP in 1983 and 1984. By 1986, one site had reduced its total recordable cases by 24 percent and its lost workday cases by one-third. The other site reduced recordable cases by 56 percent and its lost workday cases by 62 percent.

In 1986 alone direct cost savings from preventing incidents were $4.14 million at one site and $.5 million at the other.

Case 3. A rail car manufacturer in Georgia began preparing for VPP in 1989. That year it experienced a lost-workday case rate of 17.9. One year later the company had reduced their rate by two-thirds. By 1991, its lost-workday case rate was down to 5.9.

Case 4. A resident contractor at a petrochemical plant in 1989 made the commitment to begin developing a VPP quality, safety, and health program. By 1992, the year it was approved for VPP participation, the contractor had reduced its total recordable cases by 61 percent, from 18 injuries in 1989 to 6 in 1992. The company's lost workday cases dropped by 66 percent during that period. The resident maintenance contractor had workers' compensation costs of $245,543 in 1989. By 1992, these costs were down to $93,166.

Case 5. From 1985 through 1988, while participating in VPP, a Nebraska agricultural manufacturer saw its workers' compensation costs decrease by half. During the 3 years in VPP cited they noted a 13 percent increase in productivity and a 16 percent decrease in scrapped product that had to be reworked.

One plant manager testified that the adoption of a single work practice change at his 44 employee chemical plant during the first 3 years of VPP resulted in increased volume of product and savings of $265,000 per year [8].

More workplace success stories can be found on the OSHA Web site, www.vpppa.org/

SUMMARY

In the real world, management can be perceived as risk-takers, willing to do anything to be competitive with other companies to produce a profit. However, on the other side of the spectrum, there is one gamble that can be a sure loss for any manager: gambling on employee safety and the risk of incidents that can cause injury or property damage [11].

Managers are now coming to realize that the actual cost of a lost workday injury is underestimated. The problem is that we only see what is on the surface and do not always understand the hidden cost. For example, let's look at the indirect and hidden costs of one lost workday incident:

- Productive time lost by injured employee
- Productive time lost by employees and supervisors attending the victim
- Clean-up and startup of operations interrupted by the incident
- Time to hire and retrain other individuals to replace the injured employee until their return to work
- Time and cost for repair or replacement of damaged equipment or materials
- The cost of losing a valued customer due to poor performance or late delivery of goods and services

- Poor or eroded morale among employees
- Lower efficiency
- Increase workers' compensation EMR rates
- Possible penalties or other sanctions applied where the injury is determined to be caused by a violation of regulations
- The cost of completing the paperwork generated by the incident [10]

It makes good business sense to reduce the costs and risks associated with incidents, no matter whether they cause injuries. To do this, you must set a goal: to provide a service or produce a quality product efficiently without incidents. Too often, that is seen as something to be considered as time permits, over and above regular business activities. To reduce risks effectively, you must address safety just as you would production, quality control, and/or costs [10].

A safety management system must be consistent with other program requirements. A balanced program attempts to optimize safety, performance, and cost. Safety system program balance is the product of the interplay between the system safety and the other program elements of cost, schedule, and performance. Refer to Figure 1-4 to understand how the injury cost affects these elements.

Programs cannot afford incidents that will prevent the achievement of the primary goals. However, neither can we afford systems that cannot perform because of unreasonable and unnecessary safety requirements.

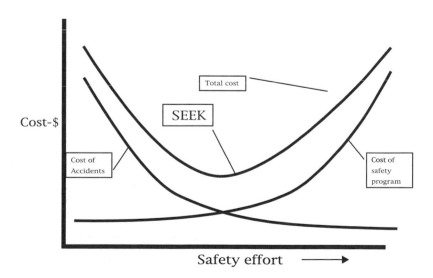

Figure 1-4 Safety efforts versus cost. Adapted from *FAA System Safety Handbook*, Chapter 3: Principles of System Safety, p. 2, Planning Principles. December 30, 2000, public domain.

Safety must be placed in its proper perspective. A correct safety balance cannot be achieved unless acceptable and unacceptable conditions are established early enough in the program to allow for the selection of the optimum design solution and/or operational alternatives. Defining acceptable and unacceptable risk is as important for cost-effective accident prevention as is defining cost and performance parameters [11].

A "visible" safety program helps to set the stage for improved employee attitude. Periodic safety related training and inspections by top management help to convince employees that the program is not merely administrative program of the month, but is an item of real concern. The employee gets involved. Once that occurs, employees participate, supervisors usually take the initiative, and the program evolves into an active force in the organization. At this stage, employees subconsciously develop the habit of planning ahead and examining the safety, production, quality, and cost aspects of the task before them. Although the physical safeguarding of the workplace is a real factor in safety, the mental attitude of the employee is the ultimate key to avoiding incidents [10].

To achieve this goal, you, as the manager, must establish a plan for eliminating employee injuries. This should be made a part of the organization's activities-based approach. The plan not only should consider the immediate needs, but also should provide for ongoing, long-lasting employee participation. Once the plan is designed, it must be followed through and used consistently. As a result, the program will let you anticipate, identify, and eliminate conditions or procedures that could result in injuries [10]. Only then will you reduce incidents and the associated cost. We will get into this in more details in later chapters.

REFERENCES

1. Bird, Frank E., Germain, George L., *Loss Control Management: Practical Loss Control Leadership*, Det Norske Veritas (USA), Inc., 4th ed., August 1996.

2. Occupational Safety and Health Administration (OSHA) Voluntary Protection Program (VPP), http://www.osha.gov/oshprogs/vpp/, public domain.

3. Oregon OSHA, The Cost of Accidents, Safety Training Module, http://www.cbs.state.or.us/osha/educate/training/pages/materials.html, public domain.

4. OSHA Web site, http://www.osha-slc.gov/SLTC/safetyhealth_ecat /comp1_mgt_lead.htm, public domain.

5. OSHA, $afety Pays 1996, OSHA Web site, http://www.osha-slc.gov/SLTC/ safetyhealth_ecat/images/safpay1.gif, public domain.

6. The Business Roundtable, Improving Construction Safety Performance Report A-3, New York, January 1982.

7. The Business Roundtable, The Workers' Compensation Crisis: Safety Excellence Will Make a Difference, New York, January 1991.

8. Proceedings of Public Information Gathering Meeting on Suggested Guidelines for General Safety and Health Programs, U.S. Department of Labor, OSHA, Docket No. C-02, p.77, October 6, 1988.

9. OSHA Web site, http://www.osha-slc.gov/SLTC/safetyhealth_ecat/mod1. htm#, public domain.

10. *The Manager's Handbook: A Reference for Developing a Basic Occupational Safety and Health Program for Small Businesses*, State of Alaska, Department of Labor & Workforce Development, Division of Labor Standards & Safety, Occupational Safety & Health, April 2000, Section 1, pp. 3–4, public domain.

11. *FAA System Safety Handbook*, Chapter 3: Principles of System Safety, p. 2, Planning Principles, December 30, 2000, public domain.

2

Defining a Value System

INTRODUCTION

Today we are bombarded with many terms that are used to help management to define and/or describe safety efforts. We use the term *behavioral-based safety* to describe why some employees engage in risk-taking behaviors. You cannot pick up a safety journal without finding an article telling you that behavioral-based safety will change your safety program. What it does not tell you is that this is only one element of a successful safety management system. "What works in one organization may not in another, there is no magic bullet" [3], as some people will lead you to believe. We will discuss behavioral-based safety in more detail in Chapter 16.

CULTURE AND SAFETY

Safety professionals pretty much ignored the concept of culture through the 1980s. As management attempted to improve culture through changing their style of leadership and through employee participation, safety efforts changed very little. Management was using the same elements in their safety programs that they had always used. I am reminded of the definition of insanity, "Doing the same thing over and over and expecting different results." Under the old traditional style, this is common. Safety programs typically consisted of the usual things: for example, safety meetings, facility inspections, and incident investigations to some degree (not getting at the root cause) (identify solutions), with little emphasis on identifying and correcting hazards. Most professionals perceived these tools as the essential elements of a safety program.

However, these tools do not create a safety culture or build a safety management system.

To understand the culture side of safety, you can review some OSHA voluntary guidelines. Several guidelines have been published since the 1980s suggesting that companies should implement specific elements of a model safety program. A number of states enacted laws requiring companies to do these same things. These elements were perceived as being a "safety program" and not building a safety culture [3, 4], Using these elements as a safety program and not a management system only supports the culture if there is top management commitment. We will discuss some of these voluntary guidelines in Chapter 3.

While OSHA and some state programs were going down the "essential element" track to safety (as was much of the safety profession), suggesting implementing a management system, a number of pieces of research began to appear that presented different answers to the safety problem. Most of the research results were consistent in saying, "There are no essential elements." As we discussed, what works in one organization may not in another. Each organization must determine for itself what will work for it. There are several major concepts. The answer seems to be clear: it is the culture of management and the employees and the organization that determines what will work in any organization [3, 4].

WHAT SETS THE CULTURE?

Certain cultures do have safety as one of their major constituent values, and not a priority. Other cultures make it very clear that safety is unimportant. In the latter, almost nothing will work; meetings will be boring, and job hazard analyses (JHAs) are perceived only as paperwork [3, 4].

The culture of an organization sets the tone for everything in the safety arena. In a positive safety culture, the culture itself says that everything you do about safety is important. In a participative environment (culture), an organization is saying to the employee, "We want and need your help." Some cultures urge creativity and innovative solutions, while others destroy them by not caring about their employees. Some cultures tap the employees for ideas and help, while others force employees to never use their brains at work [3, 4]. The following list outlines some specific elements of a culture:

- How are decisions made? Does an organization spend available funds on employees and safety? Or are these ignored in favor of other things—for example, production, quality, new equipment, or other business opportunities?

- How are employees measured? Is safety measured as tightly as production with defined activities? What is measured tightly is what is important to management.
- How are employees rewarded? Is there a greater reward for productivity than for safety? This shows management's real priorities (values).
- Is teamwork mastered? Or is it "us versus them"? In safety, is it "police versus policed"?
- What is the history? What are your traditions?
- Is your management system in place to protect employees or to comply with regulations?
- Are supervisors required to do safety tasks daily?
- Do big "bosses" (top management) walk around the facility and talk to employees?
- Are you allowed to use to use your brain or are you just a puppet (gopher)?
- Has your company downsized? Is there always a threat of downsizing?
- Is the company profitable? Too much? Too little? Is the company satisfied with its level of profit? Or do they just want more and more, never getting enough?
- Can managers and supervisors talk about safety as they can about quality and production? [3, 4]

As you can see, a culture is defined by an infinite number of things. Petersen only listed a few. It is more important to understand what the culture is than to understand why it is that way [3, 4].

We have suggested that culture dictates which program elements will work and which will not. Culture dictates results, and what the incident record will be. This is true no matter whether we look at frequency or severity [3, 4].

WHY DO CULTURES FAIL?

We have looked at what builds a safety culture, but you also need to understand how it will fail. In many cases, a culture fails based on the management style. Every management "rules" with a different style.

Douglas McGregor, who specialized in human behavior in organizations, is famous for his formulation of Theory X (authoritarian management) versus Theory Y (participative management [2]).

McGregor believed that Theory X owed its origins to the banishment of Adam and Eve from Eden into a world where they were forced to work to survive. "The stress that management places on productivity, on the concept of 'a fair day's work,' on the evils of featherbedding and restriction of output, on rewards for performance while it has a logic in terms of the objectives of enterprise reflects an underlying belief that management must counteract an inherent human tendency to avoid work" [2].

McGregor identified a series of human wants in ascending order, from the most basic physiological urges through a desire for safety and security (security in the workplace) to the "social needs" such as belonging, acceptance by one's peers, and the giving and receiving of affection. Above those again came the "egoistic needs," those that relate to an individual's self-esteem: the need for self-respect, self-confidence, autonomy, achievement, competence, and knowledge—for reputation, status, recognition, and the respect of one's peers. Ultimately in McGregor's pyramid became the needs for self-fulfillment, for realizing one's individual potential, and for continuing in self-development [2]. If management continued to focus its attention on physiological needs, providing rewards was unlikely to be effective, and the only alternative under this philosophy would be reliance on the threat of punishment [2].

Part of Theory X validates itself, "but only because we have mistaken effects for causes." McGregor continues:

> The philosophy of management by direction and control regardless if it is hard or soft is inadequate to motivate because the human needs on which this approach relies are relatively unimportant motivators of behavior in our society today. Direction and control are of limited value in motivating people whose important needs are social and egoistic. So long as the assumptions of Theory X continue to influence managerial strategy, we will fail to discover, let alone utilize, the potentialities of the average human being [2].

McGregor's Theory Y management approach was designed to tap employees' potential (participation). This was based on his observations of the way management thinking had moved a considerable way from the traditional "hard" approach and the "soft" reaction that followed the Depression years [2]. McGregor developed six basic assumptions for Theory Y:

- The expenditure of physical and mental effort on work is as natural as play or rest. The average human being does not inherently dislike work. Depending on controllable conditions, work may be a source of satisfaction (will be voluntarily performed) or a source of punishment (will be avoided if possible).

- External control and the threat of punishment are not the only means for bringing about effort toward organizational objectives. Employees will exercise self-direction and self-control in the service of objectives they are committed to.
- Commitment to objectives is a function of the rewards associated with their achievement. The most significant of such rewards, for example, the satisfaction of ego and self-actualization needs, can be direct products of effort directed toward organizational objectives.
- The average human being learns, under proper conditions, not only to accept but seek responsibility.
- The capacity to exercise a relatively high degree of imagination, ingenuity, and creativity in the solution of organizational problems is widely, not narrowly, distributed in the population.
- Under the conditions of modern industrial life, the intellectual potentialities of the average human being are only partially utilized [2].

Such assumptions, McGregor pointed out, had a deep implication for management. Where Theory X offered management an easy scapegoat for failure, because of its emphasis on the innate nature and limitations of its human resources Theory Y placed all problems "squarely in the lap of management." If employees were lazy or unwilling to show initiative or responsibility, if they were indifferent or intransigent, the fault lay in management methods. In other words, McGregor was redefining the old military adage: "There are no bad troops, only bad officers" [2].

McGregor believed that human beings had far greater potential than the industrial management of his time could understand. Theory X denied even the existence of that potential; Theory Y challenged management "to innovate, to discover new ways of organizing and directing human effort, even though we recognize that the perfect organization, like the perfect vacuum, is practically out of reach" [2]. You will understand how Theory Y applies when we discuss employee participation in Chapter 7.

WHAT ARE VALUES?

Another word that we hear is "value." What is a "value"? Can values can be defined as looking at the underlying beliefs and/or philosophies of individuals (management/employee behavior) and organizations (management support/employee participation)? When we talk about ways to change or create a proactive safety culture, we need to recognize and understand the full impact that values have on an organization. As we have discussed, there are morals and values that we will have to deal with when developing a management system.

For another way of looking at the safety culture, refer to Table 2-1.

If you view these concepts from the other side of the fence (the human side), you will find that employees bring their own set of personal beliefs and judgment to an organization. This forms the foundation of unique personal characteristics and is sometimes difficult to change because people are shaped and influenced early in life by our parents and our social environment. "Most people are a function of the social mirror, scripted by the opinion, the perceptions, the paradigms of the people around them" [1].

How about norms? Do values become norms? Is this how we operate a business? Norms have been described as unwritten rules, beliefs, attitudes, and/or practices that demonstrate proper or improper action. Norms can become expressions of personal and organizational values [1]. Many professionals will say that norms can be accomplished through observable behaviors of employees. For example, we believe that everyone can remember when a member of management knowingly permits an employee to operate equipment without proper guarding or without wearing personal protective equipment (PPE.) Why? Probably because the manager "did not have the time" to say anything to the employee or it would have slowed production. In these unspoken words, this is known as: "Production is #1 and safety is not important." The norm is to take chances if the objective is to speed up production.

Let's look at other cases: A manager sees an employee driving a fork truck too fast for conditions and does not do anything. However, employees are being praised or rewarded for production results, without any consideration given to how they achieved the results. Their risk-taking behaviors or activities may be putting themselves and others at risk. In this case, risk-taking behaviors have become the unwritten rules (norm) of getting the job done fast. If management allows this to happen, a message is sent to employees that doing something unsafe or at-risk is OK. Management has nonverbally stated that it is OK to violate a safety rule when the employee is rewarded for completing the task in record time. Given this situation, this at-risk practice will continue. You can probably think of many more examples of these types of situations in your workplace.

Table 2-1
Vision of a Safety Culture

Safety professionals use and promote the term *culture* or *paradigm shift* (an example that serves as a pattern or model) [10] as the way to promote long-term reductions of injuries. For example, one safety vision for the safety culture may be, "Safety should be integrated in the culture of every workplace." On the surface, this sounds like a great vision statement, but what does it mean? You should ask the question: What is a "safety culture" and how do we define it? In general, a safety culture can be viewed in several ways:

- Safety must be integrated into every aspect of a business just like quality, production, etc.
- All employees in the organization must understand and believe that they have a right to work safely
- All employees must accept responsibility for making sure that they protect themselves and their co-workers
- Safety is considered a value in the organization, not a priority

IS SAFETY A PRIORITY FOR YOUR ORGANIZATION?

Traditionally, most managers have viewed safety as just another priority on a growing list of things to do, just one more thing on a list of many things that should be accomplished on a daily basis. Refer to Figure 2-1 to understand how priorities change on a day-to-day basis. In this example, note how each item on the list has changed. If this is happening in your process, you need to understand why. Our life changes from day to day and we must adapt to the change. If safety is a value, it will not change.

This concept of safety as a priority is often reflected in safety slogans and speeches, for example:

- Safety starts here
- Everyone is responsible for safety
- Safety is our top priority!
- Safety is #1!
- Safety first!

Most professionals have the best intentions when setting priorities. However, the reality of trying to manage your business often interferes with the priorities. Why? Because priorities change based on the needs of the organization. To say that safety is a priority means that it will change based on the needs or urgencies of the moment, and therefore will not always be the top priority! Refer to Figure 2-1. The dictionary defines a priority as a precedence established by order of importance or urgency [10]. Think about how often your priorities change either at work or at home.

Table 2-1
Continued

SAFETY AS A VALUE

Let's look at the word value. If we say that safety is a value, what does this mean? The dictionary defines value as a principle, standard, or quality considered worthwhile or desirable: family value [10].

If you use this definition, you are sending a message that safety is important and will not be compromised. As your business needs change, so will the needs and focus of safety. However, as a value it will always be present.

Another good question that should be considered is: why does a company need a safety program? Is it because OSHA requires it, or do you just want to do the right thing? Is the basic purpose of any safety program to help prevent injuries?

If you were to look at the objective of regulatory requirements you will find that OSHA's main purpose as spelled out is: "To assure safe and healthful working conditions for working men and women." Even OSHA recognizes that physical compliance with safety requirements alone will not eliminate incidents. It is an impossible task for OSHA to attempt to write a safety regulation that will address every possible or potential hazard in the workplace. This is why some requirements are written as performance-based. You as managers have to interpret these performance-based standards and compliance initiatives. This can be done by strictly following the letter of the law or by making good management decisions that meet the requirements and providing Best Management Practice (BMP) for the business.

Why? Because *people* (employees) are in workplaces. Every organization with one or more employee has a wide variety of different backgrounds, personalities, physical characteristics, attitudes, and behaviors. We all know that humans are fallible and that we can and do make mistakes every day. This statement reminds me of an air force sergeant that I once knew. He told me that he had only made one mistake in his life. The mistake as he described it was "I thought that I had made a mistake, but really I did not make the mistake." As he put it, that was his only mistake—thinking he had made a mistake.

OSHA along with the safety community has recognized that it takes the combined efforts of BMP, physical safeguarding, training, maintenance, etc., to achieve success in an organization.

Oklahoma Department of Labor, Safety and Health Management: Safety Pays, 2000, http://www.state.ok.us/~okdol, pp. 2–3, public domain.

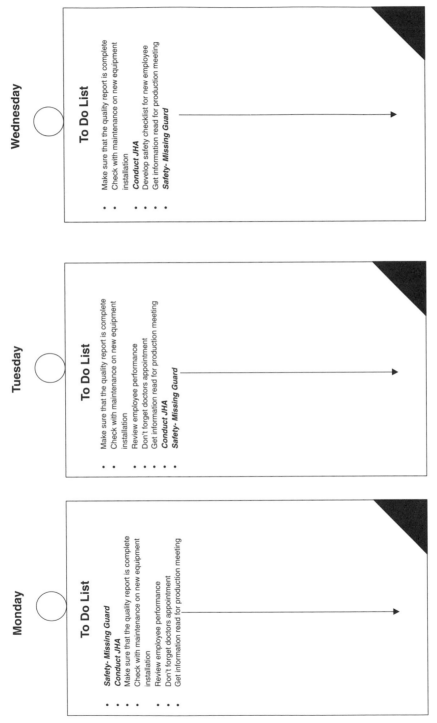

This To-Do-List represents why safety must be a value that is instilled in the minds of management and employees, instead of being on a list

Figure 2-1 Priority list.

CHANGES IN BEHAVIOR

When defining management system changes, management needs to spend time observing and evaluating indicators upstream to understand how behaviors are changing and how they affect the management system. Management needs to understand that behaviors are visible evidence of norms that are part of employee work practices. One of the best techniques that we encourage is to listen to what employees have to say. Sometimes employees just want to talk to someone about work or personal issues, etc. In these cases, you should sharpen your listening skills. This is one of the most important key concepts to understand. How many times have you seen a manager say to an employee that he/she is doing a good job and then rushes off to another project? This is a case where we need to respect the feelings of the employee and stop and listen to his/her concerns. If you do not feel that you have the time to spend with the employee, it is best not to attempt to talk to the employee. Set aside time when you can concentrate on your communication.

RECOGNIZING SUCCESS

Management must understand the indicator or warning signs and know how to react. You as part of management are no different. For example, how many times have you approached an employee to tell him that he is doing a good job, but failed to stop and listen to what he has to say because, again, "I don't have the time"? It's important to take the time. A few seconds can mean a lot to the employee.

Any system, no matter how good it is, is resistant to change. No one likes change, so we resist it. This is human nature. You must learn to deal with these changes. Behaviors need to be shaped and reinforced through praise and positive recognition.

However, we must understand the controversy surrounding safety recognition programs and how incident reporting continues to remain a hot topic among safety professionals, OSHA, organized labor, and management alike. Proponents will argue that incentives can discourage incident reporting if done improperly. In many cases, other management practices may have a negative impact on incident reporting: for example, management's overemphasis on incident-free records, punitive (disciplinary) measures taken against injured employees, embarrassing an injured employee in front of co-workers, etc. [5].

Any safety program that targets one area without addressing others is less likely to succeed. Incidents may initially decline with such a program, but soon level off or increase. An effective safety recognition program or any traditional safety program that promotes incident reporting will usually show an increase in reported incidents in the beginning of this effort. Once you have identified specific injury types, you are then able to address the factors that contributed to these incidents. Once that practice has been consistently accomplished, injuries tend to decline as the root causes of each incident are investigated, identified, and properly addressed. In some cases, this may be a better indicator that the program is working than an initial decline in number of incidents.

Why do you think this trend happens? The fact is that the more you pay attention to incidents, the more issues you will see [5]. You cannot react when you have several injuries. If your process is in place, you have to give it time to work.

As always, successful management systems involve management commitment to achieving the desired objective. It fosters a system of safety awareness and motivation at all levels of the organization. A recognition program should be rewarding, entertaining, and easy to understand, with a daily focus on safety—and not only when an incident occurs. It should generate peer group pressure; it should be visually dynamic, flexible, and involve recognition and rules enforcement. It should be administratively easy, promote employee participation, accountability, and communication, and encourage management/employee cooperation [5].

There is an increased focus on traditional performance-based incentives as potential causes of under-reporting. Although OSHA is not suggesting that companies discontinue incentive (award) programs, OSHA prefers nontraditional incentive programs where emphasis is placed on activity-based efforts such as safety suggestions, reporting hazards, incidents (near misses, loss-producing events, etc.), safety contacts, and attending safety meetings [6]. In Chapter 6 we will discuss an activity-based safety system.

OSHA's current stance places the burden of proof on management to demonstrate that incentives are not being used to manipulate incident reporting. In some cases, they have levied fines for recordkeeping violations on companies that tie bonuses to incident-free records [6].

Safety professionals are frustrated when a well-intentioned, comprehensive safety management system is put into place, yet employees continue to be injured. Safety professionals generally recognize that incentive programs that address behaviors without also recognizing performance have some limitations. Some injuries result from personal factors that cannot be traced to a specific behavior. An employee who has exemplified safe behaviors for years can have a "bad day." No matter what we believe, we have to understand that there are some things that we just cannot explain. Stress, lack of sleep, emotional strain, minor illness, substance abuse, etc., could be significant contributing factors to injuries on and off the job. An effective recognition program will capitalize on the impact of positive group dynamics as a means of shaping individual attitudes and behaviors [5]. What we need to get across is that management must feel the same way about safety as the safety professional. How many times have you, if you are a safety professional, visited a plant or work environment and employees have changed their behaviors just because they see you are there. How many times have you been called "Mr./Ms. Safety?" The question is: Does the same thing happen when managers or supervisors walk into an area? In most cases, probably not if they do not understand the culture. Somehow you need to get this point across to managers and supervisors so that they can also create the same level of awareness. This is only the first step. Of course, as the process continues building your new culture, you will see these types of changes happen. We must get to a point where employees understand and take the responsibility to do the right thing and not wait for management.

WHAT CAN WE LEARN?

Management has the controls to help reduce the number of on-the-job incidents. If you review OSHA's proposed Safety and Health Program standard, 29 Code of Federal Regulations (CFR) 1900.1, you will note that employee participation is one of the key components of the proposed requirement [9], The question is, how can management get employees involved in safety without tying rewards (recognitions) and punishments to incidents? The bottom line is this, no matter how you look at it: incident reduction—for example, reducing the number of employees getting hurt—is what counts for some managers, so why should it not be as meaningful to the employees? We will discuss these voluntary safety and heath program in more detail in Chapter 3 and continue to discuss these concepts throughout the book.

As we demonstrated in Chapter 1, insurance costs and OSHA inspections are driven by performance (number of incidents–some refer to it as OSHA incident rates [OIR]). Is it realistic to expect employees to be motivated without a performance (incident reduction) element built into their safety recognition? Some managers in a traditionally based safety system will say, "The employee gets a paycheck and that should be enough." Is this what you want to convey to your employees?

Today's environment promotes employee empowerment that translates into employee responsibility and accountability. Management needs to understand that employees are an asset to the company, not just a liability [8]. I once heard a speaker say, "We are so busy trying to please our customers that we forget the feelings of our employees." He continues to say that "you need to focus on your employees and forget the customers." In this way, your employees will be happier and more productive. In the end, more customers will come. Take this for what it is worth. I thought it was worth repeating for everyone reading this book.

The ultimate goal of any safety program or process, whether it is incentive-based, performance-based, or a management-driven system, must be to create a system that embraces specific goals and objectives. We will discuss goals and objectives in more detail in Chapter 5.

A NEW MANAGEMENT SAFETY SYSTEM BEGINS TO EMERGE

As the change process continues, personal successes become more apparent. Incidents will decrease and attitudes will become more positive. This is where more employees begin to understand the change. Observable behaviors indicate a change in the norms. Employee participation in safety increases. Companies begin to move from a reactive to a proactive state, with increasing enthusiasm for safety. One important thing to remember is that management needs to be aware that newly developed safety management systems are fragile and can be damaged easily. Step out of line one time and you will have to prove yourself again [1]. I am reminded of Stephen Covey and his emotional bank [1]. Basically what he said is that if you keep taking emotions out of the bank and never put them back in, your bank account becomes unbalanced. This is like the checkbook with all of your money—if you do not reconcile your checking account, you could be losing money. Constant attention is needed to make sure that the right values (not priorities) are being promoted and demonstrated. As this new way of thinking and acting about safety takes hold, stronger roots form. If nurtured with positive emotions, the safety management system expands, matures, and will outlive many changes, good and bad. All employees and the organization start to become a fixed asset of the organization. The system will begin to sustain itself.

SAFETY CLIMATE (CULTURE) DEFINED

Although climate is difficult to define, it is easy to see and feel. According to Petersen, "Probably the best definition I've ever heard on culture came from a worker I interviewed who stated, 'Culture is the way it is around here.' It's the unwritten rules of the ballgame that the organization is playing. Culture is what everybody knows, and therefore it does not have to be stated or written down" [3, 4].

Safety climate reflects if safety is perceived by all employees to be a "key value" in the organization. The terms "climate" and "culture" are both used here [3, 4]. The question is: has a safety climate (culture) been created that is conducive to adopting safe work attitudes and habits [3, 4]?

The concept of culture became a very popular management subject in the early 1980s, because of the popularity of a book written by Peters and Waterman, *In Search of Excellence*. That book described what it was that accounted for the economic success of a number of companies. Other books followed, delving into the concept [3, 4].

The concept of culture had been around long before *In Search of Excellence* was published. Dr. Rensis Likert wrote a book called *The Human Organization*, where he described his research on "trying to" understand the difference in "styles" of different companies, and how these "styles" affected the bottom line. Dr. Likert coined the term "Organizational Climate." We now call it culture [3, 4].

According to Petersen, Likert believed that participative management was the best kind, and the most likely to produce results. Some of his contemporary management thinkers criticized him for flatly assuming that group discussion was the only path to good decision-making and thereby abandoning or ignoring the search for better techniques of problem solving or decision-making [2].

Likert not only researched climate; he also defined it as being excellent in ten areas:

- Confidence and trust
- Interest in the subordinate's future

- Understanding of and the desire to help overcome problems
- Training and helping the subordinate to perform better
- Teaching subordinates how to solve problems rather than giving the answers
- Giving support by making available the required physical resources
- Communicating information that the subordinate must know to do the job as well as information needed to identify more with the operation
- Seeking out and attempting to use ideas and opinions
- Approachability
- Crediting and recognizing accomplishments [3, 4]

Likert invented a way to measure climate with a forced choice questionnaire that he administered to employees to find out their perception of how good the company is in the ten areas. He later took the perception survey results and ran correlation studies with things like profitability, return on investment, growth, and other bottom-line figures, invariably coming up with extremely high positive correlation. Apparently, climate determines results [3, 4]. We will discuss employee safety perception and how it will affect the operation in more detail in Chapter 17.

Likert's research method was based on detailed questionnaires, asking employees a series of questions about their supervisors. He then drew up a profile of each supervisor/manager in light of how they were viewed by the employees [2].

- *Exploitative authoritarian.* Management is by fear and coercion, where communication is top-down, decision-making is done at the top with no shared communication, and management and employees are psychologically far apart.
- *Benevolent authoritarian.* Management is by carrot rather than stick, but employees are still subservient; such information that flows upward is mainly what the manager is thought to want to hear, and policy decisions are taken at the top, with only minor ones delegated to a lower level.
- *Consultative.* Management uses both carrot and stick and does try to talk to employees; communication flows both ways, but is still somewhat limited upward; important decisions are still taken top down.
- *Participative.* Management provides economic rewards and is concerned to get employees involved in groups capable of making decisions; it sets challenging goals and works closely with employees to encourage high performance. Communication flows easily in both

directions and sideways to peers; management and employees are psychologically close. Decision-making is done through participative processes; work groups are integrated into the formal structure of the organization by creating a series of overlapping groups, each linked to the rest of the organization by a "linking pin," preferably a team leader or departmental manager, who will be a member of both group and management [2].

As culture (climate) became a popular management subject in the 1980s, top management began to look at their organizations and consider ways to "improve their culture." In many organizations, you would find new posters on the walls describing their culture. We know today that if the management of a company must write it down and make a poster of it, they are not describing their culture; they are describing what they want to see. What does it mean when the poster gets outdated, dirty, or worn out? Does this mean that the culture has suffered the same symptoms? How many cases of this do you see? The lesson here is that posters do not constitute a safety culture. Nobody needs the culture to be described to them. Everybody knows what it "is like around here" [3, 4].

AUDITS

Before we begin our journey, let discuss one more misunderstood concept: audits. How many "audits" have you performed? What does the word *audit* mean? Typically, it means that you have to find something wrong. In one author's opinion, this is the case with most auditors. We need to get over this attitude and stop looking for all the "bad" things that have happened. We encourage you to look at "audit" as a "conformance appraisal," which means looking at your management system to see how it conforms to your expectations. This will help you to keep a positive focus on conformance to practices and procedures of the management system and/or program requirements. We continually need to understand how to focus on the positive aspects of safety.

SUMMARY

One thing that we often forget that is as important as, if not more important than, employee behaviors is management behaviors. If management is trying to create or improve the safety management system, they need to make sure that they demonstrate the same behaviors they expect from their employees. Employees are watching management carefully during any process change. If management's personal behavior is not consistent with the verbal and written messages they are sending, then the process will not work, the safety culture will not be trustworthy, and the management system will fail. We will continue to discuss this subject in Chapter 4.

The management system and the safety program should be evaluated to make sure that it is effective and appropriate to specific workplace conditions. The management system must be revised in a timely manner to identify and help correct infractions during a management system evaluation. It is a living process—you must continue to feed the system.

One of the issues the authors struggle with is focusing on OSHA Incident Rates (OIR). In many cases, management has a tendency to focus on how well they are doing by using numbers to measure safety program success. We do not have a problem with presenting these statistics to management, but one must remember that people get hurt, not numbers. You must learn to focus on individuals (employees) as opposed to how the numbers line up. This chapter has offered various proven methods to help you to decide on an appropriate performance management system that works for your business—the value.

As top management, your visible commitment to safety can make a major difference in the quality of your employees' work and personal life. You can choose among a variety of formal and informal methods and styles for achieving this impact. Demonstrate to everyone that you are vitally interested in employee safety. Do this by making yourself accessible: encourage your employees to speak up about safety, listen carefully, and then follow through. Set a good example: follow the rules, make time to carry out your safety responsibilities, and insist that all managers and supervisors do the same. Make sure everyone understands that you are in charge of a business where safety will not be compromised and where hazard awareness and safe work practices are expected of everyone, including on-site contractors and their employees.

REFERENCES

1. Covey, Stephen R., *The 7 Habits of Highly Effective People*, Fireside, Simon and Schuster, New York, 1989.

2. Kennedy, Carol, *Instant Management: The Best Ideas From the People Who Have Made a Difference in How We Manage*, William Morrow and Company, Inc., New York, 1991.

3. Peterson, Dan, *The Challenge of Change, Creating a New Safety Culture, Implementation Guide*, CoreMedia, Development, Inc., 1993, Preface. Modified with permission.

4. Peterson, Dan, *The Challenge of Change, Creating a New Safety Culture, Implementation Guide*, CoreMedia, Development, Inc., 1993, Implementation Guide, Safety Climate, Category 19, pp. 90–92. Modified with permission.

5. Roughton, Jim, Marcia West, "How to Get the Most out of Safety Incentives," *Industrial Safety & Hygiene News*, June 1999.

6. "Recordkeeping Citation could Spark Ban on Safety Incentive Games," *Inside OSHA*, Volume 5, No. 21, October 19, 1998.

7. "Safety Incentive Program Claims Not Supported by Evidence, OSHA Official Says," *BNAC Safety Communicator*, Winter 1999.

8. Secretan, Lance H. K., *Reclaiming Higher Ground*, McGraw-Hill, New York, 1997.

9. U.S. Department of Labor Occupational Safety and Health Administration (OSHA), Draft Proposed Safety And Health Program Rule, 29 CFR 1900.1, Docket No. Safety and Health-0027, public domain.

10. *The American Heritage College Dictionary*, 3rd ed., Houghton Mifflin Company, 1993.

3

Voluntary Guidelines for Safety Management

INTRODUCTION

As you continue through this book, you will begin to understand why there has to be management commitment. This commitment must be strong so that a successful management system can be developed, continually improved, and be sustained. Sustainability of your process is what you want to achieve. The methods in this book will help you to get started on the right track to developing this successful management system. In addition, the OSHA voluntary program is another valuable resource. The voluntary program will provide you with a specific perspective that is suitable for a successful management system.

PROGRAM ELEMENTS

You should note that there are consistent elements associated with each OSHA Voluntary Safety and Health Program. We will discuss these elements in more detail throughout the book [2].

In the voluntary programs, OSHA outlines five elements that will help you to create a successful management system. For simplification, the authors have broken these elements into six sections. Each element will be discussed in detail throughout the book. Although management and employee participation is complementary and forms the core of an effective safety and health program, we want to make sure that you understand that there is still a clear and distinct difference between management of the operation and employee participation. As you begin to review the OSHA Model Guidelines, you will find these elements in some form or another. It will be easier for you to implement your management system if you understand what OSHA is considering a model system. The following are the Management Program Core Elements that are a common theme in all voluntary safety and health programs:

- Management commitment leadership: refer to Chapter 6
- Employee participation: refer to Chapter 7
- Hazard identification and assessment: refer to Chapter 10
- Hazard prevention and control: refer to Chapter 11
- Information and training: refer to Chapter 14
- Evaluation of program effectiveness: refer to Chapter 20

We have chosen to use a combination of each voluntary program as discussed, because they are all similar. The bottom line is that an effective system will meet the following criteria no matter if it is OSHA mandated, a best management practice (BMP), or above and beyond compliance (ABC). They should look similar in content.

OSHA has concluded that effective management of employee safety and health is a decisive factor in reducing the extent and severity of work-related injuries and illnesses. An effective management program addresses work-related hazards, including those potential hazards that could result from a change in workplace conditions or practices. In addition, it addresses hazards that are not regulatory driven [2].

OSHA encourages employers to implement and maintain a program that provides systematic policies, procedures, and practices that are adequate to protect employees from safety hazards. In other words, an effective system identifies provisions for the systematic identification, evaluation, and prevention or control of workplace hazards, specific job hazards, and potential hazards that may arise from foreseeable conditions. Compliance with OSHA standards is an important objective. However, if you develop a successful management system, this becomes a non-issue.

Whether a safety program is in writing or not is less important than how effectively it is implemented, managed, and practiced. It should be obvious that as the size of the workplace, the number of employees, or the complexity of an operation increases, the need for written guidance will increase. The program should help to make sure that there is clear communication to all employees with consistent application of policies and procedures.

Aspects of the draft safety and health program were, as mentioned, based on best practices observed at various high performing safety and health organizations, including many voluntary protection program

(VPP) sites [1]. Several practices of organizations with an effective safety culture indicate that at these sites some BMP appear as a common denominator. In general, the following sections include some of these common best management practices. Best practices will be listed throughout the contents of this book.

MANAGEMENT COMMITMENT LEADERSHIP

Management commitment leadership from the top down is the most important part of any process. "Lip service" is not going to work for you. If management demonstrates commitment and provides the motivating force and the needed resources to manage safety, an effective system can be developed and will be sustained. According to OSHA, this demonstration of leadership should include the following elements that are consistent with an effective program:

- Establishing the program responsibilities of managers, supervisors, and employees for safety and holding them accountable for carrying out these responsibilities.
- Providing managers, supervisors, and employees with the authority and access to relevant information, training, and resources they need to carry out their safety responsibilities.
- Identifying at least one manager, supervisor, or employee to receive and respond to reports about safety conditions and, where appropriate, to initiate corrective action [2].

This is the first time that OSHA has used the term "demonstrate." In reality, demonstration means "do as I do" [3]. This is an important concept no matter what you are tying to accomplish—always "walk the walk, and talk the talk."

EMPLOYEE PARTICIPATION

In any successful system, employees should be provided an opportunity to participate in establishing, implementing, and evaluating the safety program. Employee participation allows employees to develop and/or express their safety commitment to themselves and/or their fellow workers. To fulfill and enhance employee participation, management should implement some form of the following elements:

- Regularly communicating with all employees concerning safety matters
- Providing employees with access to information relevant to the safety system
- Providing ways for employees to become involved in hazard identification and assessment, prioritizing hazards, safety training, and management system evaluation
- Establishing procedures where employees can report work-related incidents promptly and ways they can make recommendations about appropriate solutions to control the hazards identified
- Providing prompt responses to reports and recommendations

It is important to remember that under an effective management system employers do not discourage employees from reporting safety hazards and making recommendations about incidents or hazards, or from participating in the safety program.

HAZARD IDENTIFICATION AND ASSESSMENT

A practical hazard analysis of the work environment involves a variety of elements to identify existing hazards and conditions as well as operations subject to change that might create new hazards. Effective management coupled with employee participation and continually analyzing the work environment to anticipate and develop programs to help prevent harmful occurrences will help to identify hazards. The following measures are recommended to help identify existing and potential hazards:

- Conducting comprehensive baseline workplace assessments, updating assessments periodically, and allowing employees to participate in the assessments
- Analyzing planned and/or new facilities, process materials, and equipment
- Performing routine job hazards analyses
- Assessing risk factors of ergonomics applications to employees' tasks
- Conducting regular site safety and health inspections so that new or previously missed hazards are identified and corrected

- Providing a reliable system for employees to notify management about conditions that appear hazardous and to receive timely and appropriate responses and encourage employees to use the system without fear of reprisal; this system utilizes employee insight and experience in safety and allows employee concerns to be addressed
- Investigating incidents and "near miss" incidents so that their causes and means of prevention can be identified
- Analyzing incident trends to identify patterns with common causes so that they can be reviewed and prevented [2]

This is a good method of defining what training is required. Hazards that employees are exposed to should systematically be identified and evaluated. This evaluation can be accomplished by assessing compliance with the following activities and reviewing safety information, for example:

- The establishment's incident experience
- OSHA logs
- Workers' compensation claims (Employer's First Report of Injury)
- Nurses' and/or first aid logs
- Results of any medical screening/surveillance
- Employee safety complaints and reports
- Environmental and biological exposure data
- Information from prior workplace safety inspections
- Material Safety Data Sheets (MSDSs)
- Results of employee safety perception surveys
- Safety manuals
- Safety warnings provided by equipment manufacturers and chemical suppliers
- Information about safety provided by trade associations or professional safety organizations
- Results of prior incidents and investigations
- Evaluating new equipment, materials, and processes for hazards before they are introduced into the workplace
- Assessing the severity of identified hazards and ranking those that cannot be corrected immediately according to their severity [2]

It is also important to evaluate other OSHA requirements that may impose additional and specific requirements for hazard identification and assessment. The hazard identification and assessment analysis should be conducted as follows:

- As often as necessary to make sure that there is compliance with specific requirements, BMP or ABC

- When workplace conditions change that could create a new hazard or there is increased risk of hazards

These elements will help identify potential safety issues. Once these areas have been identified, you can make your case for training as it applies to the operation.

Each work-related incident with the potential to cause physical harm to employees should be investigated. Records of the hazards identified, the assessments, and the actions taken or plans to control those hazards should be documented. Some recommended methods for conducting assessments will be discussed in Chapter 15, on job hazard analysis (JHA).

HAZARD PREVENTION AND CONTROL

Effective planning and design of the workplace or job task can help to prevent hazards. Where it is not feasible to eliminate hazards, action planning can help to control unsafe conditions.

Elimination or control should be accomplished in a timely manner once a hazard or potential hazards are identified. The procedures should include measures such as the following:

- Using engineering techniques where feasible and appropriate
- Establishing safe work practices and procedures that could be understood and followed by all affected employees
- Providing PPE when engineering controls are not feasible
- Using administrative controls, for example, reducing the duration of exposure
- Maintaining the facility and equipment to prevent equipment breakdowns
- Planning and preparing for emergencies, and conducting training and emergency drills, as needed, to make sure that proper responses to emergencies will be "second nature" for all persons involved
- Establishing a medical surveillance program that includes handling first aid cases on-site and off-site at a nearby physician and/or emergency medical care to help reduce the risk of any incident that may occur [2]

Once hazards are identified, an action plan should be developed to help resolve the issues that can be used to come into compliance with applicable requirements. These plans can include setting priorities and deadlines and tracking progress in controlling hazards. Action planning will be discussed in Chapter 5.

INFORMATION AND TRAINING

Safety training is an essential component of an effective safety program. This training should address the roles and responsibilities of both the management and the employees. It will be most effective when combined with other training about performance requirements and/or job practices. The complexity depends on the size and the nature of the hazards and potential hazards present.

Training is an important part of any program to make sure that all employees understand the requirements of the safety programs and potential hazards of the operation. The following section provides a brief explanation for specific areas of training. You should review your operation and expand on the brief summary. For a detailed explanation on safety training, refer to Chapter 14.

Employee Training

Employee training programs should be designed to make sure that all employees understand and are aware of the hazards that they may be exposed to and the proper methods for avoiding such hazards.

The following information and training should be provided to all employees:

- The nature of the hazards and how to recognize them
- The means to control these hazards
- What protective measures can be used to prevent and/or minimize exposure to hazards
- The provisions of applicable requirements

Anyone who has responsibilities for the information and training should be provided the level of training necessary to carry out their safety responsibilities.

Management Training

Management must be trained to understand the key role and responsibilities they play in safety and to enable them to carry out their safety responsibilities effectively. Training programs for management should include the following topics:

- Analyzing the work under their supervision to anticipate and identify potential hazards
- Maintaining physical protection in their work areas
- Reinforcing employee training on the nature of potential hazards associated with their work and on protective measures; the reinforcement is done through continual performance feedback and, as necessary, through enforcement of safe work practices
- Understanding their safety responsibilities

Note that some OSHA standards impose additional, more specific requirements for information, training, and education. Make sure that you review specific training requirements.

Evaluation of Program Effectiveness

The management system should be evaluated to make sure that it is effective and appropriate to specific workplace conditions. The system should be revised in a timely manner to correct any deficiencies as identified by any program evaluation. We will discuss the methods of evaluations in more detail in Chapter 19.

MULTI-EMPLOYER WORKPLACE

In a multi-employer workplace, the primary responsibility for the host employer is to:

- Provide information about hazards, controls, safety rules, and emergency procedures to all employers
- Make sure that safety responsibilities are assigned to specific employees as applicable

The responsibility of the on-site contractor is to:

- Make sure that the host employer is aware of the hazards associated with the contractor's work and what the contractor is doing to address any identified hazards
- Advise the host employer of any previously unidentified hazards that the contract employer identifies at the workplace

EMPLOYEE RIGHTS

Employees have the right to complain to their employers, their unions, OSHA, or other governmental agency about workplace safety hazards. Section 11(c) of the OSHA Act of 1970 makes it illegal for employees to be discriminated against for exercising their rights and for participating in other job safety-related activities. These activities include, for example:

- Complaining individually or with others directly to management concerning job safety conditions
- Filing of formal complaints with government agencies, such as OSHA or state safety and health agencies, fire departments, etc. (an employee's name is kept confidential)
- Participating in union committees or other workplace committees concerning safety and/or health matters
- Testifying before any panel, agency, or court of law concerning job hazards
- Participating in walk-around inspections
- Filing complaints under Section 11(c) and giving evidence in connection with these complaints [2]

Employees also cannot be punished for refusing a work assignment if they have a reasonable belief that it would put them in danger or cause serious physical harm, provided that they have requested the employer to remove the danger and the employer has refused; and provided that the danger cannot be eliminated quickly enough through normal OSHA enforcement procedures.

If an employee is punished or discriminated against in any way for exercising his or her rights under the OSHA Act, the employee should report it to OSHA within 30 days. OSHA will investigate the report. If the employee has been illegally punished, OSHA will seek appropriate relief for the employee and if necessary will go to court to protect the rights of the employee.

VOLUNTARY PROTECTION PROGRAMS

One of the best methods of complying with OSHA and building a successful management system is to become an OSHA Voluntary Protection Program (VPP) member. Table 3-1 summarizes what it takes to certify

Table 3-1
Voluntary Protection Program Implement Process

Pre-application Stage

1. VPP Education: Management and Employees
2. Communicate VPP to All Employees
   - Management's roles and responsibilities
   - Employees' roles, responsibilities, and rights
   - Union's roles, responsibilities, and rights
3. VPP Mock Evaluation/Assessment
   - Program deficiencies identified and corrected
   - Develop action for correcting
     Major deficiencies: 6 to 12 months
     Minor deficiencies: up to 6 months
     Basic deficiencies: up to 3 months

Application Stage

4. Correct Program Deficiencies
   - Develop action for correcting
     Major deficiencies: 6 to 12 months
     Minor deficiencies: up to 6 months
     Basic deficiencies: up to 3 months
5. Prepare Application
   - Applications to read like a "resume"
   - Every item in application to have documented proof/verification
6. Submit Application
   - Three-ring binder with tabs by subject
   - Make 5 copies
   - Send certified mail

Post-Application Stage

7. Contact OSHA Office
   - Determine evaluation date
   - Determine team
   - Determine who receives application
8. Review Program
   - Implement revisions
   - Document revisions
9. Prepare for VPP Evaluation
   - Provide conference room
   - Provide computer
   - Organize documents

Table 3-1
Continued

- Prepare list of all employees for interviews
- Provide VPP team members' escorts for plant walk-around

Evaluation Stage

10. Introduction Meeting
    - Introduce OSHA VPP team
    - Let OSHA VP team discuss the logistics of this evaluation
    - Present overview of organization
      Management structure
      Safety and Health Program
      Safety team members
    - Request daily briefing and draft pre-approval report
11. Evaluation
    - Provide all documents in conference room
    - Provide escorts
    - Provide lunch
12. Closing Conference
    - CELEBRATE that it is over
13. Post-evaluation Stage
    - Merit (with plan) approval
      Develop plan of action (POA)
      Monitor and review progress
      Document POA and activities
      Submit POA proof
      Provide for VPP evaluation
14. Star Approval
    - Monitor safety and health program quarterly
    - Conduct annual VPP safety and health program review and revise as needed
    - Prepare for VPP evaluation

Refer to Figure 3-1 for a flow diagram of the VPP process. Note that each number on Figure 3-1 corresponds to each item as listed in Table 3-1 to make it easy to read.

under the VPP program; Figure 3-1 shows a flow diagram. This tool will summarize the information as listed in Table 3-1 and help you to apply for and successfully obtain VPP status.

SUMMARY

The following is a line-by-line summary of what we discussed in this chapter. We will continue to discuss these items in a later chapter. By following the elements presented, you will be able to establish a successful safety management system:

- Provide visible top management, commitment, leadership, and involvement in implementing and sustaining the management system so that all employees understand that management's commitment is serious.
- Arrange for and encourage employee participation in the structure and operation of the management system and in decisions that affect the employee's safety. This will help to commit their insight and energy to achieving the safety program goals and objectives.
- Clearly state a policy and/or vision statement on safety expectations so that all employees can understand the value of safety activities and programs.
- Establish and communicate a clear goal for the safety program and define objectives for meeting the established goals so that all employees understand the desired results and measures planned for achieving them.
- Assign accountability and communicate responsibility for all aspects of the program so that all managers, supervisors, and employees know what performance is expected.
- Hold all managers, supervisors, and employees accountable for meeting their responsibilities so those essential tasks will be performed.
- Provide adequate authority and resources to responsible parties so that assigned responsibilities can be met.
- Review management system elements at least annually to evaluate their success in meeting the goals and objectives so that deficiencies can be identified and the program and/or the objectives can be revised when they do not meet the goal of an effective safety process.

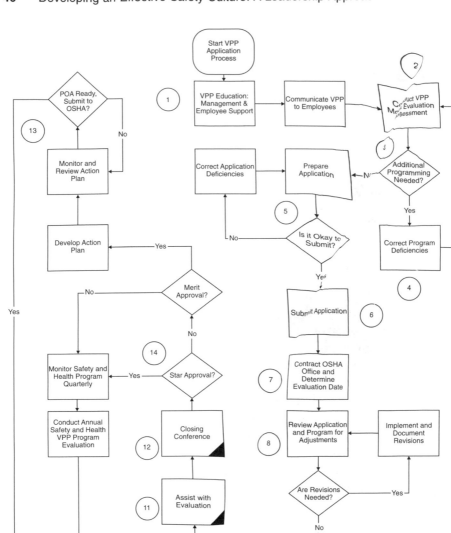

Figure 3-1 VPP process flow diagram.

Although compliance with specific OSHA requirements is an important objective, an effective management system looks beyond specific requirement and targets the development of an effective safety culture, as we described.

As we proceed through this book, you will understand the importance of effective management foundations and their impact on achieving an effective safety culture.

REFERENCES

1. OSHA Fact Sheet, January 1, 1991, Voluntary Safety and Health Program Management.

2. OSHA's Voluntary Safety and Health Program Management Guidelines published in the *Federal Register*, pp. 54 FR 3904–3916, Thursday, January 26, 1989.

3. U.S. Department of Labor Occupational Safety and Health Administration (OSHA), Draft Proposed Safety and Health Program Rule, 29 CFR 1900.1.

Part 2

Management Aspects of an Effective Safety Culture

4

Management's Role in Developing an Effective Safety Culture

INTRODUCTION

Too often safety books talk only about the technical aspects of a safety program. Certainly, some books include innovative or proactive techniques of incident prevention, while others focus on more basic, fundamental incident prevention techniques. One of the basic tenets of developing and ultimately sustaining a safety culture argues that an effective safety culture rests squarely on the shoulders of the management team, regardless of the structure. Refer to Table 4-1 for four characteristics of a safety culture. Management, starting from the individuals with the most authority, all the way through to front-line management (supervisors, superintendents, leaders, etc.), all the way to the employee must be engaged in the effort to systematically reduce and/or eliminate the exposure to hazardous situations. These situations encompass both exposures to physical hazards and work practices that put the employee at a greater risk of an injury.

First, let's review a list of principal management leadership traits we have seen effectively utilized throughout our careers in various industries and work environments. For practical purposes, we will focus only on a few of the more apparently critical attributes we have observed, which have helped to maintain a keen focus on the impact of leadership and the safety culture's management system. The list includes, but is not limited to, the following:

- Demonstrating management, leadership, and commitment to the safety program (process)

Table 4-1
Four Characteristics of a Total Safety Culture

All employees hold safety as a value.

- Each employee feels responsible for the safety of their co-workers as well as themselves
- Each employee is willing and able to "go beyond the call of duty" on behalf of the safety of others
- Each employee routinely performs actively caring and/or safety behaviors for the benefit of others

Adapted from French, Anne R., "Achieving a Total Safety Culture: Integrating Behavioral Safety into the Construction Environment," Safety Performance Solutions, Presentation June 30, 1999, AON Construction Division, Atlanta, reproduced with permission.

- Charting the course, sometimes known as creating the vision for your management system
- Defining the roles and responsibilities for all levels of management and employees in the management system
- Making sure that individuals are held accountable for their roles and responsibilities
- Creating a climate (culture) that actively fosters meaningful employee participation in the entire safety program
- Encouraging employee participation at all levels of the organization

Some readers might well take exception to these attributes, but from our perspective, if these items are performed consistently well, organizations will have a much greater opportunity for success when developing an effective safety culture and building a management system. Figure 4-1 provides a graphical view of an effective leadership model. Note that management leadership is similar to a magnet that aligns the drive force for developing a safety culture. If the management system is aligned, part of the system will not work. Therefore, the culture will fail.

As we continue, we need to examine the barriers that prevent a culture from occurring. Refer to Table 4-2 for a summary of the management attitudes and behaviors that affect a safety culture.

CHARTING THE COURSE/CREATING THE VISION

Ask yourself these questions: What would happen if an airline pilot took off from an airport without a written flight plan? How about a truck

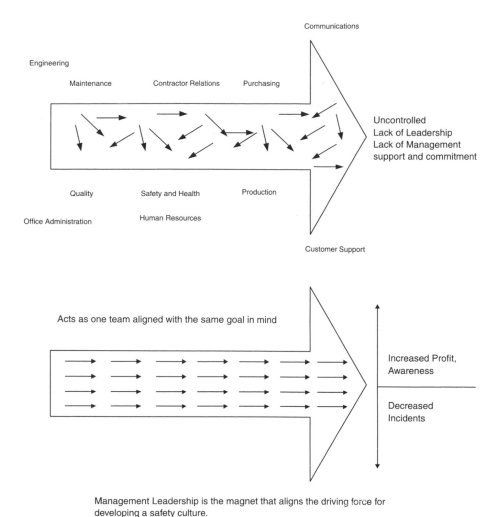

Communications

Engineering

Maintenance   Contractor Relations   Purchasing

Uncontrolled
Lack of Leadership
Lack of Management
support and commitment

Quality   Safety and Health   Production

Office Administration   Human Resources

Customer Support

Acts as one team aligned with the same goal in mind

Increased Profit,
Awareness

Decreased
Incidents

Management Leadership is the magnet that aligns the driving force for
developing a safety culture.

Figure 4-1 Leadership model.

driver who leaves for a destination without a delivery or pickup schedule? A businessperson going on a trip without any itinerary? Of course, rarely do you see or hear of such things; in every case, these individuals will have a specific plan for their destination. That is exactly what business leaders (the management system) must provide to all employees to develop a successful safety program. Top management must chart the course they wish to travel. As discussed earlier, "begin with the end in mind" [1]. Doing so allows one to maximize the use of resources while developing and implementing a new process. So, decide what your

Table 4-2
Management Attitude and Behavior Barriers

- *Inconsistency*. This week, the supervisor is intent on production at any cost, but at last week's safety meeting, she stated that safety was the company's top priority. Note: There is no mention of safety as a value.
- *Obsolete rules and procedures*. If rules are unclear, outdated, or inappropriate, employees tend to ignore them, finding ways around the requirements.
- *"Us" versus "them" thinking*. Unresolved past conflicts, combined with current stresses, are a powerful force that can encourage noncompliance.
- *Leadership deficiency*. When supervisors and other leaders fail to follow a safety protocol, they lose credibility as a role model.

Topf, Michael D., *The SAFOR Report*, A Publication from the Topf Organization, p. 2, Fall 1998. Reprinted with permission.

"vision" will be for your safety system. The vision should decide on the structure of the program—for example:

- Will we focus solely on regulatory compliance as the model for our program?
- Will we add activities that go above and beyond compliance (ABC) as part of our program?
- Do we want our program to become registered to an international safety and health standard, such as DNV's Five-Star Safety Program System or other comparable standards such as ISO 9001 or ISO 14001?
- Long term, do we want to partner with OSHA and become an approved VPP site?

By defining your destination for your safety program, you can begin a more efficient journey toward accomplishing your vision. Refer to Figure 5-1 in the next chapter for an overview representing the journey to safety excellence. It helps to define the course in generic terms. This figure describes the difference between a traditional-based safety program and a management-driven program. It highlights the difference between the two concepts and what you can achieve using each process. Note that you have several choices, either to work in the compliance state being a traffic cop or to develop your management system to achieve a sustained culture.

In addition, one thing that you must take into consideration as you develop your process is to make sure that all of the programs developed are integrated in the management process (Table 4-3).

Table 4-3
Management System: Safety and Health Integration

WHAT IS A SAFETY AND HEALTH PROGRAM, AND WHERE DOES IT FIT INTO MY MANAGEMENT SYSTEM?

If management wants to reduce incidents (injuries, illnesses, loss-producing events, and/or other related costs), everyone must place as much emphasis on safety issues as they place on other core management issues, such as production, sales, and quality control. To be most effective, safety must be balanced with and integrated in other core business processes.

"Safety First" may sound good, but in reality, safety should not be considered another program or "Flavor of the Month." It must become a basic value of your company. For example, change "Safety First" to "Safe production is our only standard." This emphasizes the idea that it's fine to produce as hard and as fast as possible, as long as it can be done safely. Refer to Figure 4-2 for a generic safety management model. This model sets the stage for our discussion on safety and health integration. You will note that this process starts in the middle of the circle and works outward integrating the entire management system.

To get an idea of where safety can fit into your organization, answer the questions below.

- Is safety an integral part of your operations?
- Is teamwork apparent in all parts of the organization?
- Are managers and supervisors on the production floor frequently, and do they always observe all company safety rules?
- Are employees encouraged to identify safety hazards and correct them on their own?
- Do employees have full and open access to all of the tools and equipment they need to do their job safely?

If you were able to answer "yes" to each of these questions, then you are on your way to developing a successful safety culture. If you want to do better, you should work on improving the safety culture in your organization.

OSHA Web site, Management System: Safety & Health Integration, http://www.osha-slc.gov/SLTC/safetyhealth_ecat/mod2.htm, public domain.

DEFINING ROLES AND RESPONSIBILITIES

Once you have decided on the type of journey you want to begin, your next major task is to outline the roles and responsibilities for each

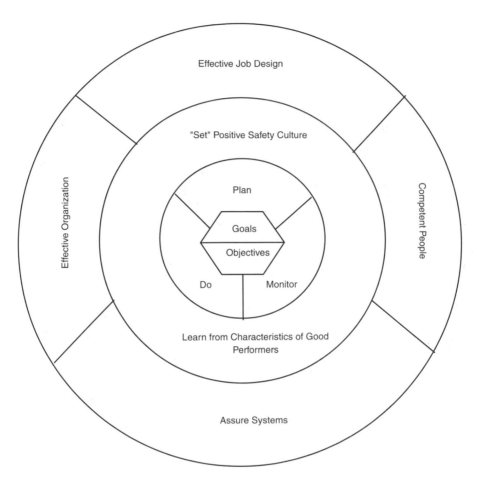

Figure 4-2 Generic safety management model. Cox, Sue, Tom Cox, *Safety Systems and People*, Figure 9-4, p. 220, Butterworth–Heinemann, Oxford. Reproduced with permission.

member of management with regard to your safety program. These roles and responsibilities will define the types of leadership skills that are necessary to support the management system. Remember our discussion on management styles in Chapter 2? For example, the role of top management may be to hold direct reports (such as managers and supervisors) accountable for their actions toward their safety objectives, provide the necessary resources time and financial support, and demonstrate leadership by actively participating in specific safety functions.

In addition, each management position, as well as all employees, should have some level of specific "responsibilities" clearly defined. Managers and supervisors must have specific safety activities that they are required to perform on a routine basis. Refer to Chapter 6 for a discussion of an activity-based safety management system. The maturity of your safety culture will determine how these activities are developing and the quality of the results demonstrated by top management and supervision. Employees should also have specific responsibilities clearly defined and well communicated—for example, complete a specific safety training program as part of a job-bid transfer [5].

For an excellent resource on this issue, refer to Dan Petersen's book *Safety by Objectives*, McGraw-Hill, 1996 [3]. We will continue to refer to Dan Petersen's concepts throughout this book.

OBSTACLES TO SUCCESSFUL SAFETY PROGRAM DEVELOPMENT

Tables 4-4 and 4-5 show obstacles that were identified by supervisors and employees, respectively, during OSHA-sponsored workshops. In the left column are identified obstacles to a successful safety program. In the right column are recommended ways to overcome these obstacles.

DEFINING ACCOUNTABILITY

Too often, a well-designed, creative safety plan, once developed and implemented, does not create the culture necessary for the accomplishment of the company vision. Why? Often the organizational culture did not include a tight (specific) accountability system, holding managers and supervisors accountable for the completion of assigned safety responsibilities. Simply stated: What gets measured, gets done. In Petersen's *Safety by Objective* book, he added one caveat: "What gets measured, and rewarded, gets done" [3]. Managers and supervisors must be positively recognized for effectively completing assigned safety responsibilities, while substandard performance should be dealt with accordingly.

These responsibilities should be treated like any other type of performance expectation and not handled as a separate function. Separating these activities from other "routine" production and related activities will lessen the value that management places on the completion of specific

Table 4-4
Supervisor-Identified Obstacles

To make a positive change in our organization, I must be willing to help eliminate the following obstacles	Therefore
Fear of losing my job	Trust in the system, do the right things, and maintain integrity
No money for needed changes	Management must support
Risk in spending money for safety	I must trust and support management
"What's in it for me?" attitude	I must take personal responsibility
Many employees want change but are afraid to take responsibility for it	I must stop worrying, trust, and risk
No support from upper management	Management must support
No time or follow-through from top management	Management must support and provide time and resources
Make it work or "can do attitude"	I must take personal responsibility and operate in the stated guidelines
Competing priorities—production is number one	Management must balance with a new value
Work orders not completed even when signed off	Management must provide resources and hold employees accountable
Overwhelmed with workload	Management must provide resources and balance competing pressures
Turnover too high	Trust that it will decrease as culture and work atmosphere improve
Double standards	Everyone must play by the same rules
Lack of trust, poor ethics in the organization	I must take personal responsibility, stop worrying, trust, and risk
Lack of open communication and listening	I must be vulnerable and trusting

OSHA Web site, Supervisor Identified Obstacles, http://www.osha-slc.gov/SLTC/safetyhealth_ecat/mod4_obstacles.htm, public domain.

Table 4-5
Employee-Identified Obstacles

To make a positive change in our organization I must be willing to help eliminate the following obstacles	Therefore
Fear and lack of trust	Trust in the system and do the right things and maintain integrity
Supervisor not willing to listen and support	Supervisors must support and be open while I must stop worrying, trust, and risk
Communication is one way (top-down)	Supervisors must risk being vulnerable and open up while I must not wait to take personal responsibility for my actions
Organization is not alignment on safety; production is number one at the line level	Management must support and provide time and resources
Supervisors not willing to hear problems and receive feedback	Supervisors must risk being vulnerable and open up, while I must not wait to take personal responsibility for my actions
Intimidation tactics	If pressures are balanced and I stop worrying, trust, and risk, my supervisor will respond
People not willing to take personal responsibility; too easy to shift blame	I must take personal responsibility and operate in guidelines, and hold my supervisor and peers accountable
Production is number one	Management must support and provide time and resources
Lack of consistency and follow-through, past efforts fade away	Management must realize and commit that this is a long-term effort. Safety excellence is a never-ending journey. I must always be willing to examine myself, receive feedback, and be willing to improve through change
"Them versus us" attitude, win–lose	Take responsibility for myself, operate from win–win [1]

OSHA Web site, Employee Identified Obstacles, http://www.osha-slc.gov/SLTC/safetyhealth_ecat/mod4_obstacles.htm, public domain.

activities. A few suggested methods to assist with accountability include the following:

- Make sure that each member of management has specific safety objectives (activities) outlined in his or her annual performance appraisal (accountability contract). It is important that you do not use the OSHA incident rates (OIR) or any number measurement system as a measurement tool. Using the OIR is counterproductive. It will force management to work to a number without regard to the specific activities. Use only specific activities that one can accomplish.
- Identify mandatory and optional activities that managers and supervisors can select to complete their list of safety objectives. Align any part of a merit compensation to the successful completion of these selected safety activities. For example, you can identify a specific number of activities. Managers and supervisors can pick and choose the ones that they are comfortable with. As time goes by and the system is working as intended, activities can be changed to meet the need. For example, if one manager/supervisor does nine activities well and fails at one activity, maybe you need to review this with the individuals and help select another activity that will work. In this way you will start to achieve success if everyone does ten activities well.
- Develop a measurement system that assists with the tracking of related activities on a periodic basis.
- Periodically review the manager's performance against the stated objectives, recognize satisfactory performance, and intervene in substandard performance [3].

Of course, the preceding list should look familiar, as performance measurement includes the basic management axiom of "Plan, Do, Check, Monitor." As we discussed in Figure 4-2, this model provides a variation of the basic management axiom as described above, but outlines all of the elements that must be controlled in a management system. In addition, we will discuss this in more detail in Chapter 8.

SUMMARY

It has become clear that basic faults in an organizational structure, climate, and procedures may predispose an organization to an incident. This background environment is being increasingly described in terms of safety culture, where culture comprises the attitudes, beliefs, and behaviors that are generally shared in an organization [4].

Various definitions have been used for safety culture. It is important to understand that safety culture is a subset of an overall culture of an organization. It follows that aspects of management that have not traditionally been seen as part of safety influence the safety performance of organizations [4].

The safety culture of an organization is the product of the individual and group values, attitudes, competencies, and patterns of behavior that describes the commitment to, and the style and proficiency of, an organization's safety programs [2].

Organizations with a positive safety culture are characterized by communications founded on mutual trust, by shared perceptions of the importance of safety, and by confidence in the efficiency of preventative measures [4].

A positive safety culture implies that the whole is more than the sum of the parts. The different aspects interact to give added effect in a collective commitment. In a negative safety culture the opposite is the case, with the commitment of some individuals strangled by the cynicism of others. There are some specific factors that appear to be characteristics of organizations with a positive safety culture:

- The importance of leadership and commitment of top management
- The safety role of line management
- The involvement of all employees
- Effective communications and commonly understood and agreed goals
- Good organizations learning and responsiveness to change
- Manifest attention to workplace safety and health
- A questioning attitude and a rigorous and prudent approach by all individuals [4]

Improving safety culture is something that must be seen as a long-term and systematic process, based on initially assessing the existing safety culture, determining priorities for change and the actions necessary to effect the change, and then going on to review progress before repeating the process indefinitely.

As the top manager, your visible commitment to safety can make a major difference in the quality of employee participation. You can choose among a variety of formal and informal methods and other styles for achieving this impact. In any type of organization, it is important to demonstrate to all employees that you are interested in safety. Do this by making yourself accessible, encouraging your employees to speak up about safety concerns, listening to them carefully, and then following up on recommendations. Set a good example: follow the rules, make time to carry out your safety responsibilities, and insist that your managers and supervisors do the same. Make sure that all employees understand that you are in charge of a business where safety will not be compromised and where hazard awareness and safe work practices are expected of everyone, including on-site contractors and their workers [4].

Setting a good example is one of the most important ways management can demonstrate and get visibly involved in safety.

Some workplaces may have rules that apply only to employees who will be in the area for several hours or who will be working with specific equipment. Make sure you know all the rules that employees are expected to follow, and then make sure you and your managers and supervisors follow them as well, even if they are visiting for only a few minutes and will not be working directly with the equipment.

In addition, if you see any infraction of the rules or safe work practices, make sure that you correct it immediately. Your support on working safely will become a model for your managers, supervisors, and employees.

Finally, make it clear to all managers and employees that you are in charge of making sure that your workplace is safe. No matter what it is, taking charge of safety means holding all managers and supervisors accountable. In addition, it means insisting that any contract work at your site be done in a safe manner.

By holding managers, supervisors, and employees accountable, you encourage positive participation in the safety program. Managers and supervisors who are held accountable for safety are more likely to press for solutions to safety issues than to present barriers to resolution. They are more likely to suggest new ideas for hazard prevention and control. Your own participation is less likely to undermined or threaten their authority [4].

In providing a complete safety management system, you must provide all managers, supervisors, and employees the tools they need to work with you in keeping the workplace safe. A complete management system addresses the needs and responsibilities of all employees including management and supervision [1].

Make sure that your supervisors know that you understand that not every safety issue can be solved at their level. Call on your managers and supervisors to help make the employee input systems work. Think of your workplace as a team trying to identify and resolve safety infractions through whatever systems are necessary. Your managers and supervisors should be team leaders, working together with you and other employees toward a common goal. You may want to recognize the teams that report hazards or suggest new control ideas. Recognition can be based on the submission of reports and suggestions or on the quality of employee input. Let your managers and supervisors know that when an employee brings a safety matter to your attention, you consider that a good reflection of the supervisor's leadership.

Encourage employees to take advantage of opportunities to become involved in problem identification, problem solving, and reporting hazards. Then, when they do become involved, make sure they get appropriate and timely feedback, including recognition for efforts.

When your management systems are working well, most safety issues will be resolved before your employees feel the need to approach you directly. Big issues may arise that the normal systems cannot handle. Your supervisors probably will understand that these problems are not a reflection on them, and that you are the proper person to address these concerns [1].

If your accountability system is going to work, any individual who continues to present barriers to an effective safety management system will have to be held accountable. It is important to try to separate any accountability activity from your immediate response to employee-raised questions, concerns, or suggestions [1].

Remember, too, that your safety systems not only encourage employee participation in identifying hazards and resolving problems, but also protect those employees from retaliatory and discriminatory actions, including reprisal. In the next sections, we will expand on all of these concepts.

REFERENCES

1. Covey, Stephen R., *The 7 Habits of Highly Effective People: Powerful Lessons in Personal Change*, A Fireside Book, 1990.

2. Ouchi, William, *Theory Z: How American Business Can Meet the Japanese Challenge*, Avon Publishers, 1981.

3. Petersen, Dan, *Safety By Objectives*, August 1996, Preface.

4. Safety, INSAG Report 75-INSAG-4—International Atomic Energy Agency, Vienna, 1991.

5. U.S. Department of Labor, Office of Cooperative Programs, Occupational Safety and Health Administration (OSHA), Managing Worker Safety and Health, November 1994.

5

Journey to a Safety Culture: Determining the Direction of Your Management System

INTRODUCTION

As we discussed in Chapter 4, when you plan your vacation you and your family have an objective in mind: where you want to go, and what you want to do when you get to your destination. You have planned for a long time to make sure that you are going in the right direction. Planning and developing a safety program is no different. You must also make sure that your objective for establishing such a program is clear to everyone.

The first step is to decide how successful you want your program to be and what you want your program to accomplish. Remember, "Begin with the end in mind" [1]. The next step is to put your plan in writing. Once you have completed these steps, you can map out the path toward your established goals and objectives. This is the method used to determine the direction of your management system (safety process) when developing your policy, goals, and objectives.

Remember: Sometimes the destination is not always as important as the journey. During the journey you keep learning. It is important to make sure that your destination is to implement each step in such a way that the end result will be able to sustain itself. You have to make sure that all of your systems are working toward getting to your final destination. Sometimes there are breakdowns along the way, as when your car breaks down when you're on a vacation. You must get it repaired before starting the journey again. This also applies to your management system.

In this chapter we will help you begin your journey by providing some guidance on how to write and communicate your safety and health policy.

In addition, this chapter will help you understand how to establish and evaluate your goal and objectives.

Before you begin on your journey, you must decide on what you want to accomplish. Refer to Figure 5-1 to review several directions. As you can see under the traditional driven safety system, management is viewed as a "Safety Cop." In this type of system, the safety program is controlled by someone who conducts audits of their workplace and finds many violations. Month after month, the same safety violations are found. The problem is that nothing ever gets fixed. Employees only do things safely when they see management coming. Management's focus is only on injuries and/or compliance. In this system, one thing works—"quick fixes"—something that can be done fast; therefore, putting a bandage on the problem. The attitude is "Ignore it and it will go away." In this type of system, you will spend a majority of time working on getting the job done due to having to deal with injured employees. This is what the authors call the injury zone; "This is the way that I have always done it."

On the other hand, the other direction as depicted in Figure 5-1 is the Management Driven Safety System. In this system you are operating below the injury zone, where the attitude is "I want to do this." Management is willing to take chances and make changes that will help to minimize incidents. It is the only right thing to do. In this type of situation, injuries only happen when something in the management system fails. In this case, management accepts the responsibilities and resolves the issues. There is a high level of management commitment and leadership with employee participation.

These concepts will be discussed in more detail later in the book.

PATHWAY TO SAFETY EXCELLENCE

Now, let's take a look at an example pathway to safety excellence (Figure 5-2). As you can see, there are many pitfalls that you will encounter on your way to a successful safety culture. In the early phases, we begin to see the desire for change take place, with employee participation, with a limited safety structure in place. At the middle phases of your journey many organizations have dealt with the trust issue and become credible in the eyes of the employees as specific safety activities have been formulated, acted upon, and maintained, thus minimizing the "program of the month" perception. In the final stage, safety is a value demonstrated by management. As decisions are made, employees and

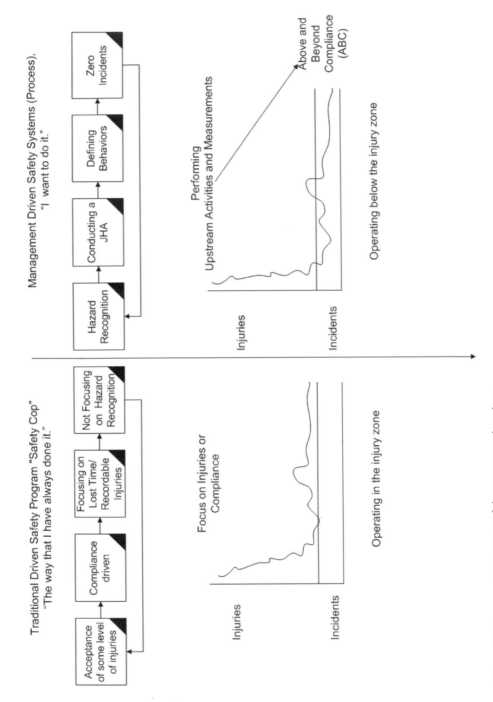

Figure 5-1 Journey to a successful management system.

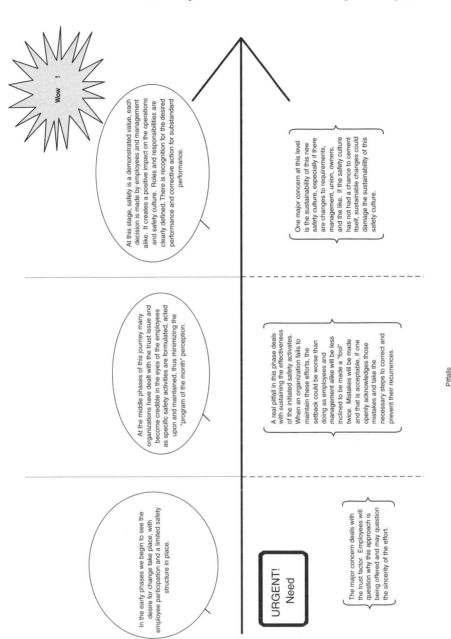

At this stage, safety is a demonstrated value, each decision is made by employees and management alike. It creates a positive impact on the operations and safety culture. Roles and responsibilities are clearly defined. There is recognition for the desired performance and corrective action for substandard performance.

Wow !

One major concern at this level is the sustainability of this new safety culture, especially if there are changes to requirements, management, union, owners, and the like. If the safety culture has not had a chance to cement itself, sustainable changes could damage the sustainability of this safety culture.

At the middle phases of this journey many organizations have dealt with the trust issue and become credible in the eyes of the employees as specific safety activities are formulated, acted upon and maintained, thus minimizing the "program of the month" perception.

A real pitfall in this phase deals with sustaining the effectiveness of the initiated safety activates. When an organization fails to maintain these efforts, the setback could be worse than doing as employees and management alike will be less inclined to be made a "fool" twice. Mistakes will be made and that is acceptable, if one openly acknowledges those mistakes and take the necessary steps to correct and prevent their recurrences.

In the early phases we begin to see the desire for change take place, with employee participation and a limited safety structure in place.

URGENT! Need

The major concern deals with the trust factor. Employees will question why this approach is being offered and may question the sincerity of the effort.

Pitfalls

Figure 5-2 Pathway to safety excellence.

management are involved. This will tend to have a positive impact on the operations and safety culture. Roles and responsibilities are clearly defined. There is recognition of the desired performance and corrective action for substandard performance.

As we continue on, there are some typical management improvements that you should be aware of.

PITFALLS TO THE PROCESS

In the beginning stages, the major concern in an organization deals with the trust factor. Employees will question why a particular approach is being offered and may question the sincerity of the effort.

A real pitfall in this phase deals with sustaining the effectiveness of the initiated safety activities. When an organization fails to maintain the safety efforts, the setback could be worse than doing nothing: employees and management alike will be less likely to be "made a fool of" twice. Mistakes will be made—and that is acceptable, if one openly acknowledges those mistakes and takes the necessary steps to correct and prevent their recurrence.

One major concern at the final level is the sustainability of this new safety culture, especially if there are changes to requirements, management, union, owners, and the like. If the safety culture has not had a chance to cement itself in the organization and has not provided employee participation, then the changes could damage the sustainability of this safety culture. Refer again to Figures 5-1 and 5-2 for a review of the process. Keep these concepts in mind as we continue our journey.

ACCIDENT PYRAMID

Before we get started, we want to discuss your journey. Let's revisit the accident pyramid. You need a good understanding of what it takes to reduce incidents. This model will provide some useful information. We will discuss this concept again in Chapter 12, "Conducting Effective Incident Investigations."

In 1969, a study of industrial accidents was undertaken by Frank E. Bird, Jr., who was then the Director of Engineering Services for the Insurance Company of North America. He was interested in the accident ratio of 1 major injury to 29 minor injuries to 300 no-injury accidents first discussed in the 1931 book *Industrial Accident Prevention* by H. W. Heinrich. Since Mr. Heinrich estimated this relationship and stated further that the ratio related to the occurrence of a unit group of 330 accidents of the same kind and involving the same person, Mr. Bird wanted to determine what the actual reporting relationship of accidents was by the entire average population of workers [6].

An analysis was made of 1,753,498 accidents reported by 297 cooperating companies. These companies represented 21 different industrial groups, employing 1,750,000 employees who worked over 3 billion hours during the exposure period analyzed. The study revealed the following ratios in the accidents reported:

- For every reported major injury (resulting in fatality, disability, lost time, or medical treatment), there were 9.8 reported minor injuries (requiring only first aid). For the 95 companies that further analyzed major injuries in their reporting, the ratio was one lost-time injury per 15 medical treatment injuries.
- Forty-seven percent of the companies indicated that they investigated all property damage accidents, and 84 percent stated that they investigated major property damage accidents. The final analysis indicated that 30.2 property damage accidents were reported for each major injury [5].

Part of the study involved 4,000 hours of confidential interviews by trained supervisors on the occurrence of incidents that under slightly different circumstances could have resulted in injury or property damage. Analysis of these interviews indicated a ratio of approximately 600 incidents for every reported major injury [5].

As you study the 1–10–30–600 ratio detailed in a pyramid (Figure 5-3), remember that this represents accidents reported and incidents discussed with the interviewers and not the total number of accidents or incidents that actually occurred.

Bird continues to say that, as we consider the ratio, we observe that 30 property damage accidents were reported for each serious or disabling injury. Property damage incidents cost billions of dollars annually and yet they are frequently misnamed and referred to as "near-accidents." Ironically, this line of thinking recognizes the fact that each property damage situation could probably have resulted in personal injury. This term is a

Serious or Major Injury

Minor Injuries

Property Damage Accidents

Incident with no Visible Injury or Damage

Figure 5-3 Safety accident pyramid. Bird, Frank E., George L. Germain, *Loss Control Management: Practical Loss Control Leadership*, revised edition, Figure 1-3, p. 5, Det Norske Veritas (U.S.A.), Inc., 1996. Adapted for use by Damon Carter.

holdover from earlier training and misconceptions that led supervisors to relate the term "accident" only to injury [5].

The 1–10–30–600 relationships in the ratio indicate clearly how foolish it is to direct our major effort only at the relatively few events resulting in serious or disabling injury when there are so many significant opportunities that provide a much larger basis for more effective control of total accident losses.

It is worth emphasizing at this point that the ratio study was of a certain large group of organizations at a given point in time. It does not necessarily follow that the ratio will be identical for any particular occupational group or organization. That is not its intent. The significant point is that major injuries are rare events and that many opportunities are afforded by the more frequent, less serious events to take actions to prevent the major losses from occurring. Safety leaders have also emphasized that these actions are most effective when directed at incidents and minor accidents with a high loss potential [5].

MANAGEMENT PROCESSES FOR IMPROVEMENT

On the other hand, we need to understand the management process that will take us to the next level. Refer to Table 5-1 for an overview of this management process as a big picture.

DEVELOPING YOUR POLICY

The hallmark of every successful management is top management's active and aggressive commitment to the safety process, where they demonstrate their involvement by following all safety rules. This commitment, in turn, influences the actions of managers, supervisors, and employees. It ultimately determines the effectiveness of your management system to minimize incidents and reduce the variations in the injuries.

Top management often states its commitment through a written and clearly communicated policy for providing a safe workplace. This policy stresses the value of providing a safe operation that will protect employees. Top management should sign the policy and have it posted and communicated to all employees. In innovative companies, management, top union representatives, and employees pledge their commitment together by signing this policy.

COMMUNICATING YOUR POLICY

Once you developed your policy, it then must be communicated. This is one of the things that we often forget. However, a successful company makes sure that workplace safety is integrated into all functional areas of the operation, for example, production, sales, and quality. If your policy statement makes this clear, it will be easier for employees to choose the correct actions when any conflict arises between safety and other functions. The following is one example of a policy statement that conveys this message [2]:

> People are our most important resource. Our company's principal responsibility is the safety and health of our employees. Every

Table 5-1
Management Processes

Define safety responsibilities for all levels of the organization; for example, make sure that safety is a line management function.

Develop upstream measures; for example, number of reports of hazards/suggestions, number of committee projects/successes, number of related specific activities, etc.

Align management and supervisors by establishing a shared vision of safety goals and objectives vs. production. Implement a process that holds managers and supervisors accountable for visibly being involved, setting the proper example, and leading a positive change for safety.

Evaluate and rebuild any incentives and disciplinary systems for safety, as necessary.

Make sure that all safety committees are functioning properly; for example, charter defined, membership, responsibilities/functions, authority, and meeting management skills.

Provide multiple paths for employees to make suggestions, concerns, or problems. One mechanism should use the chain of command and make sure that there are no repercussions against employees. Hold supervisors and middle managers accountable for being responsive to all employee concerns.

Develop a system that tracks and makes sure that there is timeliness hazard correction. Many sites have been successful in building this in with an already existing work order system.

Make sure that there are methods to report injuries; for example, first aid cases, and any near misses. Educate employees about the accident pyramid and importance of reporting minor incidents. Refer to Figure 5-2. Prepare management for an initial increase in incidents and a rise in rates. This will occur if underreporting exists in the organization. It will level off, and then decline as the system changes take hold.

Evaluate and rebuild the incident investigation system as necessary to make sure that investigations are timely, complete, and effective. They should get to the root causes and avoid blaming workers. We will discuss this in more detail in Chapter 12, "Conducting Effective Incident Investigations." [4]

OSHA Web site, http://www.osha-slc.gov/SLTC/safetyheatlh_ecat/mod4_ripe.thm, public domain.

employee is entitled to a safe work place. No job is so important it can't be done in a safe manner. If it is not safe, we will not do it [2].

To be effective, it is important that your safety policy be communicated to all employees. You can communicate your policy by words, actions, and setting an example. The following sections will discuss these options:

Communicating by Words

A new employee starts learning about the company's safety attitude from the first day on the job. By discussing the safety policy and job hazards and by providing training in safe work procedures, both one-on-one and in group meetings, you tell the employee that safety has a high value in your company. Managers' and supervisors' continuing emphasis on safety reinforces this positive attitude. In the smallest of companies, the safety policy may be easily explained and understood through spoken statements. However, for all companies, including the smallest, a carefully written policy statement is always recommended. A written statement does the following:

- Clarifies safety expectations
- Creates continuity and consistency
- Serves as a model when there is conflict with production or other functional areas
- Supports management's role regarding safety responsibilities
- Encourages employee participation in the safety program
- Describes the fundamental belief that safety is held as a business value and not a priority [2]

You want to include the written policy statement in all new-employee orientation. In addition, make sure to post a signed policy statement on the employee information bulletin boards. You can also communicate your safety policy by adding it to your company letterhead and/or posting your policy on the company's Web site, if you have one. This is an added value to the safety in your organization [2].

Keep in mind that this written statement is not the policy; it is simply one way of communicating the policy. The real policy is your attitude toward employees' safety. You demonstrate this attitude by your actions.

Communicate by Action

What you do, or fail to do, speaks louder than what you say and do not say (or write). Demonstrate your concern for your employees by committing resources to the prevention and control of unsafe conditions, to safe work practices, personal protective equipment (PPE) where needed, and training. Whenever you demonstrate a willingness to put safety before short-term production goals, your actions forcefully and clearly state and define your policy [2].

Communicate by Example

Top management, middle managers, and supervisors express the company's attitude toward safety by their own actions every day. You cannot turn this commitment on and off. It must be consistent.

For an example of a safety policy statement and guidance in writing a complete statement, refer to Appendix A, a sample policy statement worksheet [2].

DEVELOPING GOALS AND OBJECTIVES

Now that your policy statement has been developed and communicated, you are ready to define your goals and objectives. These goals and objectives provide you the direction that you can take to reach your destination. It is the results that you want your program to achieve. Refer to Table 5-2.

ESTABLISHING YOUR OBJECTIVES

Goals can only be achieved by setting objectives. Objectives are the specific paths you wish to follow to achieve your specific goal. They are statements of results or performance. They are short, positive steps along your journey to safety excellence [3].

Objectives for safety are similar to those for other organizational functions. They identify: What? When? How much? For example, "Establish

Table 5-2
Sample Guidelines for Writing Objectives

A well-defined objective can be developed using the following:

- Start with an action verb (refer to Table 5-3)
- Specify a single key result that is to be accomplished
- Specify a target date for action to be accomplished
- Specify what and when (avoid why and how)

The following are questions that should be asked when developing your objective.

- Is the objective specific and quantitative?
- Is the objective measurable and verifiable?
- Does the objective relate specifically to a manager's accountable?
- Is the objective easily understandable by those who will be contributing to its effectiveness?
- Is the objective realistic and attainable (represents a challenge)?
- Does the objective provide maximum payoff on the required investment of time and resources when compared with other objectives (quality, productivity, etc.)?
- Is the objective consistent with identified resources?
- Is the objective consistent with basic organizational policies, procedures, and practices?

and communicate a clear goal for the safety program and objectives for meeting that goal, so that all members of the organization understand the results desired and the measures planned for achieving them" [3].

IDENTIFYING YOUR OBJECTIVES

Anything can become an objective—from investigating incidents to developing a new employee orientation program. You must decide the activities that are more important for your management system to work, and the ones that will help you create an effective safety management system. The objectives you select must be consistent with your basic safety policy. Objectives should be part of the normal operations of your company, and not special projects added onto the normal workload [2]. They must be integrated into your business as we discussed.

DEFINING OBJECTIVES

Objectives should be based on performance measures, for example, indicators that tell you if you did or did not perform specific tasks as expected. When defining your objectives, keep the following points in mind:

- Objectives should relate to some part of your overall goal. For example: "Develop and implement a program to train and license fork lift truck drivers." This objective relates to the part of your goal to make sure that all employees understand the hazards and potential hazards of their work and how to protect themselves and others.
- Objectives should aim at specific areas of performance that can be measured or verified. For example: the statement "Improve safety performance next month" is too general to be an objective. It would be better to say, "Conduct weekly inspections and make sure all hazards found are corrected within 24 hours after being identified."
- Objectives should be realistic and attainable, and still present a significant challenge. For example, consider the objective "Reduce recordable cases in the upcoming year by 100 percent." This objective may be unattainable because of the extent and complexity of the measures needed to prevent all injuries. An objective that is beyond reach can soon create a defeatist attitude among all those who are working toward its achievement. On the other hand, "Reduce recordable cases by 5 percent in the next year" can destroy employee interest by presenting too small a challenge. To set a realistic injury reduction goal, you can review your pattern of injuries for the last three to five years and set a goal related to improving the best point in that pattern. For example, if you had injury rates of 5.8, 5.6, and 5.7 for the past three years, your goal for the next year could be, "Reduce recordable cases to 5.0."
- When defining objectives, solicit input from as wide a range of employees as practical. Your ideas already may influence your managers and supervisors. Nonetheless, you will find that safety objectives are most effective when you discuss them with your supervisors and employees. At the least, secure their agreement and get employee buy-in. Employees who feel they have helped set objectives will be more motivated to achieve the stated task.
- All those directly involved should understand the objectives. Use terms that have a clear meaning for everyone. Leave no doubt to

what is to be accomplished. For example: "Determine the cause(s) of all incidents" may be too abstract to be understood (and will probably not be accomplished) by those with responsibility. Be clear and specific—for example, "Investigate all incidents within 4 hours of the occurrence, determine all contributing causes (root causes), and take the appropriate corrective action within 24 hours of completing the investigation."

- Objectives need to be achievable with available resources. An objective that requires a significant financial investment or an increase in personnel in times where cost reduction efforts are under way probably won't be achieved. Defining such an objective could be perceived as futile. Rather than discard this objective, just postpone it. For the present, create an intermediate objective of working to produce the needed resources. Remember, you will move toward your goal one step at a time [2]. As the old saying goes, "You have to crawl before you can walk."

Develop objectives to include specific areas of performance that can be measured and/or verified. For example, "Improve safety performance in my department next month" is too general as an objective. A better objective may be, for example, "Reduce first aid cases by 10 percent next month." Even more measurable goals are those objectives where managers and/or supervisors have complete control—for example, "Hold 30-minute safety meetings for all employees in my area every Monday morning at 7:00 A.M." As noted previously, objectives must be realistic and attainable and should represent a significant challenge.

Appropriate authority is necessary. For example: A safety director's objective is to improve the safety record in a specific areas. Successful achievement is dependent on the performance of the manager and/or supervisor and employees. In this case, the safety director has no control over the operation. However, an objective to determine specific classroom safety training needs, and/or develop the specific safety training and notify managers of its availability, is in the control or authority of the safety director and is achievable.

Appropriate training is necessary. For example: A supervisor's objective is to investigate all incidents (near misses or other loss-producing events, etc.) that occur in his or her area and make sure that future prevention activities are accomplished. The drawback is that this objective may not be attainable if the supervisor has not received specific training in incident investigation and hazard recognition techniques. The supervisor also may need training in and access to appropriate hazard correction technology.

Adequate resources must be available. For example: A maintenance manager's objective may be "To make sure that all equipment is safe to operate by reviewing all newly purchased and installed equipment." That objective will be unattainable without an adequate budget for replacement parts and/or capital improvements. In another example, if the maintenance manager is held accountable for a clean work area at the end of each shift, but is not given the staff to complete the task and finish the cleanup, the objective of clear aisle ways and work areas will be unrealistic and not attainable.

Objectives need to be understood by all employees. When developing your objectives you must use clear, understandable language that leaves no doubt about what employees are supposed to do. For example: "Investigate incidents to determine multiple causation (root cause)." This will be unclear to almost everyone. A better example could be "Investigate incidents to determine all causes, and take corrective action in 24 hours of the incident." This objective is much clearer, more understandable, and more specific.

Employees should agree to these objectives. Even when you and your supervisors agree on most issues, you should discuss with them their safety performance objectives and secure their agreement (buy-in) and/or cooperation [2].

WRITING YOUR OBJECTIVES

The objective you achieve this year will help you to tackle a larger objective next year. Document each objective that helps you track your position at any time and determine how far along you are in accomplishing the assignment. Develop a time line. Spell out in concrete terms who will be assigned the responsibility for a specific objective, what is to be achieved, to what degree, and when. Be very specific in your wording. Focus on performance goals. You may also want to include a statement indicating the maximum amount of time or money available to accomplish the specific objective. Refer to Appendix A for sample guidelines on defining an objective. Refer to Table 5-3 for a list of action verbs for writing objectives. These guidelines will to help you write well-formulated, workable objectives. The following are some examples that can be used for safety objectives:

Table 5-3
Action Verbs for Writing Objectives

Activity	Action Verbs			
Knowledge	Define	Write	Underline	Relate
	State	Recall	Select	Repeat
	List	Recognize	Reproduce	Describe
	Name	Label	Measure	Memorize
Comprehension	Identify	Illustrate	Explain	Classify
	Justify	Represent	Judge	Discuss
	Select	Name	Contrast	Compare
	Indicate	Formulate	Translate	Express
Application	Predict	Choose	Construct	Apply
	Select	Assess	Find	Operate
	Explain	Show	Use	Demonstrate
	Find	Perform	Practice	Illustrate
Analysis	Analyze	Justify	Select	Appraise
	Identify	Resolve	Separate	Question
	Conclude	Contrast	Compare	Break Down
	Criticize	Distinguish	Examine	Differentiate
Synthesis	Combine	Restate	Summarize	Precis
	Argue	Discuss	Organize	Derive
	Select	Relate	Generalize	Conclude
	Compose	Manage	Plan	Design
Evaluation	Judge	Evaluate	Determine	Recognize
	Support	Defend	Attack	Criticize
	Identify	Avoid	Select	Choose
	Attach	Rate	Assess	Value
Skills	Grasp	Handle	Move	Position
	Operate	Reach	Relax	Tighten
	Bend	Turn	Rotate	Start
	Act	Shorten	Stretch	Perform
Attitudes	Accept	Value	Listen	Like
	Challenge	Select	Favor	Receive
	Judge	Question	Dispute	Reject
	Praise	Attempt	Volunteer	Decide

Reference: Davies, Ivor K., *Industrial Technique*, McGraw-Hill Book Company, New York, 1981.

- Conduct weekly inspections with emphasis on good housekeeping, proper use of personal protective equipment (PPE), condition of critical parts of equipment, and preventive maintenance
- Determine the root cause analysis of any incident in 24 hours
- Create a written system for documenting all incidents and near misses (loss-producing events) and all subsequent investigations and corrective actions
- Eliminate any hazard(s) identified during incident investigations within 24 hours
- Complete one job hazard analysis each month in each department, with follow-up revision of safe work procedures and employee training the following month
- Conduct and evaluate emergency drills for severe weather, including tornadoes and earthquakes (where appropriate), every 6 months
- Conduct a joint fire drill/evacuation with local emergency organizations every year [2]

Use objectives in discussions with your supervisors and employees. Make sure that everyone understands his or her assigned responsibilities and is held accountable for those responsibilities.

COMMUNICATING YOUR OBJECTIVES

To be effective, goals and objectives must be communicated to all employees. This will help to make sure that all employees understand what is expected of them. As in the policy statement, communication will also reaffirm management's commitment to safety [3].

As part of the management team you should review your objectives regularly with employees by asking the following questions:

- Are you getting the desired performance from management, supervisors, and employees?
- Are objectives being achieved?
- Are the results moving you toward your goal?

Any program or activity that you invest time and resources in on a continuing basis will prove its worth in the future. If an objective has been achieved, but there continue to be "too many injuries, too many close calls, too many unsafe acts (at-risk behavior), or no improvement in hazardous conditions," then different or additional objectives may be needed [3].

By defining your safety policy, goals, and objectives, you have now determined the reason for your journey, to establish an effective management system. Now you must choose your destination. This statement boils down to the same concept of desiring to provide working conditions that are conducive to a safe workplace [2].

When writing your objectives you must state in specific measurable terms what is to be achieved and what resources should be included for accomplishing the task. Try to keep the objective concrete and measurable. At a later time, you will need to be able to determine if the objective has been achieved [2].

Writing the objective will help you clarify your specific meaning and intent of what you want to accomplish. If questions do arise, the written objective is a document that you and others can review. The existence of written objectives will show that you are serious about meeting objectives. The following are some example objectives that can be used [2].

- Conduct weekly inspections of my assigned area with emphasis on housekeeping, employee personal protective equipment (PPE), preventive maintenance, and review of critical equipment
- Conduct an incident investigation to determine the cause(s) of any incident occurring in my area, and take corrective action in 24 hours
- Track to completion all hazards identified through hazards reported by employees, incident investigations, and weekly planned inspections
- Complete one job hazard analysis each month [2]

Provide a copy of the performance objectives to each affected employee. Refer to these objectives when discussing performance issues with employees [2].

REVIEWING OBJECTIVES

Periodically review the performance objectives to make sure that you are getting the desired performance results. For example, if a supervisor meets the objectives, but the department continues to have too many incidents (near misses, no improvement in conditions, or other loss-producing events) then the objectives need to be revised [2].

Performance evaluation can either be oral or written. However, an effective evaluation can include the following critical elements:

- Objectives should be performed at specified intervals. If performance evaluations are new to your business, short intervals will be helpful in the beginning. Unacceptable performance should be identified and changed as quickly as possible. As your employees become accustomed to working toward defined performance objectives, the intervals between evaluations can be lengthened. The evaluation can become an opportunity to provide encouragement and training and allow employee participation.
- The evaluation always should be performed against previously defined objectives. There should be no surprises to the employee being evaluated regarding what was expected. If a problem develops, it may be necessary to modify the objectives to make sure that they are understandable, realistic, measurable, and achievable. You may decide that your employee needs a more careful explanation of what is expected and that some additional training may be required.
- The evaluation is an opportunity for managers and employees to explore ways of improving both the management system and the performance of each employee. Inappropriate behavior—for example, refusal to listen to each other, animosity, blaming each other, or fear and intimidation—serves only to limit the evaluation's usefulness.
- The goal of the evaluation session should be to encourage employee responsibility and efforts toward improving the performance of the management system. For each objective that has been successfully completed, always provide positive reinforcement, such as verbal expressions of thanks or commendation or more tangible recognitions such as bonuses or raises.
- If the need arises, both parties must be able to come to some agreement on needed changes in the objectives. If the evaluation determines that performance did not meet expectations, some changes must be made. Sometimes the required changes will be obvious; in other cases, you may need to carefully explore the reasons for the objectives not being met. The key is to identify possible solutions to make the management system work. In some cases, the wrong employee may have been assigned a particular responsibility. A change in the assignments may fix the problem. As we have already discussed, the level of authority of the assigned employee may need to be increased. The objectives themselves may need to be modified and employees mentored to develop capabilities that they do not possess. In this case employees should not be held accountable [2].

The agreed-upon changes must be incorporated into the existing performance objectives. Many evaluation systems break down when managers fail to incorporate and implement changes. There must be a point where some predetermined consequence for poor performance begins.

Some task monitoring may be necessary to support the performance evaluation. For example, you may need to monitor a supervisor's incident investigation process after each incident until it is clear that the supervisor has developed the necessary investigation skills. This task monitoring can form the substance of later performance evaluations and provide mentoring to make sure that the process is being handled correctly.

Keep in mind that the complexity and formality of your evaluations should be consistent with the safety program [2].

NUMERICAL GOALS

Goals can be either numerical or descriptive. Numerical goals, as the name implies, are goals that can be measured in the form of numbers—for example, a goal of "zero hazards at any time."

Numerical goals have the advantage of being easy to measure. It is difficult to set a numerical goal that is both attainable and comprehensive enough to serve as a milestone for your journey. For example, if you set a goal of zero hazards at any time or an OIR of 5.0, it may be difficult to reach. You and your employees will become disillusioned long before you have a chance to reach your destination [3].

You could set a goal of a certain number of injuries. However, in doing so you ignore both illnesses and those existing hazards that have not yet resulted in an injury—for example, a near miss or other loss-producing events.

A goal of a certain number of injuries and illnesses may not be feasible. Illnesses often are difficult to recognize until long after employees' exposure to hazards that could have been prevented or better controlled. Similar to the example above, this goal does not address hazards that have not yet resulted in an injury [2].

DESCRIPTIVE GOALS

No numerical goal can be sufficiently inclusive and still be attainable. OSHA recommends that you adopt a broad, descriptive safety goal: a comprehensive program that assesses all existing and known potential hazards of your workplace and prevents or controls these hazards. Such a goal is neither as succinct nor as easily measurable as a numerical goal. However, it is attainable. Further, this goal will be helpful in setting objectives. It should not be difficult to evaluate objectives and program results against this goal [2].

Descriptive goals are not numerical, but can also be sufficiently inclusive and still attainable—for example, a comprehensive program that assesses all existing and known potential hazards of your workplace and prevents or controls these hazards [3].

Program goals might include the number of inspections you plan to conduct each week, and the number of safety work orders generated and completed in a specific time period. An example cited before might be the number of incident investigations completed in a set time period (24 hours, 48 hours), the number of behavioral observations conducted per active observer (contact rates), or the number of one-on-one contacts. Any number of program elements can be quantified as a means to assist you with this journey.

Many companies still use incident rates (numerical goals) as calculated from their OSHA log. Sometimes, this is all that an organization knows. Although there is widespread use of this measurement, consider these facts:

- Incident rates, both frequency (total recordable rate, that includes all OSHA recordable injuries and illnesses), and severity (the number of restricted lost days due to workplace injuries and illnesses) are considered end of the pipe (downstream) measurements. These measurements are similar to trying to inspect quality into the finished products after the fact, rather than control the production process.
- OSHA recordable injuries do not indicate if you have an effective management system, or if you were just "lucky" for that year. Figure 5-4 will give you an example of what we have discussed. There have been reports of companies that have a rather low injury rate, yet would be challenged even to demonstrate compliance with some of the most basic general industry standards [2].

If you decide to use the incident rate as one of your goals, we have seen several organizations successfully target a specific reduction to their

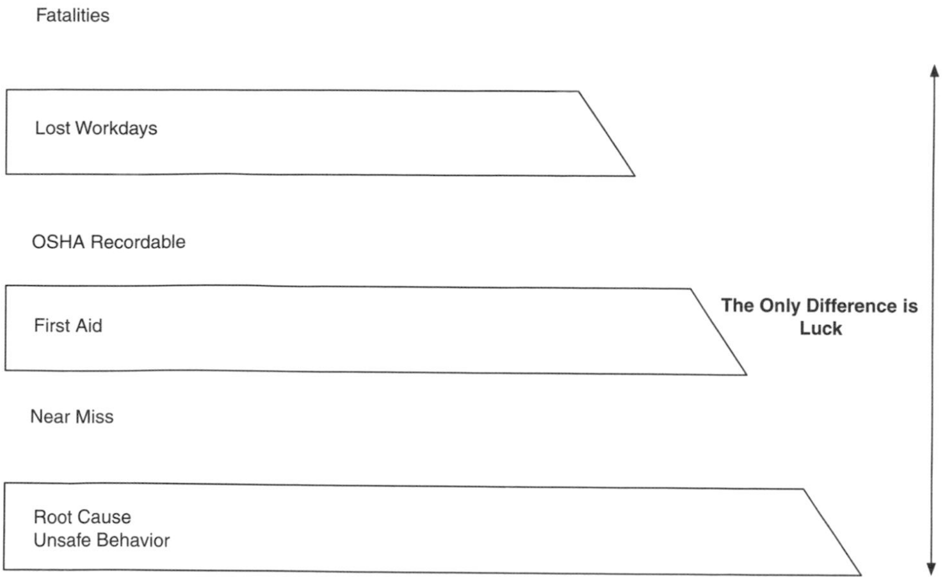

Fatalities

Lost Workdays

OSHA Recordable

First Aid

The Only Difference is Luck

Near Miss

Root Cause
Unsafe Behavior

Figure 5-4 The only difference is luck. Oklahoma Department of Labor, Safety and Health Management: Safety Pays, 2000, http://www.state.ok.us/~okdol/, *The Only Difference Is Luck*, Chapter 9, p. 55, public domain.

injury rate, coupled with specific activities they plan to complete to achieve that reduction. Remember, this approach does have its pitfalls and should only be used once the organization's safety culture understands and thus supports this type of goal setting.

We have seen in setting goals and objectives that they will determine how effective your safety process will be. You can begin to develop meaningful objectives by answering the following questions:

- Where are you now?
- Where do you want to be?
- What must be done to get from where you are now to where you want to be in the future?

The most successful goals and objectives have some of the following characteristics:

- Relates directly to management and supervision's roles and responsibilities with specific accountabilities assigned
- Is readily understandable by those who will be contributing to its attainment
- Is realistic and attainable, and represents a significant challenge

- Provides maximum payoff on the required investment of time and sources when compared with other objectives
- Is consistent with available or anticipated resources
- Is consistent with basic organizational policies and practices [3]

The following is an example of an objective that may be consistent with your basic safety program: "Conduct weekly inspections with emphasis on good housekeeping, proper use of personal protective equipment, and condition of critical parts of equipment and preventive maintenance" [3].

WHERE ARE YOU NOW?

You have established the reason behind your journey (the policy) and your desired destination (the goal). Now you are ready to decide on a travel route. The specific paths you choose in your journey are your objectives. Establishing objectives will make the difference between a journey where you get lost and a carefully planned trip where you arrive at your safety program. The latter is much more likely to get you where you want to be [2].

Before determining how to get from point A to point B, it helps to have a clear idea of the location of point A. This may seem obvious to everyone. But most of us, at one time or another, have jumped into a new project, or taken off in a new direction, without first assessing the present situation (workload). Now is the time to gather as much information as possible about the current conditions at your workplace and about practices that are already a part of your management system.

In the remaining chapters we will continue to focus on how to implement these programs that will support the management system. It will help to understand which element is missing from your workplace and what needs improvement.

This is the best opportunity to encourage employee participation. Ask employees and supervisors to help you identify both the successful and unsuccessful parts of your program. For more detailed suggestions on getting employees involved, refer to Chapter 7 [2].

Take a look at your existing safety activities. Are there some activities that work better than others? Review your records (incidents, injury or illness data, workers' compensation rates) to see what they tell you. Take a good look at your physical surroundings. What obvious physical conditions currently exist that indicate a potential hazard? In answering this question, you are beginning to identify your workplace issues and action planning to solve these issues. If you come up with an excessive number of physical problems, get these fixed before you attempt to set objectives. Further, those safety issues that are obvious to you are undoubtedly obvious to your employees. Again, correct the problems, and you begin to demonstrate your interest in employee safety [2].

HOW DO YOU GET FROM HERE TO THERE?

Now that you know where you management system stands, what do you need to accomplish the task of building your safety culture? This is another opportunity to get all employees involved in the development of your management system by allowing employees to participate in establishing the objectives. This employee participation helps to create an atmosphere of acceptance and commitment to the safety system, as will be discussed in Chapter 7 [2]. Engaging all employees in the planning stage of your journey will help employees feel as if they own part of the management system, thus aligning their efforts with the culture you are establishing.

DEVELOPING AN ACTION PLAN

At this stage you are already in the process of assessing your current situation, identifying safety program elements that present opportunities for improvement, and formulating objectives that address your program's needs. Refer to Appendix B for worksheets that can help you to develop a design for actions aimed at accomplishing defined objectives. Your action plan should address the following:

- What activities will be undertaken?
- Who has the responsibility for the stated objective?
- When should the action realistically be accomplished?
- What resources are needed, for example, employees, time, funds, and/or equipment?
- How are the corrective actions tracked and evaluated [2]?

The development of your action plan presents another opportunity to you to get employees involved. Managers and other employees can play an important role in mapping out the details they will be expected to accomplish.

Action plans can serve as a road map to move your safety program from where it is now to where it should be. An action plan defines what steps should be taken and the order in which tasks should be performed, and it identifies who is responsible for completing each task [5]. The plan must be specific and flexible enough to respond to change as the management systems and program grow and develop [5].

An action plan has two main parts. The first part is a list of major improvements needed to effect change to the management system. Each element is assigned a priority, and a target date is established for completion. In addition, an individual is identified to monitor or direct specific elements of the plan [5].

The second part of the plan involves developing a quantifiable action plan for making each change. This requires you to specify what is to be accomplished; determines specific steps required; identifies responsibilities; and assigns a target completion date [5].

The action plan will help you to monitor program improvements and to help prevent details from being overlooked. When several improvements are being made at the same time, it is easy to overlook a factor that may be an important prerequisite for the next phase of the plan [5].

For further guidance on accountability systems, refer to Chapters 8 and 9

STRATEGIC MAP FOR CHANGE

Safety cultures consist of shared beliefs, practices, and attitudes that exist at an establishment. Culture is the atmosphere created by those beliefs, attitudes, etc., which shape our behavior. An organization's safety culture is the result of a number of factors such as:

- Management and employee norms, assumptions, and beliefs
- Management and employee attitudes
- Values, myths, stories
- Policies and procedures
- Supervisor priorities, responsibilities, and accountabilities
- Production and bottom line pressures versus quality issues
- Actions or lack of action to correct unsafe behaviors

- Employee training and motivation
- Employee involvement or "buy-in" [2]

In a strong safety culture, everyone feels responsible for safety and pursues it on a daily basis; employees go beyond "the call of duty" to identify unsafe conditions and behaviors, and intervene to correct them. For example, in a strong safety culture any employee would feel comfortable walking up to any top manager and reminding him or her to wear personal protective equipment. This type of behavior would not be viewed as forward or overzealous, but would be valued by the organization and rewarded. Likewise, co-workers routinely look out for one another and point out unsafe behaviors to each other [7].

A company with a strong safety culture typically experiences few at-risk behaviors. Consequently, they experience low accident rates, low turnover, low absenteeism, and high productivity. These are usually companies that are extremely successful, excelling in all aspects of business.

Creating a safety culture takes time. It is frequently a multi-year process. A series of continuous process improvement steps can be followed to create a safety culture. Management and employee commitment are hallmarks of a true safety culture where safety is an integral part of daily operations [7].

A company at the beginning of the road toward developing a safety culture may exhibit a level of safety awareness, consisting of safety posters and warning signs. As more time and commitment are devoted, an organization will begin to address physical hazards and may develop safety recognition programs, create safety committees, and start incentive programs [7].

Top management support of a safety culture often results in acquiring a safety director, providing resources for accident investigations and safety training. Further progress toward a true safety culture uses accountability systems. These systems establish safety goals, measure safety activities, and charge costs back to the units that incur them. Ultimately, safety becomes everyone's responsibility, not just the safety professional's. The safety professional becomes a value to the organization and is an integral part of operations and not just an overhead function or cost. Management and employees are committed and involved in preventing losses. Over time the norms and beliefs of the organization shift focus from eliminating hazards to eliminating unsafe behaviors and building a management system that proactively improves employee/management relationships. Employee safety and doing something the right way takes precedence over short-term production pressures. Simultaneously, production does not suffer but is enhanced because of the level of excellence developed in the organization [7].

Any process that brings all levels in the organization together to work on a common goal that everyone holds in high value will strengthen the culture. Employee safety is a unique area that can do this. As a result, buy-in can be achieved, enabling the organization to effectively implement change. Obtaining buy-in is much easier for improving employee safety than for improving quality or increasing profitability. When the needed process improvements are implemented, all three areas typically improve and a culture is developed that supports continuous improvement in all areas.

This is intended to focus you on the process rather than individual tasks. It is common to have a tendency to focus on the accomplishment of tasks—for example, to train everyone on a particular topic or implement a new procedure for incident investigations [7].

Organizations that maintain their focus on the larger process to be followed are far more successful. They can tell the "forest" from the "trees" and thus can make midcourse adjustments as needed. They never lose sight of their intended goals; therefore, they tend not to get distracted or allow obstacles to interfere with their mission. The process itself will take care of the task implementation and ensure that the appropriate resources are provided and priorities are set [7].

OBTAINING TOP MANAGEMENT "BUY-IN"

This is the very first step that needs to be accomplished. Top managers must be on board. If they are not, safety will compete against core business issues such as production, quality, and profitability, a battle that will almost always be lost. They need to understand the need for change and be willing to support it. Showing the costs to the organization in terms of dollars lost (direct and indirect costs of accidents), as we discussed in Chapter 1, and the organizational costs (fear, lack of trust, feeling of being used, etc.) can be a compelling reason to look at needing to do something differently. Because losses due to incidents are bottom-line costs to the organization, controlling these will more than pay for the needed changes. In addition, when successful, you will also go a long way in eliminating organizational barriers such as fear or lack of trust: issues that typically get in the way of everything that the organization wants to do [7].

Continue building "buy-in" for the needed changes by building an alliance or partnership between management, the union (if one exists), and employees. A compelling reason for the change must be spelled out to everyone. Employees have to understand *why* they are being asked to change what they normally do, and what it will look like if they are successful. This needs to be done up front. If employees get wind that something "is going down" and haven't been formally told anything, they naturally tend to resist and opt out [7].

Trusting is a critical part of accepting change, and management needs to know that this is the bigger picture, outside of all the details. Trust will occur as different levels within the organization work together and begin to see success [7]. As we discussed the emotional bank [1], the bottom line is that Top Management can do 100 things right and do one thing wrong and lose the creditability of the employees.

CONDUCT SELF ASSESSMENT/BENCHMARKING

To get where you want to go, you must know where you are starting from. A variety of self-audit mechanisms can be employed to compare your site processes with other recognized models of excellence such as VPP sites. Visiting other sites to gain firsthand information is also invaluable [7].

Begin training of management, supervision, union leadership (if present), and safety committee members, and a representative number of hourly employees. This may include both safety training and any needed management, team building, hazard recognition, or communication training. This will provide you a core group of employees to draw upon as resources, and it also gets key employees on board with needed changes [7].

Establish a steering committee comprised of management, employees, union, and the safety staff. The purpose of this group is to facilitate, support, and direct the change processes. This will provide overall guidance and direction and avoid duplication of efforts. To be effective, the group must have the authority to get things done [7].

Develop the organization's safety vision, key policies, goals, measures, and strategic and operational plans to provide guidance and serve as a check. These are used to ask yourself if the decision you are about to make supports or detracts from your intended safety improvement process.

Align the organization by establishing a shared vision of safety goals and objectives versus production. Top management must be willing to support the process by providing resources (time) and holding managers and supervisors accountable for doing the same. The entire management and supervisory staff need to set the example and lead the change. It is more about leadership than management [7].

Define specific safety roles and responsibilities for all levels of the organization. Safety must be viewed as everyone's responsibility. How the organization is to deal with competing pressures and priorities, for example, production versus safety, needs to be clearly spelled out.

Develop a system of accountability for all levels of the organization. Everyone must play by the same rules and be held accountable for their areas of responsibility. Signs of a strong culture are when the individuals hold themselves accountable [7].

Develop an ongoing measurement and feedback system. Drive the system with upstream activity measures that encourage positive change. Examples include the number of hazards reported or corrected, numbers of inspections, number of equipment checks, JHAs, or pre-startup reviews conducted [7].

Although it is always nice to know the bottom-line performance of, for example, incident rates, overemphasis on these and using them to drive the system typically only drives incident underreporting. It is all too easy to manipulate incident rates, which will only result in risk issues remaining unresolved and a probability for more serious events to occur in the future [7].

Develop policies for recognition, rewards, incentives, and ceremonies. Again, reward employees for doing the right things and encourage participation in the upstream activities. Continually reevaluate these policies to make sure that they are effective and to make sure that they do not become entitlement programs [7].

It's not enough for a part of the organization to be involved and know about the change effort. The entire organization needs to know and be involved in some manner. A kickoff celebration can be used to announce it's a "new day" and seek buy-in for any new procedures and programs [7].

Implement process changes via involvement of management, union, and employees using a "Plan to Act" process as in Total Quality Management (TQM). Continually measure performance, communicate results, and celebrate successes. Publicizing results is important in sustaining efforts and keeping everyone motivated. Everyone needs to be updated throughout the process. Giving progress reports during normal shift meetings, allowing time for comments back to the steering committee, opens communications, but also allows for input. Everyone needs to have a voice; otherwise, everyone will be reluctant to buy in. A system can be as simple as using current meetings, a bulletin board, and a comment box [7].

One key requirement here is reinforcement; feedback, reassessment, mid-course corrections, and on-going training are vital to sustaining continuous improvement [7].

SUMMARY

To be effective you must periodically review your objectives and update your action plan. The following are some specific questions that can help you:

- Are you getting the desired performance from supervisors and employees?
- Are objectives being achieved?
- Are the results moving you toward your goal?

Your safety program must be carefully thought out and directed. The first step is to write and communicate your safety policy. This states your reasons for the program and your commitment to the safety of all employees. You communicate this policy by word (both spoken and written), by your actions, and by demonstrating your commitment example [2].

The second step is to establish and communicate a goal for your management system. It requires a determination of where you want to be. Your goal can be expressed either numerically or descriptively. There are advantages and disadvantages with both, but a comprehensive yet attainable goal is most likely to be descriptive.

The third step in determining the direction of your safety program is to map out your route by establishing program objectives. To do this, you first need to know where you are: take a close look at the current state of your safety program and your workplace. What more is needed to protect your employees' safety [2]?

The objectives that you develop, and the steps that you choose to take to accomplish these objectives, should be specific and measurable actions that move you toward your goal. They must be attainable and yet challenging. Use the clearest possible wording, so that your supervisors and employees understand their responsibility and accountability. Once your program is set in motion, periodically review your objectives and your action plan. Is everyone performing as expected? Are the results being achieved worth the time and resources being expended? Are you moving closer to your goal? The success of this effort depends on the commitment of top management and the participation of your employees. Involve your supervisors and employees in the setting of program objectives and the development of an action plan. The greater employee involvement in mapping the route to safety, the greater will be their acceptance of the challenges and responsibilities of the journey [2].

Make sure that your safety goals are written and that they relate directly to the safety policy or vision. The goals incorporate the essence of "a positive and supportive safety system integrated into the workplace culture" into their language. The goals are supported by top management and can be easily explained or paraphrased by others within the workplace.

Objectives exist that are designed to achieve the goals. The objectives relate to deficiencies identified in periodic assessments or reviews. The objectives are clearly assigned to responsible individuals.

A measurement system exists that reliably indicates progress on objectives toward the goals. The measurement system must be consistently used to manage specific objectives and activities. The objectives must be constructed so that all employees can easily explain them to others in the organization. Measures used to track objectives progress must be well known to all employees [3].

REFERENCES

1. Covey, Stephen R., *The 7 Habits of Highly Effective People: Powerful Lessons in Personal Change*, A Fireside Book, 1990.

2. U.S. Department of Labor, Office of Cooperative Programs, Occupational Safety and Health Administration (OSHA), *Managing Worker Safety and Health*, November 1994, public domain.

3. Oklahoma Department of Labor, *Safety and Health Management: Safety Pays*, 2000, http://www.state.ok.us/~okdol/, Chapter 2, pp. 12–15, public domain.

4. OSHA Web site, http://www.osha-slc.gov/SLTC/safetyhealth_ecat/mod4_ripe.htm, public domain.

5. "Why You've Been Handed Responsibility for Safety," *The Compass: Management Practice Specialty News*, J. J. Keller and Associates, Winter 2000, pp. 1, 4.

6. Bird, Frank E., Germain, George L., *Loss Control Management: Practical Loss Control Leadership*, revised edition, Figure 1-3, p. 5, Det Norske Veritas (U.S.A.), Inc., 1996. Adapted for use by permission.

7. OSHA Web site, http://www.osha-slc.gov/SLTC/safetyhealth_ecat/mod4_factsheets.htm#, public domain.

6

Management Leadership: Demonstrating Commitment

INTRODUCTION

Is your organization viewed as working together to create a safe working environment? On the other hand, is there a division of responsibilities between management employees?

Safety requires support and behaviors from the entire organization. Although commitment starts with top management, it is necessary to get employees to participate to make the management system work. This section is an indicator of the perception of how well the organization is working to create a safe working environment [2].

Protecting employees from hazards takes top management commitment. This commitment is essential and must be visibly demonstrated.

In this chapter, we will describe ways to provide visible leadership. Refer to Table 6-1 for a summary of management leadership traits. Ideally, this means participation in the process that demonstrates concern for every aspect of the safety of all employees throughout the organization. In addition, we have included a description of a management system for making sure that contract employees are also protected from hazards and prevented from endangering employees of the owner-company.

Successful top management uses a variety of techniques that will visibly involve them in the safety aspects of an organization. Look for methods that fit your style and that can be adapted to your workplace.

Employees in the organization will perceive that safety is supported when they see daily actions that are viewed as preventing incidents. Daily action only happens when each employee knows what to do (their roles

Table 6-1
Continuous Improvement for a Management System
through Leadership

Leadership is the keystone of any successful managing system. Safety management is integrated into all the activities of the organization with employees, customers, suppliers, contractors, and the community as key participants. All levels of management actively demonstrate commitment to safety as they carry out their responsibilities in the organization. It is the management's role to develop and nurture a culture supportive of safety values and principles. Leadership recognizes that the successful implementation of a safety management system requires active participation and meaningful employee participation.

Industrial Accident Prevention Association (IAPA), http://www.iapa.on.ca/IMS/index01. html. Presentation, slide 33. Modified with permission.

and responsibilities are clearly defined) and is required to do it (they are held accountable for their actions) [2].

In Figure 6-1, the keystone at the top of the arch represents management commitment and leadership, and the rest of the arch is the management program that must be integrated into the system. If the system does not work in the way it is intended, the system fails. Figure 6-2 shows the results of a management failure.

Let's start by looking at the role of supervisors. What drives a supervisor's performance? At this level, it is still simple. Performance is driven by the perception of what the next level of management wants done, their perception of how their manager will measure them, and their perception of how they will be recognized for their performance. According to Petersen, research shows that the answers to the following questions dictate supervisory performance [2]:

- What is the expected action?
- What is the expected recognition?
- How are the two connected?
- How are they being measured?
- How will it affect me today and in the future [2]?

Specific roles in the organization can be described as follows:

- The role of the supervisor is to carry out some agreed-upon tasks to an acceptable level of performance.
- The role of management is to make sure that:

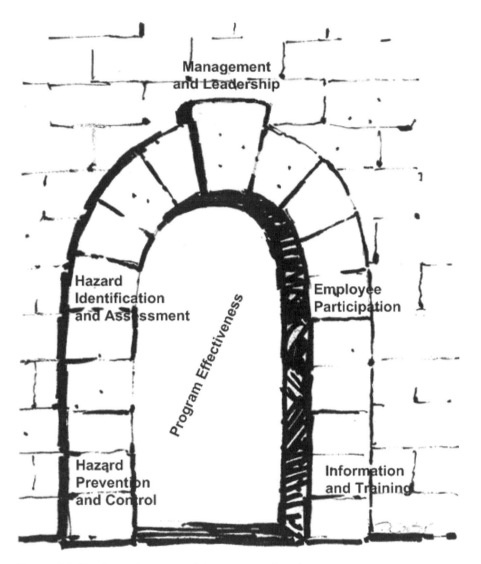

Figure 6-1 Keystone of a successful management system.

Employees perform as expected
The quality of the performance is held to a certain standard
They are personally engaged in agreed-upon tasks
- The role of top management is to visibly demonstrate the value of safety.

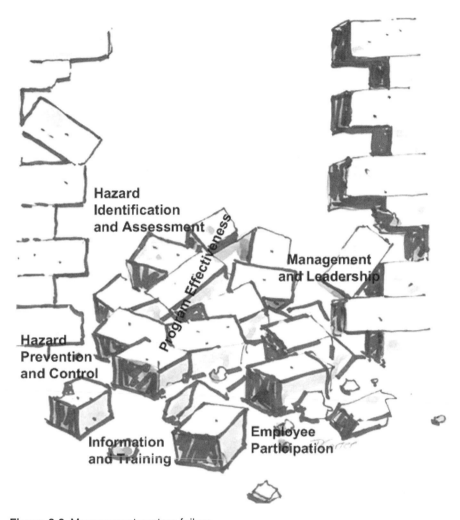

Figure 6-2 Management system failure.

- The role of the safety professional is to advise and assist each level of the organization [2]. Typically in the traditional safety approach, the safety professional becomes a "gopher," doing the bidding of management.

The supervisor's roles and responsibilities as described are simple: to carry out the agreed-upon tasks [2]. Depending on the organization, the tasks usually fall into several categories as outlined in Table 6-2.

Table 6-2
Traditional versus Nontraditional Tasks

Traditional Tasks	Nontraditional Tasks
Inspect	Give positive strokes
Hold meetings	Make sure that there is employee
Perform one-on-ones	participation
Investigate accidents	Do worker safety analyses
Do job safety analysis	Do force-field analyses
Make observations	Assess climate and priorities
Enforce rules	Crisis intervention
Keep records	

Peterson, Dan, *The Challenge of Change, Creating a New Safety Culture, Implementation Guide*, CoreMedia Development, Inc., 1993, Resource Manual, Category 17, Support for Safety, p. 79. Modified with permission.

MANAGEMENT LEADERSHIP

Management leadership is the most important part of any process. Without this leadership and support, none of the elements of a management system will work. Management's active demonstration of leadership is the key to success of any safety management system. When management demonstrates commitment, this provides the motivating force and the needed resources to manage safety where an effective process can be developed. Table 6-3 will provide you some insight into what makes a good manager.

Top management must also have clearly defined roles and responsibilities, valid measures of performance, and recognitions contingent on performance that are sufficient to get their attention. The management could use the same measurement as the supervisor, but it must be more detailed [2]. Typically the manager will develop his/her own measurement, then pass it down to the lower level of management, who would scale their measure, with the approval of their management, to a realistic goal.

Performance at this level is critical to safety success. The middle managers (the persons to whom the supervisors normally report) are more important than the supervisors in achieving safety success because they either make the system run by demonstrating commitment or allow it to fail [2]. Again, refer to Figure 6-2 for management system failure.

Petersen states: "There is probably more interest and commitment at the executive level today than we have ever seen before. Perhaps all of

Table 6-3
Traits of Management and Leadership

Effective protection from hazards takes top management leadership and commitment. Management leadership provides the motivating force and the resources for organizing and controlling activities. In an effective process, management regards employee safety as a fundamental value of the organization. Ideally, this means that that is a concern for every aspect of the safety of all employees throughout the facility.

Does your management system incorporate:

- Reasons for establishing a safety program (or the workplace policy)
- Where you want to end up (the goal)
- The path to your goal (the objectives)

These are some of the actions that can be used to make sure that you have the appropriate management leadership in place.

Other actions for management leadership include visible management involvement, assigning and communicating responsibility, authority, and resources to responsible parties, and holding those parties accountable. In addition, management must make sure that employees are encouraged to report hazards, symptoms, and injuries, and that there are no programs or policies that discourage this reporting.

VISIBLE LEADERSHIP

Successful top managers use a variety of techniques that visibly involve them in the safety of their employees. Managers must look for methods that fit their style. Some methods may include:

- Getting out where you can be seen, informally or through formal inspections
- Being accessible
- Being an example, by knowing and following the rules employees are expected to follow
- Being involved by participating on the workplace Safety and Health Committee

OSHA Web site, http://www.osha-slc.gov/SLTC/safetyhealth_ecat/comp1_mgt_lead.htm, public domain.

the above are behind this interest and commitment. Perhaps the Bhopal, Chernobyl, and *Challenger* incidents explain some of it. The safety professional that does not take advantage of this interest is missing a major opportunity" [2]. I think now that we can use the September 11,

2001, events of the World Trade center terrorist attack in the same context as Petersen used.

Although top management may be interested today, they typically do not have a clear understanding of the safety process or what to do to make safety activities successful. This is what the safety professional's job is: to outline the roles and responsibilities of management and how the management system should work [2].

When management is actively involved in safety, employees will believe it is a key value and will support the process.

In implementing an effective management system management can demonstrate their commitment and support in various ways. For example:

- Providing visible top management support in implementing the management system, so that all employees understand that management's commitment is serious.
- Encouraging employee participation in the structure and operation of the program, and in decision-making that affects their safety. This allows employees and management to commit their insight and energy to achieving the system goals and objectives.
- Stating clearly a policy and/or vision (mission) statement on safe practices, so that all employees with responsibility understand the importance of and commitment to safety awareness.
- Communicating clear goals for the management system, and defining objectives for meeting the established goals and objectives. This will allow all employees to understand the desired results and measures planned for achieving success.
- Assigning responsibility for all aspects of the management system, so that managers, supervisors, and employees know what performance is expected.
- Holding management accountable for meeting assigned responsibilities, so that they provide adequate authority and resources to responsible parties who perform essential tasks.
- Reviewing program operations periodically to evaluate successes in meeting established goals and objectives, so that infractions can be identified, action plans can be developed to address opportunities for improvement (gaps), and the program and/or the goal and objectives can be revised when they do not produce the desired effect [2].

Demonstration (showing, "doing as I do," not "as I say") is a strong motivating factor in drawing general attention to safety concerns. The active modeling of desired actions and behaviors sends a powerful message: "I am committed."

Somewhere along the line we have to establish a process to identify the tasks that management and supervision *want* to do. The problem is that most levels of management are given tasks (chores) to conduct and have no input into what they must accomplish. The attitude projected to the employee is "I have got to do this" versus "I want to do this." As we discussed before, there should be a number of agreed-upon tasks from which each manager and supervisor can pick and choose. If you do something you like, rather than something you do not like, you will accomplish more. How many times have you been given an assignment that you do not want to do? However, your assignment is to present the information to your employees. When you present this information to your employees, you say "I have got to do this, so be patient with me." What message are you sending? In this case you need to take the responsibility and project to all employees that it is the right thing to do, no matter if you agree or disagree. You will gain more creditability doing it this way.

Table 6-4 provides an overview of a management system and the specific components for a simple safety program.

The question sometimes becomes: What is management leadership in safety and health? In short, management demonstrates leadership by providing the resources, motivation, priorities, and accountability for making sure that the safety of all employees is taken into account. This leadership involves setting up systems to make sure that there is continuous improvement and that safety focus is maintained while attending to production concerns. Enlightened managers understand the value of creating and fostering a strong safety culture. Safety should become elevated so that it is a value of the organization as opposed to something that must be done or accomplished. Integrating safety concerns into the everyday management of the organization, just like production and quality control, allows for a proactive approach to incident prevention and demonstrates the importance of working safely.

You can increase employee awareness, cut business costs, enhance productivity, and improve employee morale. OSHA states that: workplaces participating in VPP have reported OSHA-verified lost workday cases at rates 60–80% lower than their industry averages. For every $1 saved on medical or insurance compensation costs (direct costs), an additional $5–50 is saved on indirect costs, such as repair to equipment or materials, retraining new workers, or production delays. During three years in the VPP, one plant noted a 13% increase in productivity, and a 16% decrease in scrapped product that had to be reworked. Bottom line, safety does pay off! Losses prevented go straight to the bottom-line profit of an organization. With today's competitive markets and narrow profit margins, loss control should be every manager's concern. Table 6-5 shows some management actions that can be taken to enhance a safety culture.

Table 6-4
Overview of Safety Management System Components

A system is an established arrangement of components that work together to attain a certain objective, to prevent injuries and illnesses. In a management system, all components are interconnected and affect each other. Using this definition, let's consider a safety system.

All elements of a safety system are like a puzzle, and all are interrelated. A flaw in one piece will probably affect all the other pieces, and therefore the system as a whole.

Management leadership and employee involvement are tied together because one is not effective without the other. Management can be totally committed, but if employees follow blindly or are not involved, problems will be solved only temporarily.

Management must provide the resources and authority so all employees can find the hazards in the worksite and, once they are found, eliminate or control them.

Training is the backbone of this system. For management to lead, for employees to analyze the worksite for hazards, and for hazards to be eliminated or controlled, everyone involved must be trained.

No parts of this system exist independently. An effective and functioning program is the sum of all the parts.

Once you know what it takes to make a safety system operate effectively, how can you tell if the system measures up (or if there is a system)?

http://www.osha-slc.gov/SLTC/safetyhealth_ecat/components.htm, public domain.

Table 6-5
Management Actions

- Establishing a safety policy
- Establishing goals and objectives
- Providing visible top management leadership and involvement
- Enlisting employee participation
- Assigning responsibility
- Providing adequate authority and responsibility
- Providing accountability for management, supervisors, and employees
- Providing a program evaluation

OSHA Web site, http://www.osha-slc.gov/SLTC/safetyhealth_ecat/comp1_mgt_lead.htm, public domain.

You should make your general safety policy specific by establishing clear goals and objectives. Make objectives realistic and attainable, aiming at specific areas of performance that can be measured or verified.

Values, goals, etc., of top management in an organization tend to be emulated and accomplished. If employees see that top management puts an emphasis on safety, they are more likely to emphasize it in their own activities. Besides following set safety rules themselves, managers can also become visible by participating in safety inspections, personally stopping activities or conditions that are hazardous until the hazards can be corrected, assigning specific responsibilities, participating in or helping to provide training, and tracking safety performance.

Everyone in the workplace should have some responsibility for safety. Clear assignment helps avoid overlaps or gaps in accomplishing activities. Safety is not the sole responsibility of the safety professional. Rather, it is everyone's responsibility, while the safety professional is a resource and mentor to help get the job done.

Accountability is crucial to helping managers, supervisors, and employees understand that they are responsible for their own performance. Reward progress and enforce negative consequences when appropriate. Supervisors are motivated to do their best when management measures their performance. Take care to ensure that measures accurately depict accomplishments and do not encourage negative behavior. As we have discussed, accountability in safety can be established through a variety of methods:

- *Charge-backs.* Charge incident costs back to the department or job, or prorate insurance premiums.
- *Safety goals.* Set safety goals for management and supervision (for example, not using accident rates, accident costs, and loss ratios).
- *Safety activities.* Conduct safety activities to achieve goals (for example, hazard hunts, training sessions, safety fairs—activities that are typically developed from needs identified based on accident history and safety program deficiencies).

Once your safety program is up and running, you will want to make sure that it is in control, just like any other aspect of your organization. Each program goal and objective should be evaluated, in addition to each of the program elements—for example, management leadership, employee participation, workplace analysis (incident reporting, investigations, surveys, pre-use analysis, hazard analysis, etc.), hazard prevention and control, and training. The evaluation should not only identify accomplishments and the strong points of the safety program, but also identify opportunities for improvement. Be honest. The audit can then become a blueprint for improvements and a starting point for the next year's goals and objectives instead of something that has to be done [1].

EXAMPLE OF MANAGEMENT SYSTEMS

When doing research for this book, I ran across the article "Safety Management: A Call for (R)evolution," written by Larry Hansen published in *Professional Safety*. This article summarizes in simple terms the styles of different companies and how they feel about safety. This may help get you thinking about your own organization. Refer to Figure 6-3 for an overview of the process.

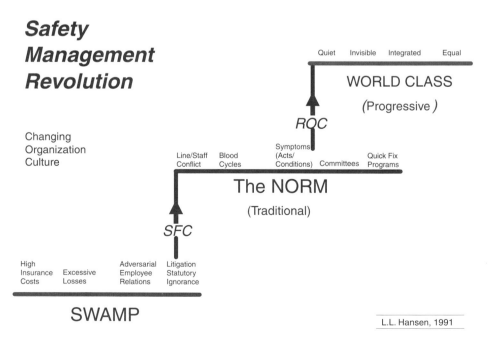

Figure 6-3 Safety management revolution. Hansen, Larry (315-383-3801), "Safety Management: A Call for (R)evolution," *Professional Safety*, March 1993, p. 20. Reproduced with permission.

Over the years I have had the opportunity as a consultant and also in the industrial environment to see all types of organizations perform. Hansen has captured the essence of these types of organizations. As you read this section, consider which type of system you may be operating in.

Safety without Any Management Process

The company without a management system in place rejects responsibility and perceives safety as a burden—another thing that "has to be done." Safety is a task that is considered overhead, not a productive value. There it is, on the priority list that is forever changing, where it never gets the emphasis it needs. Refer back to Figure 2-1. The company accepts incidents as a cost of doing business, is autocratic in its thinking, and has a heavy task focus, with safety frequently compromised in favor of production, quality, and other demands on the business. Their planning process is short-term and reactive; communication is one-way (top-down) and founded in fear. "If you get hurt, you will suffer the consequences" [5].

They employ make-do solutions (quick fixes) to equipment and facilities, often unsafe. There is minimal employee participation in decision-making. Employee–management relationships are adversarial—the traditional "them against us" [5].

This type of organization has high insurance cost driven by frequency and severity. Their EMR is typically 1.25, that is, 25 percent worse than the average. They pay a high cost in premiums to insurance (liability and workers' compensation.) They continue to operate in statutory ignorance, often in violation of recognized safety codes. The attitude is that "It is cheaper to have an OSHA fine than it is to fix the problem" [5].

Employees complain frequently. This organization becomes a target for litigation emanating from major injuries and frequently makes national headlines [5].

These companies remain in this type of environment until a significant financial crisis (SFC) occurs—normally an increase in operational cost or losses so damaging (to profit) that management finally recognizes a problem and declares, "We need a safety program" [5].

Now they start looking for a safety professional, usually low-paid, with some type of safety background (normally three to five years of experience). The safety professional is short lived because of the nature of the business. Once some of the programs are in place and everything seems to be working fine, or they see some improvements, the safety professional is no longer a value to the company [5]. Figure 6-4 shows what this stage represents.

SAFETY RESPONSIBILITY: Not Recognized/Rejected
PERCEIVED: As a "Burden"

MANAGEMENT CHARACTERISTICS:

Accidents are Accepted - C.O.D.B. - A Fluke	Communications - Fear Based
Autocratic Style	"Make Do, Make Fit" Approaches
Task Oriented - Production Compromised	Minimal EE Involvement/Interaction
Planning - Minimal; Reactive: Short-term	Bunker Mentality - Adversarial

ORGANIZATIONAL IMPACTS:

HIGH INSURANCE COST	EXCESSIVE LOSSES	POOR EMPLOYEE RELATIONS	STATUTORY IGNORANCE
E.M > 1.25	Above Avg. Incident Rates	"Blame the Union" Them vs. Us	OSHA Citations
			Complaints
	Severity in Evidence	"Antiquated" Human Resources Policies	Litigation

Figure 6-4 SWAMP (Safety Without Any Management Process). Hansen, Larry (315-383-3801), "Safety Management: A Call for (R)evolution," *Professional Safety*, March 1993, p. 20. Reproduced with permission.

Normally Occurring Reactive Management

Companies at this stage add safety activities without having an adequate understanding of the problem or the actions needed to resolve the issues: "They just have to do something." They implement a safety program patterned after what others have done—for example, they order a generic safety manual and put it on the shelf, and a safety director is assigned to create and chair the safety committees with no management support. There is a challenge to safety committees to fix: "the safety problem." There is no consideration that there is a management problem [5].

When an incident occurs, and it does, the supervisor typically will blame the employee: "It was the employee's fault" or "The employee was not trained properly." Supervisors do not recognize their key roles and responsibilities and normally embrace quick fixes. There is conflict between the safety director and the supervisor. The safety director is considered a nitpicker for looking over the supervisor's shoulder. The safety director is in the role of "Safety Cop" and is the only person fixing safety issues [5].

Safety programs have high visibility with slogans, campaigns, gimmicks, contests, and incentive programs with awards with no paybacks. Management issues policies but personally compromises most of them through their own behavior, sending a clear message to employees: "Do I say, not as I do." This type of organization must create radical organization changes (ROC), discarding the old traditional principles and adopting new process concepts. Figure 6-5 gives an overview of this type of business [5].

The World-Class Organization

When you look at the last type of company, the world-class operation, you will see a different proactive organization that is driving management processes that will be successful [5].

In this type of organization, management owns the process and drives management. Line managers perceive safety as a good business invest-

SAFETY RESPONSIBILITY: Not Understood (STAFF)
PERCEIVED: A "Cost"

MANAGEMENT CHARACTERISTICS:

Accidents are excused away - fault based

Recognizes Problems - Unwilling/Unable to solve

Is not quite sold - willing to go half way (what's easy)

Likes instant pudding (preferably vanilla)

High visibility - many labels - little results

Employees "read the actions," "see" the credibility gap, "do" nothing

Significant line/Staff authority conflicts

Programs/Campaigns – short-lived

Results (only) measured

Line "accountability" lacking.

ORGANIZATIONAL IMPACTS:

BLOOD CYCLES	SYMPTOMS (ACTS/CONDITIONS)	COMMITTEES	Q.F. PROGRAMS
E.M > 1.00 ± 25%	Inspection Heavy	Scheduled - Need it or not	Supv.'s Ignore
	Find it/Fix it	Bitch Sessions	EE's Reject
	Repetitive Conditions Problems		

Figure 6-5 The NORM, Normally Occurring Reactive Management. Hansen, Larry (315-383-3801), "Safety Management: A Call for (R)evolution," *Professional Safety*, March 1993, p. 21. Reproduced with permission.

ment with a long-term vision of positive return. Management believes that injuries are unacceptable, and accepts no excuse! Safety is not just safety; it is organizational effectiveness. The decisions management makes are time consuming and the planning is long term (three to five years.) Roles and responsibilities of line and staff are clear and communicated. There is shared vision and cooperative effort. Their employee relations policies are humanistic. Employees are empowered and rewarded, often through gain-sharing. Communications are open, informal. Feedback is encouraged. Methods to produce safely are built into job descriptions and processes. Results are closely monitored. Causes for variation are identified and corrected [5]. There are no campaigns, flashing lights, or bells and whistles; there are simply results superior to all.

In this type of organization, safety loses its identity because it is integrated into the management system; there are no "safety programs." There are few accidents; there is simply "excellent management" [5].

The inability of most organizations to attain world-class performance lies in their inability to create and effectively manage change. They fail to recognize that business enterprises exist to create change [5].

World-class effect will only become a reality when management fully integrates safety responsibility into the organization's mainstream. This will not result from safety programs that are superimposed on an organization; it will happen when safety is fully accepted as an integral part of the organization and its mission [5]. Figure 6-6 shows a summary of a world-class operation.

CONTRACTOR SAFETY

The actions of contract employees can have an adverse impact on the safety of employees at the site. Where contract employees and your own employees are intermingled, any unsafe practices or conditions of contract employees can jeopardize your own employees. But even if contract employees are removed somewhat from your normal operations, your employees will benefit from knowing that you insist on good safety practices and protection for every employee at your worksite. Hiring contractors is another issue. The bottom line is that you need to treat contractors as you would your own employees. You will need to check with the legal department to make sure that you understand the contractual requirements.

SAFETY RESPONSIBILITY: "Line Management" Owned/Driven
PERCEIVED: Good Business "Investment"

MANAGEMENT CHARACTERISTICS:

Accidents are "intolerable" - there are no excuses!

Safety isn't safety - it's "Management Effectiveness"

Decisions - time consuming and difficult

Planning - Long term - 3 to 5 years

Responsibilities and expectations - clearly defined

No "glitz" and/or "just add water" approaches

Employee centered "gain-sharing"

Communications - Informal, open, encouraged

Efforts "closely measured" - and responded to

ORGANIZATIONAL IMPACTS:

QUIET	INVISIBLE	INTEGRATED	EQUAL

E.M < .75

"Safety loses identity"

"There is no program"

"There is good management"

"Three aren't (m)any accidents"

Figure 6-6 World-class organization. Hansen, Larry (315-383-3801), "Safety Management: A Call for (R)evolution," *Professional Safety*, March 1993, p. 21. Reproduced with permission.

OTHER CORE ELEMENTS

Other companies have included additional elements into their core management system. Examples include behavioral-based safety, communication, contractor safety, and process safety management. You can add any elements to your management that will help you measure your success.

SUMMARY

Successful top managers use a variety of techniques that visibly involve them in the safety system. Look for methods that fit your style and your worksite. These methods can include:

- Getting out of the office where you can be seen by all employees
- Being accessible to all employees

- Being an example by doing all the right things
- Taking charge [3]

You may have heard the phrase "management by walking around" [4]. This phrase describes managers who walk around their workplace, getting to know employees and seeing for themselves what is working and what is not. This not only is successful as a management tool, but also sends a message to all employees that you are interested in safety. Employees who see their manager "walking around" will start to believe and trust that the manager cares about what employees are doing and how well they are doing it.

The "walk-around" management style can be demonstrated in several ways (informally or formally). For example, a manager who stops to correct hazardous conditions as he/she walks around the workplace impresses employees with the importance of safety. As you conduct your walk-around, pay attention for shortcuts in safe work procedures and/or practices. The informed manager knows that shortcuts are a form of Russian roulette. It is only a matter of time until an employee gets hurt.

If you stop occasionally to compliment employees on how well they are following safe work procedures, you can expect your comments to have a strong positive impact on the desired behavior [3]. The key is to not only praise the employee, but also stop and listen to what they are saying. You can react better to conditions.

This type of participation should be a fairly routine occurrence. If it happens only on occasion, it will not have any significant impact on management support and credibility. It only works for managers who are out in the workplace several times a week (if not several times every day). This informal style is well suited for the small business where management spends considerable time in the work area [3].

This knowledge will allow you to identify and correct hazards. As a manager you should have an understanding of what is safe and what is at-risk. If you are top management who lacks the safety expertise or is not sure of your knowledge of safety, you should not try to interfere with any lower level managers and supervisors who possess some safety expertise [3].

A formal method of demonstrating commitment is to conduct inspections with selected employees. These inspections must occur often to make a difference. One of the most common types of inspections that can be performed by top management is housekeeping. This inspection allows management to review the operation and identify some of the hazards that may exist.

During these routine inspections, some managers offer positive or negative recognition. Sometimes prizes are awarded to the best areas—for example, a rotating trophy to the department that does the best [3].

If you are part of top management, you have assigned specific responsibilities to your managers and supervisors. You want to avoid trying to undermine their authority and interfere with their accountability, since that may hurt their ability to effectively carry out their assigned responsibilities. At the same time, you want to make sure that you demonstrate your own commitment to help increase awareness. How do you walk this fine line between being the doer and being the implementer? The following are some suggestions that can help you to achieve a balance:

- Hold all managers and supervisors accountable for their activities
- Work with your employees to develop a comprehensive safety program
- Encourage employee participation to use established reporting systems
- Forge a partnership with all managers and supervisors to help encourage employee participation to speak out and use the established system [4]

If you manage your operation by walking around your workplace, you may find many opportunities to listen and respond to employees' questions and comments. However, even if you cannot spend time walking around your facility, you can still make sure that you are available in various ways.

You may want to recognize the teams that report hazards or provide suggestions on new hazard control. Recognition can be based on the submission of reports and suggestions or on the quality of the employee input. Let your managers and supervisors know that when an employee brings a safety issue to your attention, you consider that a good reflection of the supervisor's leadership [3].

If your position does not allow you to tour your workplace, an "open door" policy may be the best choice for you. If you use this approach, your office door must remain open all of the time, either continually or during regularly scheduled or communicated times. This technique may not work for all managers. If this is your policy, encourage employees to drop by and discuss their safety concerns, without fear of reprisal (only if they have been unable to get satisfactory answers through their normal chain of command). Once your policy is tested and the word gets out that your door is always open, employees will not make use of this access as long as other management systems are working [3].

As a top manager, you may not be able to spend time walking around the workplace and talking to employees. In addition, you need time for private meetings that may not allow an open-door policy. If this is the case, you should schedule periodic safety meetings (round table discussions, weekly or monthly safety committee meetings, etc.). In this type of forum you can speak directly with employees. These meetings are usually informal and open for questions, comments, and/or employee concerns. One caution: you must understand that this type of safety meeting is used only for safety concerns and not for other non-safety issues. The size of the group should be small enough that the meeting can be managed. In larger organizations, more than one meeting may be necessary to talk to all employees. Some top managers choose to hold separate meetings with supervisors and other managers with whom they do not regularly interact. You may have to try various group sizes to find out the best style [4].

The success of any safety meeting depends on how you create a climate where employees feel free to speak up, and how you handle the questions. Treat all employees' questions with respect, even if the answer seems simple or the concern unwarranted. Try to imagine how the situation looks to your employees and how you will project your image. Take the time to provide an explanation for a question until each employee understands what is being discussed. If you don't know the answer to a question, or when you need to know more about the circumstances surrounding an issue, don't be afraid to say so. Be honest, and always get back to the employee with the answer. Make sure that you follow up thoroughly, and that all employees who attended the meeting see and/or hear your response. Be consistent with your message.

Another, less formal type of scheduled meeting is where management provides a lunch for all employees with a birthday during a given time period. This type of meeting works best when you keep the group small. By grouping people by birthday, you get a random selection of employees from all parts of the worksite [4]. Try to steer the conversation to questions or concerns that your employees may have. In small groups, some people may be afraid to discuss perceived safety issues.

REFERENCES

1. OSHA web site, http://www.osha-slc.gov/SLTC/safetyhealth_ecat/comp1_mgt_lead.htm, public domain.

2. Peterson, Dan, *The Challenge of Change, Creating a New Safety Culture, Implementation Guide*, CoreMedia Development, Inc., 1993, Resource Manual, Category 17, Support for Safety, pp. 78–79. Modified with permission.

3. U.S. Department of Labor, Office of Cooperative Programs, Occupational Safety and Health Administration (OSHA), *Managing Worker Safety and Health*, November 1994, public domain.

4. Ouchi, William, *Theory Z: How American Business Can Meet the Japanese Challenge*, Avon Publishers, 1981.

5. Hansen, Larry (315-383-3801), "Safety Management: A Call for (R)evolution," *Professional Safety*, March 1993, pp. 16–21. Modified with permission.

7

Employee Participation

INTRODUCTION

The success of any business depends on the employees. That is the bottom line. As we have discussed, protecting employees from hazards not only makes good business sense, but also is the right thing to do. As part of management, you do not have to face this task alone. In this chapter, we will outline how employee participation can strengthen your management system [7].

This chapter takes a look at some of the reasons behind employee participation, and some of the ways you can implement a successful program [7].

This is one of the things that we as management do not do well—getting employee inputs. Why do you think this is so hard? The one reason that the authors have dealt with is simple. Management in many cases is afraid of what they hear. If they listen to employees, they may have to solve problems and really do not want to deal with it—"I do not have the time." This is a common statement from many managers. After all who needs more problems? The bottom line is that if you want true employee participation you must ask the question and accept what you hear, solve the issues, and continue to go about your business. Once the culture is developed and trust begins to build, issues will solve themselves because employees will start to solve issues on their own and only come to management when they cannot get the situation resolved.

Note to this section on "I do not have the time." As I said, this is a common response of many supervisors and management. I find that this is just an excuse. I recall when I worked for a major manufacturer, that common response was also my favorite line. It was easy to say and I got used to saying it. On a 3×5 card we had to stop every 15 minutes and write down what we did. We all protested, but had to conform to what management wanted. You know, after a couple of days, I began to realize that I could not put anything constructive on the card. I had to write something down so I did. Since that point in time, I do not use these words anymore—"I do not have the time." What I learned is that you have the time to do what you want to do when you want to. Think about what I have just described. Try it; you may be surprised to find the same thing that I did.

Refer to Table 7-1 for an overview of guidelines for employee participation.

Many of these activities require training to make sure that each employee can perform these functions proficiently. The training does not need to be elaborate and can be conducted at the workplace by employees who are appropriately trained [7].

WHY SHOULD EMPLOYEES BE INVOLVED?

Involving all employees in the management system is one of the most effective approaches you can take to develop an effective safety culture. The advantage to this participation is that it promotes employees awareness, instills an understanding of the comprehensive nature of a management system, and allows employees to "own" a part of this system. Employees make valuable problem solvers because they are closest to the action. No one knows the job better than the employee does. The following outline provides some ideas on how to deal with employee participation:

- Employees are the individuals that are in contact most with potential safety hazards every day. They have a vested interest in making sure that there is an effective management system.
- Group decisions have the advantage of the wider range of experience.
- Employees who participate in safety programs' development are more likely to support and use the programs.
- Employees who are encouraged to offer their ideas and whose contributions are taken seriously are more satisfied and more productive [7].

In the beginning of a process, employees do not understand what is going on and are afraid to get involved. Therefore, the typical response from employees is, "I don't have a clue" (Figure 7-1). They do not want any part of what you are trying to do. They just want you to go away.

**Table 7-1
Guidelines for Employee Participation**

Employee participation provides the means by which employees can develop and express their own commitment to safety, for both themselves and their co-workers.

You need to recognize that the value of employee participation and the increasing number and variety of employee participation arrangements can raise legal concerns. It makes good sense to consult with your human relations department to make sure that your employee participation program conforms to any legal requirements.

Why should employees be involved? The obvious answer is: because it's the right and smart thing to do:

- Employees are the individuals most in contact with potential safety hazards. They have a vested interest in an effective management system that supports safety.
- Group decisions have the advantage of the group's wider range of experience.
- Employees are more likely to support and use programs that they have input (buy-in).
- Employees who are encouraged to offer their ideas and whose contributions are taken seriously are more satisfied and productive on the job.

What can employees do to be involved? The following are some examples of employee participation:

- Participating on joint labor—management committees (as applicable) and other advisory or specific purpose committees (short-term projects)
- Conducting workplace inspections
- Analyzing routine hazards in each step of a job or process (JHA), and preparing safe work practices or controls to eliminate or reduce exposure
- Developing and revising the workplace safety rules
- Training both current and newly hired/transferred and seasoned employees
- Providing programs and presentations at safety meetings
- Conducting and participating in incident investigations
- Reporting hazards and fixing hazards under their control
- Supporting co-workers by providing feedback on risks and assisting them in eliminating hazards
- Performing a pre-use or change analysis for new equipment or processes in order to identify hazards up front before use

OSHA Web site, http://www.osha-slc.gov/SLTC/safetyhealth_ecat/comp1_empl_envolv. htm, public domain.

I Have No Clue!

Figure 7-1 I have no clue. Created by Damon Carter.

Have you ever had a manager that you did not want to talk to? Every time he/she would call you, they would want you to do something that you just did not want to do.

CLOSE CONTACT WITH HAZARDS

As a manager you have an understanding of how your overall business operates. You provide the vision. However, employees who work in the organization have a more detailed knowledge of each operation and task than anyone else because they do these tasks day in and day out. No matter how educated you think you are, unless you have recently performed a particular task over and over, your employees will know the job task much better than you do.

Employees who understand the hazards that are associated with the workplace will realize that they have the most to gain from preventing or controlling exposure to identified hazards. Employees who are knowledgeable and aware tend to be safe employees and also good sources of ideas for better control [7].

IMPROVED SUPPORT

Many managers complain that they cannot get employees to comply with safety requirements, whether it is wearing the appropriate personal protective equipment (PPE) or following safe work procedures. So the question becomes: How do you change employee attitude [7]? From our experience, there is one problem that management as a whole must address. Consider this: When you walk through an area, what is the first thing that you discuss with your employees? The only real thing that many levels of supervision and management know is personal protective equipment. Therefore, the first thing that you may say is, "Put on your safety glasses." "Where are your earplugs?" Why? Because safety glasses and earplugs are visible and easy to recognize. It is like the police setting up roadblocks to see if you are wearing your seatbelt. It's easier than catching speeders.

In a management system, supervision managers must understand that personal protective equipment is only one aspect of the safety system. They must know how to recognize other elements of a management system such as machine guarding and housekeeping, among other elements of the process. PPE is only one part of the process.

If you were to review any Japanese company, you would find that that there is an approach to employee participation. They have used their employees to help solve various types of quality-related issues. Some American companies are finally starting to realize the value of employee participation. Employee input in the United States is most common in the area of quality control. We now must take a broader approach on safety issues [5].

It is a well-known fact that no one wants ideas "forced down their throats." We as humans naturally resist change—but we have a tendency to support ideas that we help to develop and implement. Employees who are allowed to participate in the development and implementation of safety rules and procedures have more of a personal stake in making sure that these rules are followed [7].

As we start to develop our systems and get employees more involved, we have a tendency to provide too much information, too many meetings (sometimes called information sharing but is really information dumping), etc. In this case, you go through the phase, "Don't bombard me with all of this stuff" (Figure 7-2). Leave me alone and let me do my job.

Don't Bombard Me With
All of This Safety Stuff!

Figure 7-2 Don't bombard me with all of this safety stuff. Created by Damon Carter.

MORE PARTICIPATION, MORE AWARENESS

Using safety committees may not always be the best approach to reach a decision. However, these group decisions are often the best buy-in from employees on safety matters. The benefit is that many viewpoints with varied experiences can help produce better decisions.

Employee participation can be used to help identify and solve safety issues. Employees who participate enjoy their work more than those who simply do what they are told. When employees enjoy their work, they take a greater responsibility in their jobs and tend to produce a better quality product. Reduced turnover is a benefit of increased employee participation [7].

HAWTHORNE STUDIES

This section will describe one example, the Hawthorne Studies, of how employee participation can be used. This was a well-known study in the

1920s. Although it was not called "employee participation," we believe that you will understand what the final results concluded.

The Hawthorne Studies of 1924 came about as the results of experiments conducted at Western Electric that involved changes in the workplace conditions and produced unexpected results in employee performance. Two teams of employees took part in these experiments, in which the lighting conditions for only one team were improved. Production rose dramatically [3]. The interesting thing that happened was that production also improved in the group where the lighting remained unchanged.

The studies were undertaken in an effort to determine what effects such things as hours of work and periods of rest might have on employee fatigue and productivity. As these studies progressed, it was discovered that the social environment could have an equal if not a greater effect on productivity than the physical environment [2].

The studies revealed the influence that informal work groups can have on the productivity of employees and on their response to such factors as supervision and financial incentives. It also represented the beginning of nondirective counseling with employees [2].

Baffled by the results, the researchers called in a team of psychologists from Harvard University lead by Elton Mayo. Over the course of five years, hundreds of experiments involving thousands of employees were conducted. In these experiments, the researchers altered variables such as wage payment, rest periods, and length of the workday. The results were similar to those obtained in the lighting experiments: production increased in no obvious relationship to the environment. After much analysis, the psychologists concluded that other factors besides the physical environment affected worker productivity [6].

Elton Mayo took the experiments further, making as many as ten changes in working conditions—for example, shorter hours, varied rest breaks, and a number of incentives. Mayo's research team spent a great deal of time with the work groups, each consisting of six women. Changes were discussed before they were put into effect. Output increased each time a change was made. The interesting thing was that when the teams were asked to return to their original working conditions, with a 48-hour work week, no incentives, and no rest breaks, output rose again. This time it rose to the highest ever recorded for the Hawthorne plant. There were other significant benefits such as a decline of 80 percent in absenteeism [3].

These studies emphasized the impact of human motivation on production and output. When Mayo began the first phase the experiments he believed that every social problem was "ultimately individual." Mayo soon found that group rather than individual psychology was the key

Table 7-2
Elton Mayo and the Hawthorne Studies: Summary

Prejudgments	Findings	Safety Culture
Job performance depends on the individual employee	The group is the key factor to employee's job performance	Being involved in activities and providing input to management
Fatigue is the main factor affecting output	Perceived meaning and importance of the work determine output	Daily leadership from management through specific activities, employee ownership of process elements, several keys to employee safety
Management sets production standards	Workplace culture sets its own production standards [1]	Top management shows employees visible commitment to safety and is the driving force

Adapted from Bone, Louis E., David L. Kurtz, *Management*, 3rd ed., Figure 2–4, p. 41. Mayo's contribution to management philosophy was important to the field of managing employees because it revealed the importance of human emotions, reactions, and response to managing others. It also pioneered the concept of good communication between management and employees [3].

factor in the production performance of the workers [1]. Table 7-2 summarizes the Hawthorne Study. In addition, we have added a section to show how it related to safety.

Mayo identified the importance of the findings by specifying that the relationship of working groups to management was one of the fundamental problems of large-scale industry. Organizing teamwork, developing, and sustaining cooperation had to be a major preoccupation of management. Management needed to think less about what "we" wanted to get across to "them" than about listening to what "they" wanted to know and would be receptive to [3].

LESSONS LEARNED

The Hawthorne Studies helped management to understand that although an organization is a formal arrangement of functions, it is also a social system.

When the methods used and the conclusion reached by the Hawthorne researchers have been questioned, they have generated interest in human problems in the workplace and focused attention on the human factor [6]. These studies added much to our knowledge of human behavior in organizations and created pressure on management to change the traditional ways of managing human resources. The human relations movement pushed managers toward gaining participatory support of lower levels of the operation in solving organization problems. The movement also fostered a more open and trusting environment and a greater emphasis on groups rather than just individuals [8].

The research also found that employees reacted to the psychological and social conditions at work, such as informal group pressures, individual recognition, and participation in making decisions. In addition, the significance of effective supervision to both productivity and employee morale was identified and emphasized.

The bottom line: paying attention to employees will have a great impact on the outcome that you are trying to accomplish.

COMMITTEE PARTICIPATION

Joint labor–management committees are a popular method of employee participation. Other types of committees also have been used successfully to allow employee participation. At many unionized facilities, employee safety committees (with members selected by the union or elected by employees) work alone, with little direct management participation, on various tasks. In other workplaces, employees participate on a central safety committee. Some worksites use employees or joint committees for specific purposes, such as conducting workplace surveys, investigating incidents, training new employees, and implementing behavioral-based safety (BBS) systems, etc. [7].

As the system continues to develop and mature, employees get more involved. If this involvement is not handled properly, there is still a thought in the employee's mind that this is the "program of the month." After all, they have seen this type of increased awareness before, particularly after a major injury. Then it dies down. As the old saying goes, "If you do not like it, wait around, it will change again." Hence, "Ignore it and it will go away" (Figure 7-3).

**Ignore it and It
will Go Away!**

Figure 7-3 Ignore it and it will go away. Created by Damon Carter.

GETTING EMPLOYEE PARTICIPATION STARTED

One of the keys to getting employees involved is to meet with them. Here are some proven techniques for encouraging employee participation:

- Meet with employees in one large group, as appropriate (by shift, department, one-on-one, etc.) depending on the nature of the business.
- Explain the safety policy and the goals and objectives that you then want to achieve.
- Explain how you want your employees to help with the safety efforts. Ask for their input, suggestions, etc. Try to use as many of the reasonable suggestions as possible in some visible way [7].

FORMING A COMMITTEE

Form a committee that is suited for your business. It should be large enough to represent different parts of the organization. All committees must be integrated. For example, a committee designed for one part of a facility should integrate with other parts of the facility. If there is no integration, then there is division between different departments.

Try to have equal numbers of management and non-supervisory employees on the committee. Choose middle management members who can get things done. One method is to pick one employee, have that employee pick another employee, and so on. With this method, or if employees are picked at random, there is no perceived bias from management.

Make sure that the safety professional serves as a resource for the committee and is not the leader of the committee.

If your workplace has a union, allow them to choose non-supervisory members. If your workplace is not unionized, you may want to consult with a qualified labor relations professional on the best way to obtain employee participation if you decide to use a committee [7].

One word of caution: Each committee must have a written description of its intent charter. This description will define exactly what the committee will be doing. In addition, it will allow you to perform some level of measurement on work that is accomplished.

HOW TO USE EMPLOYEES IN THE PROCESS

You can involve your employees in the management system and safety program by having them conduct regularly scheduled, routine physical surveys using a checklist. Make sure that employees have adequate and appropriate training; also, employees should be expected to help with decisions about hazard correction as well as hazard identification.

You may also choose to ask the safety committee to study one or two difficult safety problems that management has been unable to resolve. Once the committee is established and functioning successfully, it will be in a position to suggest other ways to involve your employees in the safety program. One last committee could not be a short-term committee to solve a specific issue. Employees will probably get more involved if they can see the "light at the end of the tunnel" (one or two month assignments).

Always remember that it is the employer who has ultimate legal responsibility for making sure that the workplace is safe for all employees [7].

JOINT LABOR–MANAGEMENT COMMITTEES

These types of committees usually have equal representation of labor and management. The chairperson may alternate between an employee representative and a management representative as appropriate. There usually are quorum requirements and formal voting. The power of this type of committee is worked out through negotiation. Although tasks depend on the outcome of these negotiations, the committees typically conduct the following activities [7]:

- Site evaluations with oversight of hazard recognition
- Investigating hazards reported by employees
- Incident investigations
- Safety awareness program development [7]

Sometimes the committees only receive reports from other committees on selected activities and monitor hazard correction and program effectiveness [7]. This will depend on your workplace.

OTHER JOINT COMMITTEES

In other joint committees, there may be either more employee participants (for example, at construction sites where several different trade unions represent employees) or more management participants (especially where safety, medical, and industrial hygiene professionals are counted as management). The top management at the workplace frequently chairs these committees. In other cases these committees are employees who are elected by the committee itself [7]. They work by consensus and do not take formal votes. Their usual functions are similar to those of the joint labor–management committees.

Employee Safety Committees

These committees are usually union safety committees, with membership determined by the union. Some workplaces with more than one union will have more than one union safety committee. The committee operates without management and meets regularly with management to discuss safety issues. At these meetings, the committee raises concerns,

and management provides the necessary responses to help fix the identified issue. The committee may conduct inspections, investigate hazards
reported by employees, and bring safety issues to management for corrective actions. The committee also may design and present safety awareness programs to employees [7].

Central Safety Committee

At non-union sites the central safety committee may consist of the site
manager and a member of the top management. In recent years, some
companies have discovered that it is helpful to have employees on the
committee. Some sites rotate employees on this committee so that all
employees can participate in safety planning. At other sites, management
selects the employees for their experience and achievements in other
safety management systems [7].

The central safety committee is an oversight committee with an interest in every part of the safety program. It sometimes serves as the hazard
correction tracking management system. It is sometimes known as a
follow-up committee. In this case, the committee receives reports of all
inspections, incident investigations, and hazards reported by employees,
and makes sure that all reported hazards are tracked until they are
resolved [7].

Specific-Function Committees

Some companies use single-function standing committees. In these
committees employees are given the opportunity to volunteer for
membership. These committees may consist of employees only, with management support; or there may be a joint membership with some management and/or the safety professional (including on-site nurse/physician
or medical provider, as applicable). Each committee has a single responsibility, such as incident investigation, site safety inspections, development
of safety rules, providing training, or creating safety awareness programs.
The company provides committee members with the necessary training
and resources [7].

Safety observers usually work with the area supervisor to get hazards
corrected. Normally, they do their inspection alone. Some companies
periodically bring together their safety observers to brainstorm problems
or ideas that extend beyond the individual work areas. For your safety

observers' participation to be fully effective, they should also be involved in correcting the hazards that they spot [7].

CONDUCTING SITE INSPECTIONS

Employee participation is common in workplace inspections. A joint committee, an employee committee that performs several functions, or a single-function inspection committee can conduct inspections, or an individual employee acting as safety observer can do so, as discussed [7].

Whatever method you choose, employees must be trained to recognize hazards. They also should have access to a safety professional and written references. For meaningful participation, the committee or safety observer should be able to suggest methods of correcting hazards and to track corrections to completion [7].

ROUTINE HAZARD ANALYSIS

Employees can be helpful in analyzing jobs, processes, or activities for hidden hazards and in helping to design improved hazard controls. Employees and supervisors frequently are teamed up to accomplish these activities. For complicated processes, an engineer probably will lead the team [7].

Employees are more likely to accept the changes that result from these analyses if they are involved in the decisions that revise practices and processes. For more information on routine hazard analysis and job hazard analysis in particular, refer to Chapter 15.

DEVELOPING OR REVISING SITE-SPECIFIC SAFETY RULES

Giving employees responsibility for developing or updating the site's safety rules can be profitable. Employees who help make the rules are more likely to adhere to them and to remind others to adhere to them. Employees possess an in-depth knowledge of their work environment and their co-workers and can contribute significantly in improving and strengthening the overall safety rules [7].

TRAINING OTHER EMPLOYEES

Use qualified employees to train other employees on safety matters— for example, rules and procedures. This technique can be effective and can improve your ongoing training efforts. Many companies have seen excellent results from delegating responsibility for training to employees. For more information on safety training, refer to Chapter 14 [7].

Employee Orientation

In many cases, employees can make excellent trainers for new employees. You will want someone in management to present the personnel/employee relations portions of the orientation, but trained employees can handle other topics. The trainer who provides this introduction to the job can follow up by acting as "buddy" and watching over the new employee, providing advice, and answering questions that a new employee might be afraid to ask a supervisor [7].

There is one problem that must be addressed at this time. Seasoned employees may train new employees with old procedures. These procedures may not be correct and could in effect cause injuries, some due to a shortcut. You need to be careful that you are not training employees on the "bad habits" that may exist in your workplace.

DIFFERENT APPROACHES: UNION AND NON-UNION SITES

Employee participation at a unionized facility is usually achieved differently than at non-union facilities. Neither type of workplace is more conducive than the other to enhancing safety awareness through employee participation. At both union and non-union worksites, employee participation is characterized by a commitment to cooperative problem solving. Just as important, employee participation relies on: respect among employees and representatives of organizations [7].

Unionized Work Sites

A reduction in incidents is clearly beneficial to management and employees. This goal should lend itself to joint union–management effort. The union will need to participate from the beginning [7].

The most common form of cooperative, participatory effort is the joint labor–management safety committee, as discussed. Sometimes an all-employee safety committee will be used [7].

Committee duties can range from reviewing hazards reported and suggesting corrections, to conducting site surveys, to handling incident investigations and follow-up of hazards. Some committees are advisory; others have specific powers to correct hazards and, in some circumstances, to shut down unsafe operations [7]. Table 7-3 presents a sample employee participation case study for union sites.

Non-union Work Sites

Sometimes there is employee hesitation (resistance) when trying to implement employee participation in a non-union work environment. The key is to be careful not to force or impose "voluntary" employee participation on any employee who does not want to get involved. In some cases, you may need to convince employees that their participation is necessary to help develop an effective safety culture. One method is to conduct buy-in meetings where employees with management support present the benefits of participation. Once employees feel comfortable with management support, the program will have an improved chance of success. Management support includes protecting employees from reprisal when they get involved in safety activities [7].

Table 7-4 presents sample employee participation case studies for non-union sites.

During the final phases of the culture building process where employee participation is important, if you are not careful, you will wear your employees down and their feeling will be "Okay, you win" (Figure 7-4). If you keep pushing the "safety stuff" to your employees, they will give up and only do what they have to do and nothing else. You do not want

Table 7-3
Sample Employee Participation Case Studies for Union Sites

Case 1. Employee participation at a paint manufacturing facility with 72 employees works primarily through the Safety Committee. Three members of the committee are hourly bargaining unit employees selected by the union, and three are salaried non-bargaining unit employees selected by management. Members participate in committee meetings, hold monthly plant inspections, and recommend safety- and health-related improvements to management.

Case 2. An oil refinery with almost 400 employees involves its employees in a variety of ways. Employees act as safety and health monitors assigned to preventive maintenance contractors. Employees developed and revised safe work procedures and are part of the team that develops and reviews job hazard analyses. They serve as work group safety and health auditors.

Case 3. A chemical company with 1,200 employees has found numerous ways to include its employees in the site's safety and health program. For example, the Safety and Health Committee, which includes equal labor and management representatives, has responsibility for a variety of activities, such as monthly plant inspections, incident investigations, and examination of any unsafe conditions in the plant. Employees are also involved in process and operations review, safety inspection, and quality teams. Two hourly employees work full-time at monitoring the safety and health performance of on-site contractors.

Case 4. An electronics manufacturer with almost 5,800 employees has established a joint committee consisting of seven management and eight hourly employees. They conduct monthly inspections of preselected areas of the facility, maintain records of these inspections, and follow up to make sure that identified hazards are properly abated. They investigate all incidents that occur in the facility. All committee members have been trained extensively in hazard recognition and incident investigation.

U.S. Department of Labor, Office of Cooperative Programs, Occupational Safety and Health Administration (OSHA), *Managing Worker Safety and Health*, November 1994, public domain.

this to happen. You need buy-in from your employees. You need to look beyond this and use employees as a resource to help you understand your process better. Be careful in your journey to take small steps that will lead to larger steps. Do not push too hard. Focus on small steps and let it happen naturally.

Table 7-4
Sample Employee Participation Case Studies for Non-union Sites

Case 1. A textile manufacturer with more than 50 plants and employee populations of 18 to more than 1,200 has established joint safety and health committees on all shifts at its facilities. All members are trained in hazard recognition and conduct monthly inspections of their facilities.

Case 2. A small chemical plant with 85 employees involves employees in safety activities through an Accident Investigation Team and a Safety and Communications Committee consisting of four hourly and three management employees. The team investigates all incidents that occur at the facility. The committee conducts routine site inspections, reviews all accident and incident investigations, and advises management on a full range of safety and health matters.

Case 3. Employee participation at a farm machinery manufacturer of 675 employees includes active membership on several committees and subcommittees. Members change on a voluntary, rotational basis. Committees conduct routine plant-wide inspections and incident investigations. Employees also are involved in conducting training on a variety of safety and health topics. Maintenance employees are revising the preventive maintenance program.

Case 4. A large chemical company with 2,300 employees implemented a dynamic safety and health program that encourages 100 percent employee participation. Its safety and health committee is broad and complex, with each department having its own committee structure. Subcommittees deal with specific issues, for example off-site safety, training, contractors, communication, process hazard analysis, management, and emergency response. The plant-wide committee, which includes representatives from all departmental committees, is responsible for coordination. All committee members are heavily involved in safety and health activities, for example, area inspections and incident investigations. They also act as channels for other employees to express their concerns. Members receive training in incident investigations, area assessments, and interpersonal skills.

U.S. Department of Labor, Office of Cooperative Programs, Occupational Safety and Health Administration (OSHA), *Managing Worker Safety and Health*, November 1994, public domain.

USUAL FORMS OF EMPLOYEE PARTICIPATION

At many non-union facilities, employee participation is rotated through the entire employee population. Programs receive the benefit of

Okay You Win!

Figure 7-4 Okay, you win. Created by Damon Carter.

a broad range of employee experience. Other employees' benefits include increased safety knowledge and awareness. At other non-union facilities, employee participation relies on volunteers, while at some facilities employees are appointed to safety committees by their supervisors [7].

The best method for employee participation will depend on what you want to achieve and the direction that you want your program to go. If improved employee awareness is a major objective, rotational programs are a good choice. If high levels of skill and knowledge are required to achieve your safety objectives, volunteers or appointees who possess this knowledge may be preferable [7].

WHAT MANAGEMENT MUST DO

Management sets the tone. Refer to Chapter 6 on management commitment and leadership. If management is not supportive of getting employees involved, and unless your employees believe you want their participation, getting them to participate will be difficult and may not be successful [4].

Employees often do not believe management actually wants their input on serious matters, so the participation is minimal. Sometimes this is because managers claim that safety committees only want to talk about "trivial" things such as cafeteria menus. Management may decide from this evidence that employees are either unwilling or unable to address the serious issues of the worksite. It is essential that mistrust and miscommunication between management and employees be corrected. You can accomplish this by demonstrating visible management commitment and support and providing positive feedback [7].

The following are some recommended solutions that you can use to encourage employee participation:

- Show your commitment through management support and leadership. This helps your employees to believe that you want a safe workplace, whatever it takes.
- Communicate clearly to your employees that a safe workplace is a condition of their employment.
- Tell your employees what you expect of them. Document the requirements. Refer to Chapter 8 on roles and responsibilities. Communicate to all employees specific responsibilities in the safety program, appropriate training, and adequate resources for performing specific activities that were assigned.
- Get as many employees involved as possible: brainstorming, inspecting, detecting, and correcting.
- Make sure that employee participation is expected as part of the job during normal working hours or as part of their assigned normal jobs.
- Take your employees seriously. Implement their safety suggestions in a timely manner, or take time to explain why they cannot be implemented.
- Make sure co-workers hear about it when other employee ideas are successful.

SUMMARY

Employee participation has been shown to improve the quality of safety programs. Your employees are equipped to provide assistance in a wide variety of areas. What employees need are opportunities to participate. This can be shown by clear signals from management through leadership, training, and resources.

Employee participation differs at unionized and non-union sites. No matter what forms of participation you choose to establish your program, you have the opportunity and responsibility to set a management tone that communicates your commitment to safety, thereby eliciting a high-quality response from your employees. Remember, no matter what recommendations management may glean from employee participation groups; make sure safety always remains the legal and moral obligation [7].

The objective of employee participation is to provide for and encourage employees to help in the structure and operation of the safety program and in decisions that affect their safety. If this is done properly, it will help employees commit their insight and energy for achieving the safety program's goal and objectives.

The following are some general guidelines for forming committees and making them successful:

- Include equal numbers of management and non-supervisory employees.
- Choose management members who can get things done.
- If your workplace is not unionized, you may wish to solicit employee suggestions on how to select non-supervisory members of the committee.
- One important consideration is to consult with your human resources professionals before holding an election for Safety Committee members. These employees may volunteer and be put on a rotating basis to extract as much information and knowledge from as many employees as possible. If you are not sure where you stand on these issues, it would be advisable to consult with your company attorney.

The following are general guidelines for involving employees. The key is to provide employees:

- The opportunity to participate
- Clear signals from management
- Needed training and resources
- The knowledge that they are taken seriously
- The understanding that a safe environment is a condition of employment

- Implementation of employee suggestions in a timely manner, or a clear explanation of why they cannot be implemented
- A policy statement that employees are protected from reprisal resulting from safety program participation
- The knowledge that all employees hear about the success of other employees' ideas
- Opportunities and mechanism(s) for employees to influence safety program design and operation—make sure that there is evidence of management support of employee safety interventions
- The certainty that employees have a substantial impact on the design and operation of the safety program

There are multiple avenues for employee participation, and these avenues are well known, understood, and utilized by employees. The avenues and mechanisms for involvement are effective at reducing accidents and enhancing safe behavior.

REFERENCES

1. Bone, Louis E., David L. Kurtz, *Management*, 3rd ed., Random House Business, New York, 1987, p. 677.

2. Chruden, Herbert J., Arthur W. Sherman, Jr., *Personal Management: The Utilization of Human Resources*, 6th ed., South-Western Publishing Co., Cincinnati, OH, 1980.

3. Kennedy, Carol, *Instant Management,* revised 1993, Williams Morrow and Company, Inc., New York, 1991.

4. Oklahoma Department of Labor, Safety and Health Management: Safety Pays, 2000, http://www.state.ok.us/~okdol/, Chapter 4, pp. 21–23, public domain.

5. Ouchi, William, *Theory Z: How American Business Can Meet the Japanese Challenge*, Avon Publishers, 1981.

6. Rue, Leslie W., Lloyd L. Byers, *Management: Theory and Application*, 5th ed., Irwin, Homewood, IL, 1989.

7. U.S. Department of Labor, Office of Cooperative Programs, Occupational Safety and Health Administration (OSHA), *Managing Worker Safety and Health*, November 1994, public domain.

8. Wertheim, Edward G., Historical Background of Organizational Behavior, College of Business Administration, Northeastern University, Boston, MA 02115, Adapted from Web site, http://www.cba.neu.edu/~ewertheim/introd/history.htm#HR

9. OSHA Web site, http://www.osha-slc.gov/SLTC/safetyhealth_ecat/comp1_empl_envolv.htm, public domain.

8

Assigning Safety Responsibilities

INTRODUCTION

Management is responsible for establishing the purpose of the management system, determining measurable objectives, and taking the actions necessary to accomplish those specific objectives [3].

As a member of management, you must understand the importance of taking the accountability for the safety of your employees—new employees, transferred employees (temporary employees), contractors, etc. To be successful you must learn to assign specific accountabilities to responsible individuals in the organization. It is important to understand the difference between assigning responsibilities and delegating responsibilities. You cannot delegate your responsibilities. However, if you assign your responsibilities, you will still have control.

What you can do is to expect other individuals in the organization to share the responsibility for certain elements of the management system. Many managers question why they should be assigned the responsibility for safety. In most cases, you have a working knowledge of your business issues and you should be close to your employees. However, as a business grows and the numbers of employees increase, being responsible for all of the details of an effective safety management system may become less feasible. It is important to understand how to have a management system in place for assigning some of the safety responsibilities to others [2].

In larger organizations, managers must develop a management system for clearly assigning safety responsibilities and authority, and providing the necessary resources to get this accomplished.

One method that can be used is through written job descriptions. These job descriptions can include the following elements:

- Clarify the specific safety responsibilities and authority of individuals
- Distribute responsibilities evenly among supervisors and employees
- Establish a charter for safety committees [1] (Make sure that this charter is written)

In this chapter we will focus on how to develop job descriptions that can be used to help you to achieve safety responsibilities. To understand the need for the job description, you will have to accomplish the following:

- Review your existing organizational structure. How does this influence the operation?
- Determine what part each position should play in your management system and what level of responsibilities you want to assign to each position
- Determine the level of authority that the selected position should play and what resources will be needed

Once you have decided what you want to do, communicate what has been established to all employees [2].

There is one caution that must be considered at this point. You must not confuse the job descriptions with JHAs. JHAs are not job descriptions; rather, they are lists of job tasks. We will discuss JHAs in detail in Chapter 15.

THE VALUE OF WRITTEN JOB DESCRIPTIONS

A written job description describes the structure of the organization, including the safety management system and other associated elements—for example, production and quality. An individual job description describes the most important characteristics and responsibilities of a position [1].

Some small companies do not understand the value of job descriptions. OSHA believes that, with respect to safety responsibilities, written statements are preferable to oral assignments [2].

Regardless of what OSHA believes, any organization should always do what is right for the business. Each company must understand that a carefully written job description accomplishes the following:

- Removes any doubt about the responsibilities and authority for each position

- Helps to determine that all responsibilities have been accounted for in the organization
- Identifies how new tasks and responsibilities should be assigned
- Helps to develop job performance objectives and establishes performance measurements
- Improves and enhances communication throughout the organization [2]

REVIEW THE EXISTING ORGANIZATION

In every business, there are individuals who should be assigned and involved in developing and implementing the management system. Refer to Figure 8-1 for a sample safety program responsibilities worksheet. This will help you with developing descriptions of the basic positions normally involved in a safety program. List all the positions in your business [2].

DETERMINE ROLE OF EACH POSITION

One of the keys in determining the safety roles of individuals is to determine the following:

- What role do you want each position or group to play in your management system?
- What level of authority will the person holding these positions need to accomplish the goals and objectives [2]?

The method to accomplish this is to write a general statement of responsibility and authority for each position. Although authority is built into managerial and supervisory roles and responsibilities, you may want to make changes as they relate to your management system. This can be accomplished by:

- Clearly stating the scope of authority by indicating supervisory relationships
- Establishing a budget that the position can use
- Any other measures that describe what a person in this position can do without obtaining further approval [1]

All employees are responsible for implementing the provisions of the safety policy that pertain to specific operations. The responsibilities listed are minimum expectations.

Job Title

General policy statement

Limits of authority and resources (expenditures, reporting hazards, authority to shut down equipment)

List specifically how the employee will be responsible and held accountable:

Figure 8-1 Sample safety program responsibilities worksheet. U.S. Department of Labor, Office of Cooperative Programs, Occupational Safety and Health Administration (OSHA), *Managing Worker Safety and Health*, November 1994, public domain.

At this stage, do not attempt to describe each job's specific tasks in detail. This is suggested wording for job descriptions; which of these responsibilities fit into your safety program, and at what authority level and specific positions in your business should these responsibilities be assigned? The following are some examples of some assigned safety roles for each category.

- Top management establishes and provides the leadership and resources for carrying out the stated company safety policy. Table 8-1 shows sample responsibilities of top management.

Table 8-1
Sample Responsibilities of Top Management

- Establish a policy to make sure that the worksite stays in compliance with all applicable regulatory requirements and best management practices
- Provide a safe work environment and working conditions for all employees
- Provide the leadership and resources to accomplish safety policies and procedures
- Resolve any conflicts of production priorities where safety is a concern
- Establish goals and objectives
- Support the safety professional and employees in their requests for information, training, other professional services, facilities, tools, and equipment needed to develop an effective safety program
- Assign clear and understandable responsibilities for the various aspects of the safety program
- Make sure that employees with assigned responsibilities have adequate resources and authority to perform their assigned duties
- Hold managers, supervisors, and employees accountable with assigned responsibilities by checking to make sure they are meeting their responsibilities and by recognizing them, as appropriate
- Evaluate the effectiveness of managers and supervisors in regard to the safety program
- Keep in touch with employees and other individuals who perform safety activities
- Assist in providing direction and authority for specific activities and visibly demonstrate involvement
- Set an example by demonstrating commitment (following safety rules and safe work practices)
- Make sure that all vendors, customers, contractors, and visitors comply with the company safety policy
- Thoroughly understand all hazards and potential hazards that employees may be exposed to; make sure that a comprehensive program of prevention and control is established and operating
- Provide a reliable system for employees to report hazardous conditions and other situations that appear hazardous; make sure that responses to employees reports are appropriate and timely (Chapter 11)
- Encourage employees to use established hazard reporting system(s); guarantee a strict prohibition of reprisal for all employees, supervisors, and managers
- Establish an inspection system, for example, self-inspections, and review the results periodically to make sure that the proper and timely hazard corrections are made

Table 8-1
Continued

- Develop a preventive maintenance program to make sure that there is proper care and functioning of equipment and facilities
- Review incident reports to keep informed of causes and trends
- Provide medical and emergency response systems and first aid facilities adequate for the size and hazards at the worksite
- Require periodic drills to make sure that each employee knows what to do in case of an emergency
- Develop training programs for all employees, managers, and supervisors to help them recognize and understand hazards and protect themselves and others

U.S. Department of Labor, Office of Cooperative Programs, Occupational Safety and Health Administration (OSHA), *Managing Worker Safety and Health*, November 1994, public domain.

- Managers and supervisors help to maintain safe working conditions in their respective areas of responsibility. Table 8-2 gives sample responsibilities for plant/site superintendents/managers, and Table 8-3 for supervisors.
- Employees exercise care in the course of their work to prevent injuries to themselves and their coworkers (Table 8-4).
- Visitors, vendors, customers, and contractors comply with all safety requirements and/policies and procedures while at the workplace.
- Engineering helps to make sure that all equipment that could affect the safety of employees is selected, installed, and maintained in a manner that eliminates or controls potential hazards.
- Purchasing (sometimes called procurement) helps to make sure that safety equipment and materials are purchased in a timely manner; that new materials, parts, and equipment are analyzed for potential hazards so preventive measures or controls can be implemented; and that such materials, parts, and equipment are obtained in accordance with all applicable safety requirements.
- Safety professionals help in assessing safety issues by working with management to resolve identified problems. In addition, the safety professional is a consultant to management [2]. Table 8-5 shows sample responsibilities for safety and health managers/coordinators.

Table 8-2
Sample Responsibilities of Plant/Site Superintendents/Managers

- Provide the leadership and appropriate direction essential to maintain the safety policy as the fundamental value in all operations
- Hold all supervisors accountable for all assigned safety responsibilities, including their responsibility to make sure that employees under their direction comply with all safety policies, procedures, and rules
- Evaluate the safety performance of supervisors, taking into account indicators of good performance—for example, safety activities leading to reducing injuries; housekeeping; a creative, cooperative participation in safety activities; a positive approach to safety problems and solutions; and a willingness to implement recommendations
- Make sure that the physical plant is safe—for example, structural features, equipment, and the working environment
- Insist that a high level of housekeeping be maintained, that safe working procedures be established, and that employees follow procedures and apply good judgment to the hazardous aspects of all tasks
- Participate in regular inspections to observe conditions and to communicate with employees
- Provide positive reinforcement and instruction during inspections, and require the correction of any hazards
- Actively participate in and support employee participation in safety program activities
- Provide timely and appropriate follow-up to recommendations made by employee or joint labor–management committees
- Make sure that all new facilities, equipment, materials, and processes are reviewed for potential hazards. This must occur during design and before purchase so that potential hazards are minimized or controlled before their introduction into the work environment, that tools and machinery are used as designed, and that all equipment is properly maintained
- Make sure that JHAs are conducted and reviewed periodically for all jobs, with emphasis on specific tasks, so those hazards can be identified and minimized or controlled
- Make sure that employees know about the management system and are encouraged to use the systems for reporting hazards and making safety suggestions. Make sure that employees are protected from reprisal, their input is considered, and that their ideas are adopted when helpful and feasible
- Make sure that corrective action plans are developed where hazards are identified or unsafe acts are observed

Table 8-2
Continued

- Make sure that all hazardous tasks are covered by specific safe work procedures or rules to minimize injury
- Provide safety equipment and protective devices, as appropriate, and make sure employees understand how to use the equipment properly
- Make sure that any employee's injury, no matter how minor, receives prompt and appropriate medical treatment
- Make sure that all incidents are promptly reported, thoroughly investigated, and documented, and that safety recognition programs do not discourage reporting of incidents
- Stay on top of incident trends; take proper corrective action, as appropriate, to reverse these trends
- Make sure that all employees are physically qualified to perform their assigned tasks
- Make sure that all employees are trained and retrained, as applicable, to recognize and understand hazards and to follow safe work procedures
- Make sure that supervisors conduct periodic safety meetings to review and analyze the causes of incidents and to promote free discussion of hazardous conditions and possible solutions
- Use the safety manager to help promote aggressive and effective safety programs
- Help develop and implement emergency procedures; make sure that all employees have the opportunity to practice their emergency duties
- Participate in safety program evaluation
- Refer to the corporate level and conflicts between productivity and safety
- Refer to the corporate level any major concerns and capital needs for safety

U.S. Department of Labor, Office of Cooperative Programs, Occupational Safety and Health Administration (OSHA), *Managing Worker Safety and Health*, November 1994, public domain.

DETERMINING AND ASSIGNING SPECIFIC RESPONSIBILITIES

Now that you have determined who should participate in your safety program, you need to develop written statements that specify what each position must do to help meet any stated goals and objectives. This corresponds to the last entry on the worksheet presented in Figure 8-1 [2].

When developing responsibilities for non-supervisory employees, be careful not to confuse these responsibilities with specific work rules and

**Table 8-3
Sample Responsibilities of Supervisors**

- Demonstrate your commitment. Be thorough and conscientious in following the safe work procedures and safety rules that apply. Practice what you preach
- Evaluate employee performance, for example, safe behavior and work methods
- Encourage and actively support employee participation. Provide positive reinforcement and recognition and/or group performance
- Maintain up-to-date knowledge and skills required to understand specific hazards, for example, improperly functioning tools and/or equipment
- Maintain good housekeeping
- Make sure that the preventive maintenance program is being followed and that any hazards are found and tracked to completion
- Conduct frequent inspections, using a checklist, to evaluate physical conditions
- Investigate incidents thoroughly to determine root cause and how hazards can be minimized or controlled
- Hold employees accountable for their safety responsibilities and actions
- Consistently and fairly enforce safety rules and procedures
- Discourage shortcuts
- Provide on-the-job training in safe work procedures and the use and maintenance of PPE
- Make sure that each employee knows what to do in case of an emergency
- Refer to top management any resource needs you cannot resolve

U.S. Department of Labor, Office of Cooperative Programs, Occupational Safety and Health Administration (OSHA), *Managing Worker Safety and Health*, November 1994, public domain.

safe work practices. A brief, general statement about the employee's responsibility to understand and follow rules and safe work practices is more appropriate [2].

You should assign the details of carrying out your management system to the same individuals who are responsible for plant operations, the office environment, production, and other areas that should be included. In this way, you build safety into the management structure as firmly as into the production environment. Be sure that each assigned responsibility comes with the authority and resources needed to fulfill the requirements [2].

Table 8-4
Sample Responsibilities for Employees

- Learn the safety rules. Understand and follow specific rules. Avoid shortcuts
- Review all safety educational material posted and distributed. If you do not understand something, ask questions
- Take personal responsibility for keeping yourself, your co-workers, and equipment free from hazards
- Make sure that you understand work instructions before starting a new task
- Avoid taking shortcuts through safe work procedures
- Seek information on hazardous substances that you may work with so that you understand the associated dangers that may exist and how to protect yourself
- If you have any doubt about the safety of a task, stop and get instructions from your supervisor before continuing. Be sure that you completely understand instructions before starting work
- If you have a suggestion for reducing safety risks, discuss it with your supervisor. It is your responsibility to get involved and help resolve these issues
- Take part in the employee participation system and support other employees in their assigned safety roles (Chapter 7)
- Make sure that you understand what your responsibilities are in emergency situations
- Know how and where medical help can be obtained
- Report all incidents and unsafe conditions to your supervisor or use the management system to report safety conditions

U.S. Department of Labor, Office of Cooperative Programs, Occupational Safety and Health Administration (OSHA), *Managing Worker Safety and Health*, November 1994, public domain.

SUMMARY

After you have developed the safety responsibilities and specific activities for each position, you must communicate this information to all employees. You may find it useful to combine all these written statements of safety responsibility into one document. Then post it or circulate it to all employees involved. Discuss the job descriptions and responsibilities in one-on-one (face-to-face) meetings with the employees who will be responsible for carrying out the responsibilities. Keep a copy of this document and always refer to it when meeting with employees, no matter whether you are in a general meeting or performance reviews [2].

Table 8-5
Sample Responsibilities for the Safety and Health
Manager/Coordinator

- Keep top management informed of all significant safety developments—for example, the status and results of hazard evaluations, inspections, and incident investigations, and any other serious safety issues or opportunities for improvement in the management system and safety program
- Maintain safety knowledge through attending training seminars, reading safety literature, networking, selected conferences, and use of other outside professionals
- Stay abreast of and be able to interpret specific requirements dealing with employee risk
- Act as the eyes, ears, and "conscience" of top management where employee safety is concerned
- Working with managers, supervisors, employees, and other professionals as needed, aid in developing an inventory of hazards and potential hazards, and planning a program of prevention and control
- Evaluate the preventive maintenance program's effectiveness to make sure that the workplace is safe
- Conduct a hazard analysis—for example, a hazard detection, prevention, or control plan—when new equipment, facilities, or materials are designed, purchased, and/or used, and when new processes are designed
- Provide technical assistance and support to supervision and employees
- Assist management in making sure that the appropriate safety rules are developed, communicated, and understood by all employees
- Assist in or oversee the development of a system for consistent and fair enforcement of rules, procedures, and safe work practices
- Assist management in providing adequate equipment for personal protection, industrial hygiene, safety, fire prevention, etc.
- Inspect and/or assist in inspection of facilities to identify hazards that may have been overlooked during prevention and control mechanisms and to identify any previously undetected hazards
- Investigate or oversee investigation of hazards
- Respond to employee safety suggestions
- Assist supervisors in investigating incidents such as property damage and near misses
- Provide technical assistance to supervisors and employees in the performance of their duties under the safety program
- Assist in developing and providing safety training to all employees, so that they will understand the hazards of the workplace and their responsibility to protect themselves and others
- Oversee, analyze, and critique periodic emergency drills to improve worksite emergency readiness
- Mentor management on the appropriate direction of the management system

U.S. Department of Labor, Office of Cooperative Programs, Occupational Safety and Health Administration (OSHA), *Managing Worker Safety and Health*, November 1994, public domain.

Many managers can be perceived as risk-takers, willing to put their business against others in a competitive world. However, there is one gamble that is a true loss for management to consider: a gamble on safety and the risk of incidents that cause injuries to employees or damage to company property [1].

To reduce risks effectively, you must address safety along with production, quality control, and costs. After all, the cost involved, as discussed in Chapter 1, comes off the bottom line. To accomplish this task you must set specific goals and objectives to provide a service or produce a quality product efficiently without an injury. Too often, that is seen as something to be considered as time permits, over and above regular business activities.

For your management system to be successful, you need to assign responsibility to specific positions, departments, and staff levels in your organization. The following steps can help you to make sure that your safety program elements are communicated properly:

- Review your current structure. Understand what you want it to look like
- Determine what part each job position should have in the safety program
- Determine what authority and resources are necessary to carry out these roles and responsibilities
- Determine and assign safety responsibilities, and write responsibilities into each employee's job description
- Communicate with all employees by discussing the responsibilities and authority in face-to-face meetings [2]

In the next chapter, we will discuss accountability.

REFERENCES

1. *The Manager's Handbook: A Reference for Developing a Basic Occupational Safety and Health Program for Small Businesses*, State of Alaska, Department of Labor and Workforce Development, Division of Labor Standards and Safety, Occupational Safety and Health, April 2000, Section 1, pp. 3–4, public domain.

2. U.S. Department of Labor, Office of Cooperative Programs, Occupational Safety and Health Administration (OSHA), *Managing Worker Safety and Health*, November 1994, public domain.

3. Crosby, Phillip B., *Quality Is Free*, McGraw-Hill Book Company, New York, 1979.

9

Developing Accountability

INTRODUCTION

Why is it important to develop an accountability management system? Imagine a sports organization with a coach (manager) and players (employees, contractors, temporary employees). Each player is assigned specific tasks and responsibilities that are critical to the success of the team [1].

A clearly defined accountability system is the mechanism to make sure that employees successfully fulfill their assigned responsibilities. When players (employees) fail to show up for practice (work) with no reasonable cause, they are fined (penalized). If they perform poorly, no matter the reason, they do not make the starting lineup (promotion). Player contracts (accountability contracts) reflect trends in poor performance or relative value to the team (co-workers), thus creating a form of employee accountability for performance. We have all heard of coaches (managers) fired (reassigned) at the end of the season (after a series of injuries), sometimes even in the middle of a season because of their performance. The potential for dismissal creates a sense of employee accountability among coaches (managers) and players (employees). For the organization, profit and loss statements (management's salary) are strong motivators to do the job well [1].

Based on this example, one should see how important an accountability management system is for a sports club (plant environment), and the purposes of a good management system in the organization. Reputation and public approval are strong motivators for all team members and companies. As we discussed, if we take this team concept (employee participation) and apply it to an organization, we find that business consists of managers, supervisors, and leaders (coaches) and the employees (the players). Employees on the team have their own areas of responsibility. Unfortunately, in business these areas are not always clearly defined as on a sports team. Management may not understand that each employee must perform at top efficiency to create a successful team (develop a management system and a safety culture).

In many cases, supervisors are asked to double as managers or as production employees, as the need arises. These kinds of flexible and undefined organizational structures can lead to breakdowns in communication that lead to lack of accountability. Other programs may suffer or fail [1]. The bottom line, no matter how you look at it, is that management sets priorities, and what is on top of the list gets done. As we have discussed, safety must be a value that is built into the management system. As all of us know, priorities change over time, but a value, once established, will be less apt to change with those priorities. Review our discussion on a to-do list. This is one of the most important concepts that management must understand. This is how the culture is created and will continue to function and improve.

Responsibilities frequently become so complex that some activities are neglected because of the number of duties that must be performed. As new responsibilities and business initiatives are created, accountability breaks down when responsibilities are assigned and the authority or resources are not provided [1].

The purpose of an accountability system is to help employees understand how critical their role is, and to help them understand what to do to take responsibility for their performance and actions. In the present context, accountability makes sure that your safety program is not just a "paper program," a book or manual that is developed and then sits on the bookshelf collecting dust, with no real purpose in achieving its objectives. The following sections will help you assess and clearly define your safety accountability [1].

ESTABLISHING CLEAR GOALS AND ASSIGNING RESPONSIBILITIES

Before you can hold management/employees accountable for their actions, you must make sure that they know what is expected. They must have goals set for their performance. Employee goals for safety stem from the overall company goal and objectives. The method for defining your company goals was discussed in Chapter 5. You need to understand these guidelines before you can establish your safety goal, the objectives leading to the specific goal, and a set of job descriptions that clearly define

Table 9-1
Responsibility, Authority, and Accountability

When you have authority or responsibility, your performance is not necessarily measured. However, when you are held accountable, your performance is measured in relation to goals or objectives that may result in certain positive or negative consequences.

Top management assigns specific responsibilities to managers or supervisors. If top management undercuts this authority, it will interfere with the manager's/supervisor's ability to carry out those responsibilities. At the same time, the manager wants to demonstrate their own commitment to reducing safety hazards and protecting employees.

When managers and employees are held accountable for their responsibilities, they are more likely to press for solutions to safety problems than to present barriers. By implementing an accountability system, positive involvement in the safety program is created.

Do you have a safety accountability system in place at your workplace? If so, can it be improved?

http://www.osha-slc.gov/SLTC/safetyhealth_ecat/comp1_responsibility.htm, public domain.

Table 9-2
Essential Components of an Effective Accountability System

- Established formal standards of behavior and performance—for example, policies, procedures, or rules that clearly convey standards of performance in safety to employees
- Resources provided to meet those standards—for example, a safe workplace, effective training, and adequate oversight of work operations
- An effective system of measurement
- Appropriate application of consequences, both positive and negative
- Consistent application throughout the organization

Oregon OSHA Web site, http://www.cbs.state.or.us/external/osha/educate/training/pages/materials.html. Continuous Safety Improvement, Training program 100w, public domain.

safety responsibilities [1]. Table 9-1 gives an overview of a model application of an effective responsibility, authority, and accountability system.

An effective management system has five essential components of an effective accountability system. Refer to Table 9-2 for the essential components.

Table 9-3 provides an example of accountability for a behavior-based system. We will discuss this in more detail in Chapter 16.

Table 9-3
Examples of Measuring Safety Behavior Responsibility, Authority, and Accountability

The following are some examples of measuring safe behavior at various levels:

- *Top/mid-level managers.* Measurement at this level includes personal behavior, safety activities, and statistical results, for example, following company safety rules, enforcing safety rules, arranging safety training, and monitoring workers' compensation costs.
- *Supervisors.* Measurement should include personal safety behavior and safety activities that they are able to control, for example, making sure employees have safe materials and equipment, following and enforcing safety rules, and conducting safety meetings, one-on-one contacts, etc.
- *Employees.* Measurement usually includes personal behavior, for example, complying with safety rules, participating in safety processes, and reporting injuries and hazards.

http://www.osha-slc.gov/SLTC/safetyhealth_ecat/comp1_responsibility.htm#, public domain.

The next step is to establish performance objectives for all employees with assigned safety responsibilities. These objectives must be realistic, understandable, measurable, and achievable. It is your job as a manager to clearly establish who is responsible for performing specific tasks. Evaluate your assignments of responsibility to make sure that they specify who does what and how they do it, and that they are realistic and reasonably attainable. When objectives are unclear, they can be misunderstood. In this case, it will be hard to determine whose performance is lacking—the manager or the employee [1].

When you assign responsibilities to employees, it is essential that you also delegate the necessary authority and/or commit sufficient resources. Only a few things can be more demoralizing to a motivated employee than being given an assignment without being provided the means to accomplish it. By providing the resources, you will help to make sure that each employee accomplishes the desired goals and objectives [1].

ESTABLISHING EMPLOYEE OBJECTIVES

Objectives for employees should be based on performance measurements. These indicators tell you whether the employee performed the desired objective as expected [1]. Refer to Chapter 5 for a discussion on developing objectives.

ESTABLISHING CONSEQUENCES FOR FAILURE TO PERFORM ADEQUATELY

As the management system is being developed and the employees acquire new skills and start to change behavior patterns, there should be minimal or no consequences for poor performance. In this case, use positive reinforcement during the initial phase of performance evaluation. This will encourage employees' natural desire to do what is right and to be recognized [1].

Although the goal of any accountability system should be to develop a sense of employee accountability for actions, employees must understand that there are negative consequences for poor performance. Consequences reinforce the importance of meeting stated objectives. Make sure that each manager and supervisor understands when the consequence will occur. There should be no surprises [1].

However, it is important that consequences be appropriate to the situation. For example, termination of a supervisor for his first poorly conducted incident investigation is an obvious example of overreacting to a problem. Gradually the consequences of poor performance should be increased to some specified maximum severity [1]. How many times have you heard of this situation? Table 9-4 illustrates the difference between responsibility and accountability.

DEFINING ACCOUNTABILITY

Most supervisors today know that they are responsible for safety, and they know what they should be doing, yet they do not do it. Why? Because they usually are not held accountable. That is, they are not measured in safety [3].

Accountability is nothing if you cannot measure what you have established. To hold someone accountable, you must know if they are performing their job functions correctly. To understand these accomplishments, you must measure their performance. Without mea-

Table 9-4
Responsibility and Accountability: Are they the same?

Responsible

- Expected or obliged to account for; answerable to
- Able to distinguish between right or wrong
- Dependable, reliable

Responsibility

- Being responsible; an obligation (Webster)
- A thing or person one is responsible for
- Establishes an obligation to perform assigned duties
- An antecedent, activator, initiating behaviors under behavior-based safety

Accountable

- Responsible; liable, explainable
- Liable; legally bound, subject to or bound by standards and subject to consequences
- Perception of being subject to consequences based on performance

Accountability

- The state of being accountable, liable, or answerable
- Confers the obligation to perform assigned duties to a desired standard, or else
- Fixes liability through measurement and consequences
- A system of performance measurement, evaluation, and consequences

OR OSHA 116, Responsibility and Accountability: Are They the Same?, p. 20, http://www.cbs.state.or.us/external/osha/educate/training/pages/materials.html, public domain.

surement, accountability becomes an empty, meaningless, and unenforceable concept [2].

Starting with what we propose as the single most important factor in getting good true management accountability, we find that we are really talking about ways to measure management better. Measurement has been a major downfall in safety for many years, because many managers do not understand the intended measurements and how to apply them to safety activities [2].

According to Dan Petersen, "For the line manager, to measure is to motivate." Although this statement might have sounded a little ridiculous 20 years ago, Petersen believes that it expresses a profound truth, at least in terms of the safety performance of top management. Managers react to the measures used by the "boss"; they perceive a task to be important only when the boss thinks it is worth measuring [2]. This is typical in any organization.

Once you understand the importance of performance measurement in obtaining good performance, we then hit our biggest snag. The questions we should ask: "What shall we measure?" "Should we measure our failures as demonstrated by incidents that have occurred in the past?" This is known as the OSHA incident rate (OIR). If this is a good measurement, as has historically been believed (for that is what we usually measure), then what level of failure should be measured? We can measure the level of failure we call "fatalities." Is this a "good" measure? Perhaps if we are assessing the national traffic safety picture, but it would be a little ridiculous in the case of a supervisor. A supervisor who never does anything to promote safety may never experience a fatality in his or her workplace. In this case, measuring fatalities makes no sense [2].

Although this may sound a little ridiculous, it accurately describes what is going on in many safety programs today. A supervisor can do nothing related to safety for a year and attain a zero frequency rate with a small bit of luck. By measuring and recognizing any part of supervision using the OIR, we are reinforcing nonperformance in safety [2].

> If fatalities (or frequency rates) are a poor measurement of supervisory performance, what is a good measurement? What is wrong with fatalities as a measurement? Measuring our failures is not the best approach to use in judging safety performance. This is not the way we measure employees in other aspects of their jobs. We do not, for example, measure line managers by the number of parts the employee in their departments failed to make. We do not measure the worth of sales professionals by the number of sales they did not make. In cases like these we decide what performances we want and then we measure to see if it is getting done [2].

What would be a good measure for supervisory safety performance? More important, what set of criteria can we develop for measuring supervisory safety performance? Or the safety performance of the corporation? Or our national traffic safety performance—or anything else related to safety [2]?

Taking a look at the issues of measurement shows us that we need different measurements for different levels of an organization, for different functions, and for different types of management style. What is a good measurement for one supervisor of ten employees may not be a good measure for another, much less for a plant manager of one plant or the general manager of seven plants and 10,000 people [2].

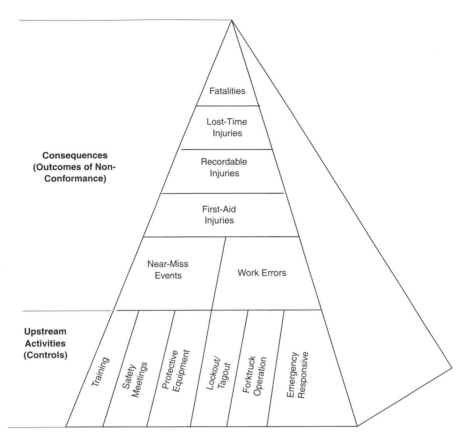

Figure 9-1 Safety performance pyramid. Manzella, James C., "Measuring Safety Performance to Achieve Long-Term Improvement," *Professional Safety*, September 1999, p. 35, Figure 1. Reproduced with permission.

MEASUREMENT TOOLS

Refer to Figure 9-1 for a safety performance pyramid. This pyramid is a graphical overview of consequences of outcomes of nonconformance and upstream activities (controls). Basically, if you do proactive activities, you will reduce your near misses, work error, etc. This is because there is more involvement of employees as we discussed in Chapter 7, "Employee Participation."

Measurement can classified into two categories:

- Activity (or performance-based) measurements
- Results measurements

Table 9-5
Activities and Results Measures

Activity Supervisor: Objectives Met	Manager: Objectives Met	System-Wide
# Inspections	Use of media	Audit
# Quality investigations	# Job hazard analyses (JHAs)	Questionnaires
# Trained	# Job safety observations	Interviews
# Hazard hunts	# One-on-ones	
# Observations	# Positive reinforcement	
# Quality circles	# Group involvement	

Results Supervisors	Managers	System-Wide
Safety sampling	Safety sampling	Safety sampling
Inspection results	Inspection results	Safety performance indicator
	Safety performance indicator	# First aid or frequency
	Estimated costs	# Near misses or frequency
	Control charts	Property damage
	Property damage	Frequency–severity index
		Estimated cost control charts

Peterson, Dan, *The Challenge of Change, Creating a New Safety Culture, Resource Guide*, CoreMedia Development, Inc., 1993, Activity/Results Chart (Activities and results measures for supervisors, managers and system-wide safety programs), p. 110. Modified with permission.

As a rule of thumb, in selecting measuring devices, use only activity measurements at the supervisory levels, and primarily activity measures (with some results measurements) at the middle to upper management levels. Reserve the pure results measurements for top management [2].

Table 9-5 provides details on specific activities and the results that can be achieved from them.

Table 9-6 gives an overview of the types of measurement that can be used to measure supervision.

According to Petersen, we can use either activity measurements or results measurements to determine performance. These measurements can be used at the supervisory level, middle and top managerial, or system-wide, provided that we use caution in identifying measurement selections. At the supervisory level the activity measurements are more

Table 9-6
Examples of Activity Indicator Measure Functions Carried
Out by Supervision

- One-on-one meetings
- Safety meetings conducted
- JHAs conducted
- Safety inspections conducted
- Employees recognized
- Hazards reported
- Incident investigations
- Safety observations
- Safety training conducted
- Any activity objectives determined jointly by supervisor and manager

OR OSHA 100w, Activity Indicators, p. 38, http://www.cbs.state.or.us/external/osha/educate/training/pages/materials.html, public domain.

appropriate because the supervisor is close to the employee. The key is to use the number of inspections, number of employees trained, or number of observations made. These activity measurements are equally appropriate at the managerial level, and can even be used at the system-wide level [2]. For an overview of the types of measures that can be used to measure supervision, refer to Figure 9-2.

Results measurements can also be used at all levels, as long as caution is used at the supervisor level. The traditional safety measurements—for example, frequency rate or severity rate—cannot be used at the supervisor level except over long periods of time, and then probably only as a quality check [2].

PERFORMANCE MEASUREMENTS FOR SUPERVISORS

We need to begin with the evaluation of a supervisor's performance by reviewing what he or she does to get results, and by determining whether the stated objectives have been accomplished.

Performance measurements have certain distinct advantages:

- They are flexible and take into account individual supervisory styles. It is important that you understand that you do not have to use the

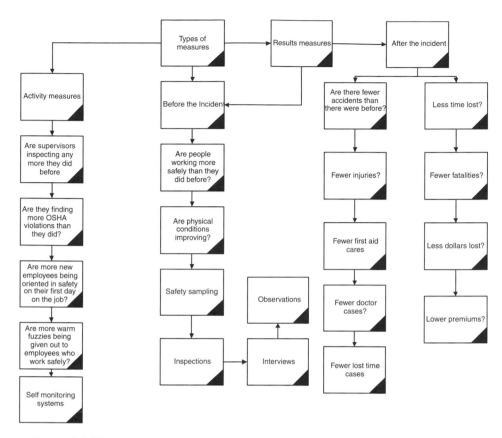

Figure 9-2 Types of measures. Peterson, Dan, *The Challenge of Change, Creating a New Safety Culture, Resource Guide*, CoreMedia Development, Inc., 1993, Type of Measures, p. 110. Modified with permission.

same measurement for all supervisors. As we discussed, you can let each supervisor select performance objectives and levels of performance that they want to see and then measure these performances and levels of achievement. Performance measurements are excellent for use in the MBO approaches presented by Peterson [3]. Let the supervisor pick an activity that she can do well, and she will exceed her own expectations.

- They all measure the presence, rather than the absence, of safety activities.
- They are usually simple and thus administratively feasible.
- They are clearly the most valid of all measures.
- They provide immediate feedback, since most supervisors are required to report their level of performance to their manager [3].

```
Supervisor _____ Department _____
Date _____ Reporting Period _____ to _____
Inspections Made
Date _____ # Hazards Corrected _____ # Recommendations _____
Meetings Held
Date _____ Type _____ # Attending _____ Subject _____
Accidents Investigated____
# Investigations _____ # Hazards Corrected _____ # Recommendations _____
Employee Contacts
# New Employee Orientations _____ # Performance Recognitions _____
# Enforcement Actions _____ # Safety Training: OJT ___ Formal _____
Comments: _____
_____
_____
```

Figure 9-3 Sample report of supervisor's safety activities. OR OSHA 100w, p. 39. http://www.cbs.state.or.us/external/osha/educate/training/pages/materials.html, public domain.

For examples of a self-monitoring system, refer to Figures 9-3 and 9-4. Table 9-7 provides specific activities that can be measured to get the desired results.

Figures 9-5 and 9-6 provide an overview of an activity-based safety system that will be successful if implemented properly. It allows the supervisor to choose selected topics for discussion with their employees. I have seen this type of system develop a successful culture, because it puts the supervisor in tune with the employees.

MIDDLE AND UPPER MANAGEMENT RESULTS MEASUREMENTS

At the middle management level we concentrate on results measurements. Here again you can work with before-the-fact (upstream) measurements or with failure measurements. Refer to Figures 9-1 and 9-2. The validity and reliability of your failure measures may depend on the size of the unit you are working with. Before-the-fact measurements used to judge middle managers' performance would include, for example, safety sampling and management inspections. Failure measurements can be used a bit more extensively at this level. Here our traditional indicators become more useful [2].

You may find that positive consequences produce the desired result. You can experiment with a variety of consequences as long as your employees are informed of your intentions.

Month	Date of Report
Department	Manager/Supervisor

Methods for Improving Safety Culture	
"What have I done to communicate & train Supervisors on the key elements of the Safety management system?"	
Management Commitment & Leadership	
Employee Participation	
Hazard Recognition & Control	
Education & Training	
Incident Investigation & Analysis	
Performance & Measurement	
Leadership Commitment and Support	
"What have I done to understand & implement the key elements of the Safety management system?"	
Management Commitment & Leadership	
Employee Participation	
Hazard Recognition & Control	
Education & Training	
Incident Investigation & Analysis	
Performance & Measurement	
Daily Safety Topic	
"What percent of required Daily Safety Tips were accomplished this week?"	
Shift Safety Reviews	
"What percent of required Shift Safety reviews were accomplished this week? Did I have a daily group safety review with each employee?"	
Safety Walk-Through	
"How many Safety Action Plans were developed this month?" _____ "How many remain open?" _____ "Why?"	

Figure 9-4 Sample supervisor monthly/weekly activity safety report.

Safety Meetings
"How many Safety Meetings were held?" _____ "How many other Manager/supervisors attended?" _____

Safety Training
"What Training Topics were covered this week?"

Housekeeping and Department Reviews

"List the score of your quarterly Housekeeping & Audits next to month it was conducted with upper management"	
January	February
March	April
May	June
July	August
September	October
November	December

Incidents/Near Misses/Investigations			
"Were Incident, Near Miss, & Investigation Reports completed & submitted to the Safety Dept. within 72 hours?" Yes or No. "If not, why?"			
Did I participate in any Incident Reviews? List Injury in my Department and a brief description of the incident			
Name	Injury Type Description	Root Cause	Action Plan

Figure 9-4 *Continued.*

You may eventually conclude that the employee is not capable of handling the assigned responsibilities, especially when sufficient training and mentoring through the accountability system have been documented, and poor performance still continues. At this point, the reason for the problem (inadequate capabilities, improper attitudes, etc.) should not be the issue. The maximum degree of consequence must be enforced. Otherwise, other employees will conclude that the appropriate consequences are not taken seriously or do not apply equally to everyone. This belief among employees will destroy any chance for an effective accountability management system [1].

Table 9-7
Accountability for Activities: Management Measures What Supervisors Are Doing

(1) Safety meetings
(2) Tool box meetings
(3) Activity reports on safety
(4) Inspection results
(5) Incident investigation and reports
(6) Job hazard analyses
(7) Any defined tool

Management Measuring Tools

(1) Safety sampling
(2) Statistical controls
(3) Critical incident techniques
(4) Safe-T-scores
(5) Regular reports
(6) SCRAPE (ABSS)*
(7) Performance ratings
(8) Objectives met Safety By Objective (SBO)
(9) Audits

Accountability for Results

(1) Charge incidents to departments
    (a) Charge claim costs to the line
    (b) Include incident costs in the profit and loss statements
(2) Prorate insurance premiums
(3) Put safety into the supervisor's appraisal
(4) Have safety affect the supervisor's income

Peterson, Dan, *The Challenge of Change, Creating a New Safety Culture, Resource Guide*, CoreMedia Development, Inc., 1993, Accountability for Activity Chart, p. 112. Modified with permission.
*Systematic method of counting and rating accident prevention efforts (Activity Based Safety System) [3].

UPPER MANAGEMENT PERFORMANCE MEASUREMENTS

While we begin to emphasize results performance measurements more at the middle management level, some performance measurements

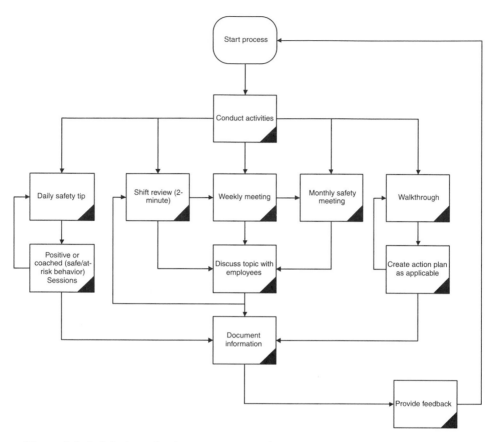

Figure 9-5 Activity-based safety systems overview, part 1.

should be retained. Performance measurements at this level are simple measures of what middle managers do, if they perform the tasks that are necessary for them to perform. Everything we have discussed about the supervisor performance measurement also applies to the middle management level. There are different activities at the middle to top management levels [2].

As we discussed, it is important to note that the supervisor and middle management performance should be different, necessitating some changes in the measurements used at this level. Normally, we want the middle managers to get their supervisors to do something in safety. Thus, we can measure managers to see if they meet with their supervisors and review activities or monitor the quality of the supervisors' work [2].

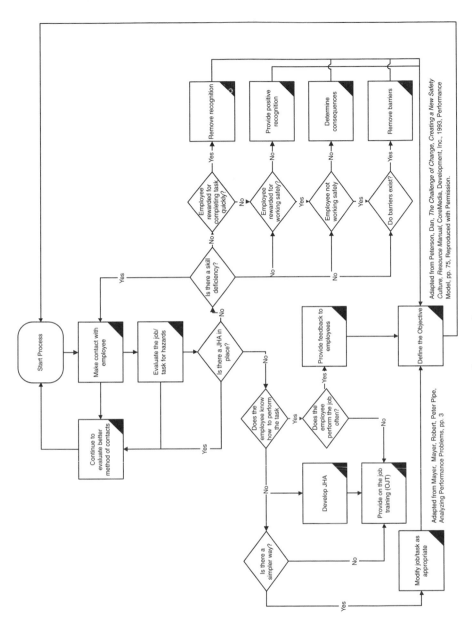

Figure 9-6 Activity-based safety systems overview, part 2.

FAILURE MEASUREMENTS

Failure measures tend to be generated from the injury review system. If injury rates are to be used for measuring the results of line management's safety performance, they should be set up so that:

- They are broken down by unit
- They provide some insight into the nature and causes of incidents
- They are expressed eventually in terms of dollars by unit
- They conform to any legal and insurance requirements [2, 3]

Beyond these broad outlines, each organization can devise any system that seems right for them. The "dollar" criterion is included because of the belief that a dollar measuring stick is much more meaningful to employees than any safety measuring stick (such as the frequency, severity rates, and/or experience modification rate [EMR]) [2, 3].

BEFORE-THE-FACT MEASURES

Before-the-fact measurements measure the results of supervisor actions before an incident occurs. For example, a periodic inspection is made of the supervisor's work area to measure how well he/she is maintaining physical conditions. This is a measure of whether things are wrong and, if so, how many are wrong. We can also measure how well a supervisor gets through to the people in the department by measuring the employee's work behavior [2].

One of the more recent and one of the best measurement tools is the perception survey.

Traditionally "safety programs" dealt with the physical environment. Later we looked at management and attempted to build management principles into our safety programs. Today we recognize the need to look at the behavior environment—the climate and culture in which the safety system must live [2].

THE OVERZEALOUS COMPANY

This is the kind of company where a great amount of safety equipment must be worn, and machines are so guarded as to make them difficult to get near and work with. The tone of the safety program is heavy. Such a company is likely to impose harsh punishment for minor infractions of safety procedures. There seem to be endless meetings, videos, manuals, and preaching about safety, often not involving the employees in any direct personal way. Such a company's response to safety education material is not lively, and employees feel overexposed to it [2].

THE REWARDING COMPANY

This is the kind of company that might offer prizes for safety records or for safety slogan competitions. Although the prizes are relatively small, there's a sense of competition. Such companies feel that safety programs are important, and the company generally feels responsible about safety matters [3].

THE LIVELY ORGANIZATION

This type of organization has a program that stipulates competition among its various plants, offers plaques, has boards to record the number of hours passed without incidents, or posts a continuing safety record at the plant entrance. Such a company teaches the employees to identify the safety goals, and the employees are proud of the record. Safety in this company becomes one of the lively aspects of the job and is more than simply avoiding risks or accidents [2].

THE REACTIVE ORGANIZATION

This is the kind of organization that seems to develop programs only after the fact. This company gets busy about safety only after a major incident occurs. Employees feel that the company does not care. The company passes out safety equipment and safety education materials, but only because this is currently the custom, and all in order to protect itself. This is similar to the SWAMP we discussed in Chapter 6 (Figures 6-3 to

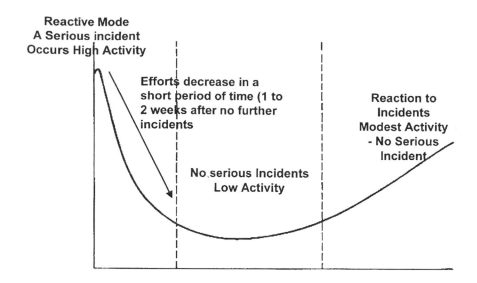

Note: This cycle will continue if a safety culture does not exist.

Figure 9-7 Safety Bathtub on Incidents. Kolb, John, Steven S. Ross, *Product Safety and Liability: A Desk Reference*, McGraw-Hill, 1980. Adapted from "The stylized bathtub curve," Figure IV-18, p. 127.

6-6). In addition, I have modified the "bathtub effect" diagram in Figure 9-7 to highlight the reactive company.

In the overzealous company, safety personnel usually seem loud, harsh, overly watchful, and too quick to criticize. In plants where programs for safety are livelier and more rewarding, safety professionals are usually seen as friends, people much like the employees. In the more lax organizations, the employees complain that there are not enough safety professionals around, and they doubt that reporting to them does much good, since little happens [2].

SUMMARY

There is more to an accountability system than enforcing punishment for employees. The accountability system aims to methodically help managers, employees, and supervisors to take responsibility for their actions and the subsequent effect of these actions on the team as follows:

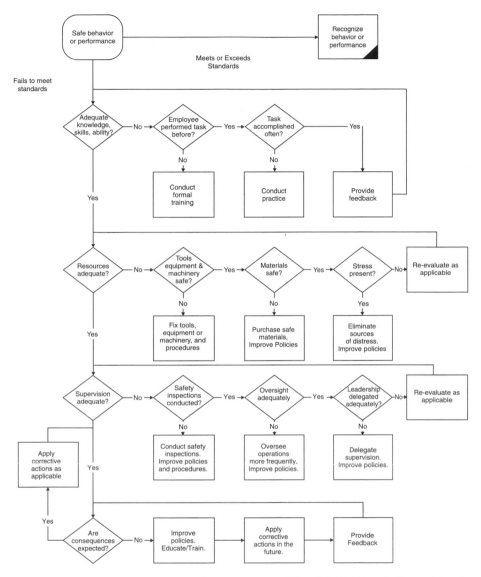

Figure 9-8 Model application for effective accountability. OR OSHA 119, Essentials of Effective Safety Accountability. http://www.cbs.state.or.us/external/osha/educate/training/pages/materials.html, public domain.

- Clearly defining expected performance in written performance objectives
- Periodically evaluating this performance jointly with employees
- Allowing employees the freedom to learn and develop in a positive, non-threatening manner

- Enforcing negative consequences only when training and mentoring have not been effective [1]

Your employees deserve to have a clear understanding of the nature, severity, and timetable of consequences. [1]

The communication between management and employees provided by an effective accountability system allows employees to choose their challenge for themselves. They can:

- Change their performance
- Attempt to change but ultimately acknowledge an inability to perform adequately
- Choose to ignore your expectations and endure the consequences [1]

An accountability system is essential. All of the hard work and effort you spent in developing your safety culture does not end with positive consequences. Do not let it fail. You are the leader and must make sure that the system works and will stay the course. As a review, refer to Figure 9-8 for a model application for effective accountability. This will help you to understand how a properly functioning accountability system works.

REFERENCES

1. U.S. Department of Labor, Office of Cooperative Programs, Occupational Safety and Health Administration (OSHA), Managing Worker Safety and Health, November 1994, public domain.

2. Peterson, Dan, *The Challenge of Change, Creating a New Safety Culture, Implementation Guide*, CoreMedia Development, Inc., 1993, Resource Manual. Modified with permission.

3. Petersen, Dan, *Safety by Objectives: What Gets Measured and Rewarded Gets Done*, 2nd ed., Van Nostrand Reinhold, 1996.

Part 3
Safety and Health Programs That Support the Safety Culture

10

Developing a Hazard Inventory

INTRODUCTION

Do you know all of the potential hazards that are associated with your type of industry and your site-specific workplace? If you do, we would be surprised. Yet if you want to protect your employees from workplace hazards, you must understand what those hazards are and how to control them [2].

A means of identifying hazards should be systematic. The following activities form the basis for a good hazard recognition, prevention, and control program and outline the three major actions needed to control hazards [2]:

- *Comprehensive survey*. Periodic, comprehensive safety and industrial hygiene surveys
- *Change analysis*. Potential hazards associated with new facilities, new/relocated equipment installation, materials, and processes; refer to Chapter 15
- *Job hazard analysis*. Routine hazard analysis, such as job hazard analysis, process hazard analysis, or task hazard analysis [2]; refer to Chapter 15

After hazards have been identified and controls are put in place, additional worksite analysis tools can help to make sure that the controls stay in place and that other hazards do not appear. Refer to Chapter 11 for detailed discussions of specific tools—for example, inspections, methods for employees to report hazards, incident investigations, and pattern analysis.

First, you must understand the existing and potential hazards that may exist in your workplace. Table 10-1 lists some methods of preventing hazards.

Table 10-1
Preventing Hazards

Continually review the work environment and work practices to control or prevent workplace hazards.

Some ways to prevent and control hazards include:

- Perform regular and thorough equipment maintenance
- Make sure that hazard correction procedures are in place
- Make sure that everyone knows how to use and maintain PPE
- Make sure that everyone understands and follows safe work procedures
- Make sure that, when needed, there is a medical surveillance program tailored to your facility to help prevent workplace hazards and exposures

http://www.osha-slc.gov/SLTC/safetyhealth_ecat/comp3.htm#, public domain.

COMPREHENSIVE SURVEYS

One thing that you must remember is that a comprehensive survey is not the same as an inspection. Employees often do inspections on a routine basis as part of a safety committee, etc. Other individuals who can bring fresh ideas and a knowledge of safety, health, or industrial hygiene should perform comprehensive surveys. Because there are few professionals equipped to do comprehensive surveys in all three areas, the best approach is to use a multiple-discipline team consisting of several specialists: a safety professional, an industrial hygienist, and a medical provider. In other cases, it could be comprised of a safety professional or someone from operations (management and an employee from another part of the operation) [2]. Figure 10-1 is a flowchart of a risk toolkit.

To conduct a comprehensive survey you may need some level of professional knowledge. The following is a summary of those professionals you should consider when conducting any type of comprehensive survey for your organization.

The medical provider can be a physician or a registered nurse with specialized training and experience in occupational medicine. He/she can assist the safety or industrial hygiene professional or can do a separate health survey, depending on the makeup of your workplace. Refer to Chapter 13 for suggested criteria for the occupational health professionals and/or medical provider. Many workers' compensation carriers and insurance companies offer loss control professional services to help

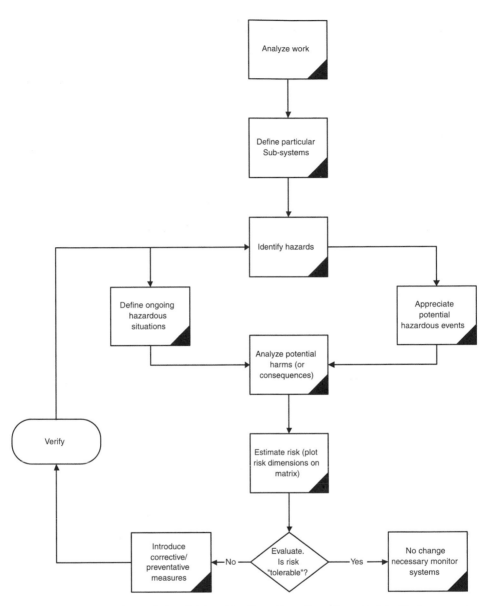

Figure 10-1 Flowchart of the risk toolkit. Cox, Sue, Tom Cox, *Safety Systems and People*, p. 41, Figure 2.6. Butterworth-Heinemann, 1996. Modified with permission.

clients evaluate safety hazards. Larger businesses may find the needed expertise at the company or corporate level or in an independent consultant.

If you use a professional from within your company, be wary of "tunnel vision," which can lead to failure to identify specific issues [2]. As the old saying goes: Sometimes you cannot see the forest for the trees. It's possible to stay in a position so long that you cannot see what is really going on around you.

For the industrial hygiene survey you should, at a minimum, want to inventory all chemicals and hazardous materials at the site, review the hazard communication program, and conduct air sampling in various areas to analyze the air for hazardous contaminants. For specific applications, a noise level survey and a review of the respirator program could be required [2].

EVALUATING THE PROFESSIONAL RESOURCE

If you hire a professional and you want to make sure that your worksite will be evaluated properly, you may want to ask the consultant specific questions:

- What type of training and experience have you had?
  How recent is it?
  What is its scope of the training? Is it limited to a specific industry?
  Does it consist of only practical experience, without formal training [2]?
  If he/she has a professional certification, is it valid, or has it lapsed for lack of maintenance points—for example, training, presentations, seminar attendance, activity participation in a local society [2]?
  Check references where surveys have been recently completed.
  Determine if any OSHA inspections occurred after the survey. Were any serious hazards found that the consultant missed?
  What tools did the consultant use for the survey?
  How much knowledge or transferable skills does the consultant have [2]?
- What kind of information will he/she need in advance?
  A professional who is planning an in-depth survey should prepare by learning as much as possible about the workplace and processes that exist
  Both safety and industrial hygiene professionals will usually want to see a layout of the operations
  The industrial hygienists may ask for a list of chemicals used or the Material Safety Data Sheets (MSDS) you received from your suppliers and the types of processes where you use them [2]

- What type of test equipment will the consultant use? Each professional will usually bring several items to the survey. You should expect the safety professional to bring a tape measure; a ground loop circuit tester to test electrical circuits; a multimeter (for 220 and/or 440 volts only); a tic tracer (or similar equipment) to check wires or electrical equipment to see if they are energized; and a ground fault circuit interrupter tester [2].
- The industrial hygienist may bring noise monitoring equipment. Depending on the chemicals or other contaminants expected, sampling pumps or grab sampling devices may be used.
- How long will the review take?
  It should take several times longer than a routine inspection of your workplace. It will also depend on the complexity of the workplace For example, if the industrial hygienist does air sampling, it should be time-weighted, 8-hour or full-shift sampling to measure the overall exposure to employees [2].

EVALUATING THE SURVEY

The following are some signs of a thorough survey:

- The consultant (safety professionals, industrial hygienists, and/or medical provider) should start with your injury and illness logs and look for patterns.
- The safety professional also may want to see program documentation.
- The industrial hygienist and health professional may want to see your hazard communication program, if applicable, your hearing conservation and/or respirator program.
- The medical provider may want to see your records of employee visits to clinics, first aid stations, and other workplaces where treatment is provided for work-related injuries.
- He/she will want to examine records of employee training in first aid and CPR or blood-borne pathogens. In addition, they may want to see a baseline and follow-up testing records.
- The safety professional should start at the beginning of your process, where raw materials are brought in, and carefully go through all processes, inspecting each operation and talking to employees, until the point where the product is finished and shipped to the customer. The following areas should be evaluated or inspected:

How materials are handled from the time they are received and how they are moved and stored, checking the stability of storage racks and the safe storage of flammables/explosives

The openings that expose moving parts for pinch points, machine guarding, and other hazards

Hand tools and equipment and wiring in the maintenance shop and other areas as appropriate

Operations on every shift and any after-hours operations, such as cleanup or forklift battery recharging

How you manage your hazard prevention and control program

Outside of buildings, for example: chocks for trucks at the loading/unloading docks, fork lift ramps, outdoor storage of flammables/explosives, oil, and any fueling areas

- The professional should also do the following:

Open every door and look in every corner of the facility

Suggest target tasks for job hazard analysis, especially those tasks that might involve ergonomic-related hazards

Assist in developing and/or improving your injury reduction program

- The industrial hygienist and medical provider should start at the beginning of your production operation, observe all processes, talk to employees, and follow the production flow to the point of shipping [2]. They may do the following:

Review a list of all hazardous substances that can be found in the worksite (as required by 29 CFR 1910.1200(e), OSHA's Hazard Communication standard), or the accuracy of the existing inventory list

Determine what metals are used in any welding operations

Check any production areas where eating or smoking is allowed

Check for the possible presence of asbestos, lead, mercury, silica, other carcinogens, etc.

If respirators are used, determine whether employees are using each brand properly, how to fit test each employee, whether pulmonary function testing is done, and how the respirators are cleaned, maintained, and stored

Perform full-shift sampling of contaminants thought to be present to understand employee exposures

Watch the movements that employees make in performing their tasks to see if there are existing or potential cumulative trauma disorders (CTDs) or other ergonomic issues

Suggest processes for routine process hazard analysis

Help you set up or improve regular monitoring programs for any contaminants or other health hazards found to be present [1, 2].

The items listed do not constitute an exhaustive list of activities you should expect [2].

This hazard inventory should be expanded and improved by what you learn from periodic surveys, change analyses, and routine hazard analyses. The baseline survey should provide the basic inventory of hazards and potential hazards. However, the foundation of your inventory is the baseline comprehensive survey. It is important that this initial survey be done well [2].

FOLLOW-UP SURVEYS

You will need periodic follow-up surveys if you are to apply the rapidly growing scientific and engineering knowledge about hazards, their prevention, and control. These follow-up surveys help to identify hazards that may have developed after the processes and/or procedures have been in place. The frequency of follow-up surveys depends on the size and complexity of the organization.

INVOLVING EMPLOYEES IN ESTABLISHING THE INVENTORY

Always use the knowledge that your employees have gained from their close involvement with equipment, materials, and processes. You should encourage your employees to communicate openly with the professionals who do the comprehensive surveys [2].

Do not make the mistake of limiting your employees' participation to what they can tell the professionals. Make sure that employees participate in the various types of hazard analysis. It makes good sense to involve employees in change analysis of new equipment and/or processes because of their valuable insights into how things really work. As indicated in Chapter 15 on job hazard analysis, many companies regularly include employees in this activity. Employees can play a similar role in process hazard analysis. Greater understanding of hazards, prevention, and control helps employees do a better job of protecting themselves and their co-workers.

CHANGE ANALYSIS

Before making changes in the worksite, analyze the changes to identify potential hazards. Anytime you change your worksite, whether you are adding a piece of equipment, different materials, or a new process, relocating equipment, or building, you may unintentionally introduce new hazards. If you are considering a change in your workplace, you should analyze it thoroughly before making these changes. You need a good baseline (starting point) and independent review. This change analysis is cost effective in terms of the human suffering and financial loss it helps to prevent. Stopping problems before they develop is less expensive than attempting to fix them after the fact [2].

An important step in preparing for a worksite change is to consider the potential effect on all employees. Individuals respond differently to changes, and even a clearly beneficial change can throw an employee temporarily off balance and increase the risk of an incident. You want to make sure that you communicate all changes to affected employees as soon as possible, provide training as needed, and pay attention to employee work activities until everyone has adapted to the new change [2]. Refer to Chapter 15 for a description of change analysis.

Building or Leasing a New Facility

Something as basic as a new facility needs to be reviewed carefully to identify hidden hazards—for example, the fire system, security, and environmental issues. This is often overlooked. A design that seems to enhance production of your product and appears to be a marvel to the architect may be a harmful or even fatal management decision. Safety professionals should take a careful look at all design/building plans prior to final approval [2].

When you lease a facility that was built for a different purpose, the risk of acquiring safety problems is even greater. This a true unknown. You should make a thorough review of the facility, by reviewing the blueprints and/or plans for any renovations. One of the most obvious concerns in acquiring an existing facility is whether any environmental issues such as lead paint, asbestos insulation, mercury, or PCBs may be present. But you also may discover that something as easy to fix as a loose stair railing has gone unnoticed in the rush to renovate production areas. Save frustration, money, and lives: Make sure that you have a safety professional involved in the planning of any facility construction, purchase, or lease [2].

Installing New Equipment

Equipment manufacturers do not know how their products will be used at your facility. Therefore, you cannot rely on the manufacturer to have analyzed and designed guarding, controls, or safe procedures for the equipment. Another issue to take into consideration is if the equipment is produced in a foreign country. It may not meet U.S. requirements and laws. Therefore, you should also involve a safety professional in the purchasing decision, as well as in installation plans and startup [2].

Many companies also provide a period to test newly installed equipment. The company assigns its most experienced employees to watch for hidden hazards in the operations before full production begins. As with new facilities, the sooner hazards are detected, the easier and cheaper the corrections will be [2]. One of the best methods is to conduct a JHA (Chapter 15).

Using New Materials

Before introducing any new materials into your production processes, you must research the hazards that the materials may present. Also try to determine any hazards that may appear as a result of the processes you plan to use with the materials [2].

When you review new material, the place to start is the manufacturer's Material Safety Data Sheet (MSDS). An MSDS is required for all materials containing hazardous substances. It should arrive with each shipment. The MSDS should provide hazard information. You should have someone knowledgeable to analyze the hazard a chemical presents and to prevent or control any associated hazard [2].

Starting Up New Processes

New processes require employees to perform differently. Consequently, new hazards may develop even when your employees are using familiar materials, equipment, and facilities. Carefully develop safe work procedures for new processes. After the employees have become familiar with these procedures, perform routine job hazard analysis to identify any hidden hazards [2].

ANALYZING MULTIPLE CHANGES

Often a big change is composed of several smaller changes. When you begin producing a new product, chances are you will have new equipment, materials, and processes to monitor. Make sure that each new change is analyzed not only individually, but also in relation to the other changes [2].

Once you have analyzed the changes at your worksite, add this information to your inventory of hazards. This inventory is the foundation of the design of your hazard prevention and control program.

WHEN EMPLOYEES CHANGE

When an employee changes position there can be huge safety ramifications. In staffing changes, one employee now is performing a task previously done by another. The new employee may bring to the position a different level of skill. He/she will possess a different degree of experience performing the tasks, following the specific work rules and procedures, and interacting with other employees. Especially in high hazard situations, these differences should be examined and steps should be taken to minimize any increased risk, both to the new employee and to anyone affected by their presence. Refer to Chapter 14 for a variety of training and job orientation methods that will help you to make sure that there is a safe employee transition [2].

The change may be related to temporary or chronic medical issues, a partial disability, family responsibilities and/or crisis and other personal problems, alcohol or drug abuse, aging, or the employee's response to workplace changes. An analysis of this change, followed by physical and/or administrative accommodations to ensure safe continued performance, sometimes may be appropriate—for example, when an incident affects an employee's ability to function. At other times, a less formal response may be more suitable [2].

Workplace hazards do not exist in a vacuum. The human element is always present, and the human condition is a source of change that you must understand. An effective manager is sensitive to these changes and their potential effect on the safety of the individual and the company as a whole [2].

ROUTINE HAZARD ANALYSIS

To combat this issue, a routine hazard analysis should be conducted. One of the most common reasons for injuries are tasks that are performed over and over many times daily, weekly, monthly, year after year, etc. These become a hazard because we get used to doing the same task over and over. It becomes a natural routine. We loose the ability to think of the consequences. For example, driving a car becomes second nature to us. How many times have you passed a landmark and do not remember passing it? This is the same as an employee doing a specific task over and over without thinking.

- Managers and employees are made aware that hazards can develop within existing job processes and/or phases of activity.
- You will be able to determine if one or more hazard analysis systems designed to address routine job, process, or phase hazards are in place at the facility.
- All jobs, processes, or phases of activity are analyzed using the appropriate hazard analysis system.

For example, different management measures may be needed for hazards that do not stay corrected, compared to those that have never received attention/correction. For hazards that cannot be quickly corrected, a record is required so that final correction is not forgotten. Figure 10-2 presents a hazard identification flowchart [1].

The importance of interim protection cannot be overemphasized: There is no way to predict when a hazard will cause serious harm, and there is no justification for continuing to expose employees to risk unnecessarily. These are suggested guidelines for establishing a hazard reporting system:

- Develop a policy for hazard reporting
- Make sure that all employees know about the policy for reporting hazards and that they understand that policy.

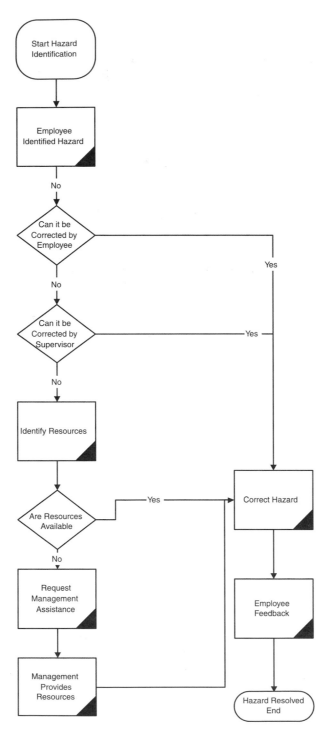

Figure 10-2 Hazard identification flowchart. Oklahoma Department of Labor, Safety and Health Management: Safety Pays, 2000, http://www.state.ok.us/~okdol/, Chapter 7, p. 41, public domain.

- See that the policy is written, distributed to all employees, posted, and discussed during regular safety meetings
- Demonstrate that the policy for reporting hazards is real: Involve employees in creating the policy, share and discuss the policy with all employees, and follow through with the policy
- Protect employees who report hazards from harassment or reprisal
- Use the information gained through the hazard reporting system to improve the overall safety and health program

It is better to have some non-hazards reported than to risk even one real hazard that was not reported because it was believed that the organization would not respond.

The system provides for data collection and display as a means to measure the success of the system in resolving identified hazards [1].

SUMMARY

The authors would like to offer one caution here. This survey does not constitute a safety culture. You cannot implement inspections into a culture. These inspections are only to identify issues/hazards that you must fix to enhance your culture.

A practical hazard analysis of the work environment involves a variety of elements intended to identify existing hazards and conditions, and operations subject to change that might create new hazards. Effective management systems continually analyze the workplace to anticipate and develop policies and procedures to help minimize the introduction and reoccurrence of hazards. The following are recommended measures that can be used to help identify existing and potential hazards:

- Conducting comprehensive baseline assessments, updating assessments periodically, and allowing employees participation in the review
- Analyzing planned new facilities, processes, materials, relocated equipment, and any plant, machine, and process modifications
- Performing standard job hazards analyses, to include assessing ergonomic risk and other related factors with regard to employees' tasks
- Conducting regular site inspections so that new or previously missed hazards can be identified and corrected

- Providing a reliable management system for employees to notify management about conditions that appear hazardous, to receive timely and appropriate responses, and to encourage employees to use the management system without fear of reprisal
- Investigating incidents (near miss/hit and significant loss-producing events) so their causes and means of prevention can be identified
- Analyzing trends to identify patterns with common causes so they can be reviewed and corrected [2]

These measures provide a good method of defining specific hazard training. Hazards that employees are exposed to should be systematically identified and evaluated. Evaluating other requirements that may impose additional and specific requirements for hazard identification and assessment is also important.

Hazard identification and assessment analysis should be conducted:

- As often as necessary to make sure the facility is in compliance with specific requirements, best management practices (BMP) or above and beyond compliance (ABC)
- When workplace conditions change that could create a new or increased risk of hazards

Each work-related incident with the potential to cause physical harm to employees should also be investigated. Keep accurate and complete records of the assessments and hazards identified. Develop action plans to control identified hazards.

Establishing a complete hazard inventory is not as complicated as it may sound. It begins with having a professional conduct a comprehensive survey to determine existing and/or potential hazards of your worksite. Periodic surveys, conducted at intervals that make sense for the size and complexity of your worksite, will help to identify any new engineering or scientific knowledge of hazards and their prevention. These subsequent surveys also can help find new hazards that have evolved as a result of changing work procedures.

Change analysis prevents expensive problems before they occur. Individuals who are knowledgeable in employee safety can help to design and plan for changes at the worksite. Change analysis uses elements of routine hazard analysis appropriate to the type of change being contemplated.

Routine hazard analysis also adds hazards to your inventory. It will help you to control hazards that develop in work procedures or in processes, or that occur because of changes in the phases of the operation.

The tools and approaches used in the various types of hazard analysis tend to overlap. This overlap helps ensure total coverage and a more comprehensive inventory on which to base your prevention program.

When assessing workplace hazards, do not overlook the human element. Whenever one employee replaces another employee, the difference in skill level and experience can increase risk both to new employees and to their co-workers. Changes in the individual employee's health, ability to function on the job, and personal life, no matter whether these changes are sudden or gradual, can affect workplace safety. A manager needs to be sensitive to these changes and willing to provide training and orientation, physical and administrative adjustments, and/or other accommodations.

Refer to Chapter 11 for a detailed review of this program. You should minimize hazards by substituting less hazardous materials or equipment whenever possible and engineering controls that distance the worker from the hazard. For the remaining hazards, design safe work practices, train your employees adequately in these practices, and enforce the practice consistently. In some cases, you may also need to establish other administrative controls, such as employee rotation or more frequent work breaks where needed.

You need to use the surveys and analyses that we have described to plan and accomplish a program of hazard prevention and control. Remember, the first step in protecting your employees is recognizing the hazards.

REFERENCES

1. Oklahoma Department of Labor, Safety and Health Management: Safety Pays, 2000, http://www.state.ok.us/~okdol/, public domain.

2. U.S. Department of Labor, Office of Cooperative Programs, Occupational Safety and Health Administration (OSHA), *Managing Worker Safety and Health*, November 1994, public domain.

11

Developing a Hazard Prevention and Control System

INTRODUCTION

A management system that uses inspections, observations, etc., identifies the hazards and potential hazards to which employees could be exposed and has designed a system to minimize and control identified hazards. However, some hazards may not be detected during the hazard inventory; other hazards may have a way of slipping out of control. You need to find ways to identify hazards that were not identified during the initial inventory. In other words, you must develop a formal method to help control hazards before employees get injured [2].

MANAGING OR CONTROLLING HAZARDS

After hazards are identified, all current and potential hazards must be prevented, corrected, or controlled. The following are some systems that can be used to prevent and control hazards:

- Engineering controls
- Administrative controls
- Personal protective equipment (PPE)
- Safe work practices
- Work rules

- Training
- Emergency plans
- Medical programs

These systems are used to help track hazards: corrective actions; preventive maintenance systems; emergency preparation; and medical programs [2]. Refer to Figures 11-1 and 11-2 for an overview of how a hazard control system works.

Hazards take many forms: for example, air contaminants (microorganisms, toxic liquids), tasks involving repetitive motions (CTD), equipment with moving parts or open areas (poor guarding) that can catch body parts or clothing, extreme heat or cold, or noise. The terms that we use here to describe the principles of engineering control may sound strange when applied to some of these hazards. You may find that, in other discussions of hazard control, the terms are used differently. There should be agreement about the concepts that the terms describe [3].

In the next sections we will discuss these methods of controlling hazards in more detail. In addition, refer to Table 11-1 for an overview of hazard prevention and controls.

ENGINEERING CONTROLS

One important feature is that the work environment and the job or task should be designed to minimize or reduce employee exposure to hazards [3].

Engineering controls (Table 11-2) focus on the source of the hazard. This is unlike other types of controls that generally focus on the employee exposed to the hazard. The basic concept behind engineering controls is that the work environment and the job itself should be designed to minimize employee exposure to hazards. Although this approach is called engineering control, it does not mean that an engineer is required to design the control. Some of the solutions are "just plain common sense" [2].

Engineering controls can be very simple in some cases. These controls are based on some general principles: for example, as practical and feasible, design the facility, equipment, or process that will help to remove the hazard and/or substitute a non-hazardous or less hazardous condition [2, 3].

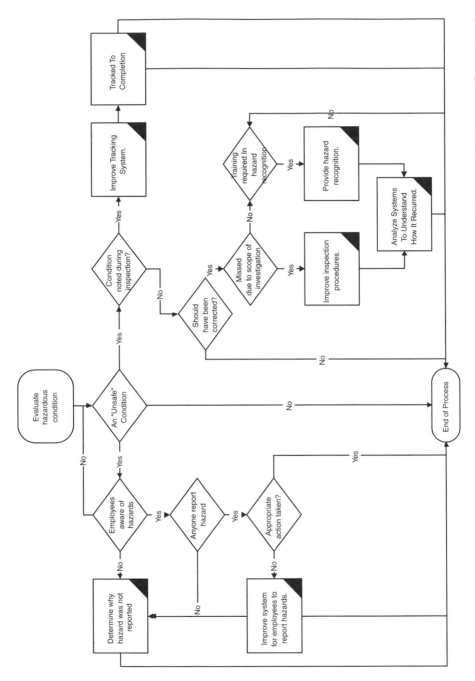

Figure 11-1 Hazard analysis flow diagram, part 1. U.S. Department of Labor, Office of Cooperative Programs, Occupational Safety and Health Administration (OSHA), *Managing Worker Safety and Health*, November 1994, public domain.

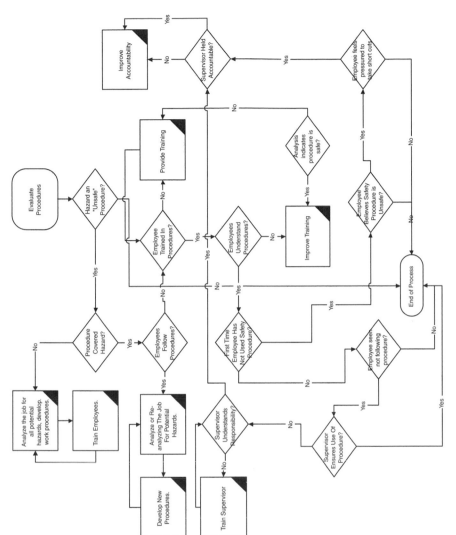

Figure 11-2 Hazard analysis flow diagram, part 2. U.S. Department of Labor, Office of Cooperative Programs, Occupational Safety and Health Administration (OSHA), *Managing Worker Safety and Health*, November 1994, public domain.

Table 11-1
Hazard Prevention and Controls

ENGINEERING CONTROLS

The first and best strategy is to control the hazard at its source. Engineering controls are the best choice, unlike other controls that focus on the employee exposed to the hazard. The basic concept behind engineering controls is: to the extent feasible, the work environment and the job itself should be designed to eliminate hazards or reduce exposure to hazards. Engineering controls can be simple in some cases and are based on the following principles:

- If feasible, design the facility, equipment, or process to remove the hazard or substitute something that is not hazardous.
- If removal is not feasible, enclose the hazard to prevent exposure in normal operations.
- Where complete enclosure is not feasible, establish barriers or local ventilation to reduce exposure to the hazard in normal operations.

SAFE WORK PRACTICES

Safe work practices include general workplace rules and other site-specific rules. For example, even when hazards are enclosed, exposure can occur when maintenance is necessary.

ADMINISTRATIVE CONTROLS

While safe work practices can be considered forms of administrative controls, OSHA uses the term administrative controls to mean other measures aimed at reducing employee exposure to hazards. These measures include additional relief employees, exercise breaks, and rotation of employees. These types of controls are normally used in conjunction with other controls that more directly prevent or control exposure to the hazard.

PERSONAL PROTECTIVE EQUIPMENT (PPE)

When exposure to hazards cannot be engineered out of normal operations or maintenance work, and when safe work practices and other forms of administrative controls cannot provide sufficient protection, a supplementary method of control is the use of PPE. PPE may also be appropriate for controlling hazards while engineering and work practice controls are being installed. For specific OSHA requirements on personal protective equipment, see OSHA's standard, 1910 Subpart I.

PPE HAZARD ASSESSMENT AND TRAINING

The basic element of any management program for PPE should be an in depth evaluation of the equipment needed to protect against the hazards at the

Table 11-1
Continued

workplace. The evaluation should be used to set a standard operating procedure for personnel, then train employees on the protective limitations of the PPE, and on its proper use and maintenance.

Using PPE requires the appropriate hazard awareness and training. Employees must be aware that the equipment does not eliminate the hazard. If the equipment fails, exposure will occur.

To reduce the possibility of failure, equipment must be properly fitted and maintained in a clean and serviceable condition.

TRACKING HAZARD CORRECTION

An essential part of any safety system is the correction of hazards that occur despite the overall prevention and control procedure. Documentation is important so that management and employees have a record of the correction. Many companies use the form that documents the original discovery of a hazard to track its correction. Hazard correction information can be noted on an inspection report next to the hazard description.

Employee reports of hazards and reports of accident investigation should provide space for notations about hazard correction.

PREVENTIVE MAINTENANCE

Good preventive maintenance programs play a major role in making sure that hazard controls continue to function properly. It also keeps new hazards from arising due to equipment malfunction. Reliable scheduling and documentation of maintenance activity are necessary. The scheduling depends on knowledge of what needs maintenance and how often. The point of preventive maintenance is to get the work done before breakdown occurs and/or repairs or replacement is needed. Documentation is not only a good idea, but is a necessity. Certain OSHA standards require that preventive maintenance be done. How many injuries do you know that have occurred because equipment was jammed and employees reached into the nip point (to stop downtime) to remove the material? Most of this can be traced to poor management leadership practices and the perception of employees to keep the product moving. Historically, we as managers push production. This is natural because this is what makes the money for the business. But in the same context as we discussed in Chapter 1, there are underlined costs that management must understand that affect the bottom line. Sometimes it is less expensive to shut a piece of equipment down, than it is to take a chance on an employee getting hurt. I think that this has been explained in detail in this book. In addition, these issues can also be tracked to poor maintenance, not repairing equipment for the same reason. This is what is called in some circles as breakdown

Table 11-1
Continued

maintenance. A good preventive maintenance program will enhance the safety aspects of a job and also provide an increase in production. I recall one incident where the employee tried to unjam a piece of equipment without shutting it down. During the investigation, it was determined that the belts were worn. With this condition, the machine keeps malfunctioning. The equipment was shut down and repaired and the jam ups stopped.

EMERGENCY PREPARATION

During emergencies, hazards appear that normally are not found in the workplace. These may be the result of natural causes (floods, earthquakes, tornadoes, etc.), events caused by humans but beyond control (train or plane accidents, terrorist activities, etc.), or within a firm's own systems due to unforeseen circumstances or events. You must become aware of possible emergencies and plan the best way to control or prevent the hazards they present. Some of the steps in emergency planning include:

- Survey of possible emergencies;
- Planning actions to reduce impact on the workplace;
- Employee information and training;
- Emergency drills as needed;
- Medical programs.

A company's medical program is an important part of the safety and health system. It can deliver services that prevent hazards that can cause illness and injuries, recognize and treat illness and injury, and limit the severity of work-related injury and illness. The size and complexity of a medical program will depend on many factors, including the:

- Types of processes and materials and the related hazards,
- Types of facilities,
- Number of workers,
- Characteristics of the workforce, and
- Location of each operation and its proximity to a health care facility.

Medical programs consist of everything from a basic first aid and CPR response to sophisticated approaches for the diagnosis and resolution of ergonomic problems. Depending on the size of the site, this may be in-house or through arrangements made with a local medical clinic. Whatever the type of medical program, it is important to use medical specialists with occupational health training. Refer to OSHA standard 1910.151(b) for first aid requirements.

http://www.osha-slc.gov/SLTC/safetyhealth_ecat/comp3.htm#SafeWorkPractices, public domain.

Table 11-2
Engineering Controls

- Address hazards
- Can eliminate the hazard; should be considered first
- Redesign, substitute, enclose, replace equipment, tools, or workstation
- Need to consider feasibility, cost, ease of implementation
- Major strength: usually do not rely on human behavior to be successful

OR OSHA 100w, Hierarchy of Hazard Control Strategies, p. 16, http://www.cbs.state.or.us/external/osha/educate/training/pages/materials.html, public domain.

Minimizing Hazards through Design

Designing facilities, equipment, or processes so that the hazard is no longer a threat is the best employee protection. Some examples may include:

- Redesigning, changing, or substituting equipment to remove the source of the hazard—for example, excessive temperatures, noise, or pressure
- Redesigning a process to use less hazardous substances
- Redesigning a work station to relieve physical stress and/or remove ergonomic hazards
- Designing general ventilation with sufficient fresh outdoor air to improve indoor air quality [2]

Guarding Hazards

Engineering controls could include guards placed on equipment to prevent body parts from coming in contact with hazard areas—for example, pinch points that can cut or pinch body parts or catch clothing, hair, or body parts and pull them into the machinery. These guards should be reviewed during regular inspections to make sure they are in good working order. All guarding must remain in place while equipment is in operation [2].

When you cannot remove the hazard and cannot replace it with a less hazardous alternative, the next best control is to guard the exposure. Guarding the exposed hazard means reducing an employee's exposure during normal operations. However, there is potential exposure during maintenance activities or if the guarding system breaks down. In this case, additional controls such as safe work practices or personal protective equipment (PPE) may be necessary to limit exposure [2].

Some examples of guarding designs include:

- Guarding of moving parts of equipment
- Containment of toxic liquids or gases from the beginning of the process, safe packing for shipment, or safe disposal of toxic waste products
- Glove box operations to enclose work with dangerous microorganisms, radioisotopes, and/or toxic substances
- Containment of noise, heat, or pressure producing processes with materials especially designed for those purposes [2]

ADMINISTRATIVE CONTROLS

Workplace rules and safe work practices are considered administrative controls (Table 11-3). We use administrative controls to decide other measures that are aimed at minimizing exposure to hazards. Such measures can include longer rest breaks, additional substitute employees, exercise breaks to vary body motions, or rotation of employees through different jobs to reduce or "even out" exposure to hazards. Administrative controls are normally used in conjunction with other controls that more directly prevent or control exposure to hazards [2].

You also can control hazards with specifically designed work practices and procedures. For example, what precautions are taken by an employee who starts equipment after a jam? The supervisor should watch out for any unauthorized modifications of work practices [2].

**Table 11-3
Administrative Controls**

- Address exposure, behaviors
- Reduce the duration, frequency, and severity of exposure
- May include job rotation and adequate rest breaks
- Design safe work procedures
- Key elements that must be managed: training, compliance, reporting
- Major weakness: Relies on appropriate behavior
- Fix system weaknesses (root causes)

OR OSHA 100w, Hierarchy of Hazard Control Strategies, p. 16, http://www.cbs.state.or.us/external/osha/educate/training/pages/materials.html, public domain.

The negative part of administrative control is just what its name indicates. Administrative control is "paperwork" that tells someone how to be safe, but does not monitor activities. What happens after a couple of days or a couple of weeks? Employees forget to do the task as described. Supervisors forget to enforce the rules of the workplace. Then the employee gets hurt. Controls must be engineered into the site to prevent this type of control.

PERSONAL PROTECTIVE EQUIPMENT

Another method of control is the use of personal protective equipment (PPE). Personal protective equipment can consist steel-toed shoes and boots, safety glasses with side shields, ear plugs, goggles, face shields, etc. Personal protective equipment should be used only as a last resort [3].

One must remember that when using specific PPE, you must make sure that you comply with requirements addressing the use of specific items [3].

The use of PPE should be considered in conjunction with other control efforts, or when no other method or practice is available. The supervisor may examine the equipment to make sure that it is in good condition. In addition, the supervisor should see if employees are wearing the required personal protective equipment and if the equipment is being used properly [2]. For example, employees may place disposable hearing protection only partway into their ears and leave most of the plug protruding.

Do not overlook areas outside of the production areas. Your search for common hazards and regulatory requirements or infractions should cover the entire workplace, including the office area [2].

PPE Hazard Assessment and Training

The basic element of any management program for PPE should be an in-depth evaluation of the equipment needed to protect against the hazards at the workplace. The evaluation should be used to set a standard operating procedure for personnel, then train employees on the protective limitations of the PPE, and on its proper use and maintenance.

Using PPE (Table 11-4) requires hazard awareness and training on the part of the user. Employees must be aware that the equipment does not

Table 11-4
Personal Protective Equipment

- Equipment provides a barrier between the employee and the hazard
- Major weakness: Relies on human behavior and vigilant supervision

OR OSHA 100w, Hierarchy of Hazard Control Strategies, p.16, http://www.cbs.state.or.us/external/osha/educate/training/pages/materials.html, public domain.

eliminate the hazard. If the equipment fails, exposure will occur. To reduce the possibility of failure, equipment must be properly fitted and maintained in a clean and serviceable condition [1].

Limitations of Personal Protective Equipment

Employees need training to understand why personal protective equipment is necessary and how to use and maintain it properly. They must also understand its limitations. Personal protective equipment is designed for specific uses and is not always suitable in all situations. For example, no one type of glove or apron will protect employees against all solvents or physical hazards. To pick the appropriate glove or apron, employees should refer to recommendations on the material safety data sheets of the chemicals they are using or any manufacturers' data sheets [3].

Your employees need positive reinforcement and fair, consistent enforcement of the rules governing personal protective equipment usage. Some employees may resist wearing personal protective equipment according to the rules, because the PPE is uncomfortable and places additional burden of stress on the body. This stress can make it unpleasant or difficult for employees to work safely. This is a significant limitation, particularly when heat stress is a factor in the work environment. Ill-fitting or improperly selected PPE is hazardous, since PPE is only used when other feasible controls have failed to minimize a hazard [2].

Bearing the Cost

OSHA standards require employers to provide personal protective equipment that fits individual employees. Most employers provide the required PPE, with the exception of safety shoes and safety glasses. But even when employees must provide their own safety shoes, safety glasses, or other PPE, employers usually pay part of the cost [2].

INTERIM PROTECTION

When a hazard is identified, the preferred corrective actions and/or control cannot always be accomplished immediately. Interim measures must be taken to minimize risk—for example, taping down loose wiring that poses a tripping hazard, shutting down a piece of equipment temporarily, etc. The importance of taking these interim protective action steps cannot be overemphasized. There is no way to predict when a hazard will cause serious harm, and no justification for continuing to expose employees unnecessarily [2].

GENERAL SAFETY RULES

General work rules that you develop and make part of your organization are important parts of a hazard prevention and control program. These rules play a major part in identifying acceptable and unacceptable behaviors. For example, you may have rules concerning horseplay or violent behavior on company property, or requiring your employees to wear personal protective equipment. The following are some sample workplace rules that can be used as a guide:

- Procedures regarding the use of equipment or materials
- Safe acts or behaviors tied with behavior-based systems
- Lockout/tagout procedures
- Personal protective equipment procedures
- Good housekeeping practices [2]

You must understand that these work rules are generic and need to be expanded or condensed based on your site-specific needs. These categories can be further divided into specific terms that are more specific to your operation. For example:

- Report unsafe conditions to your supervisor immediately.
- Report all injuries promptly to your supervisor.
- Dress properly. Wear appropriate work clothes, gloves, and shoes and/or safety boots.
- Loose clothing and jewelry should be not be worn except in designated areas.
- Never operate any equipment unless all guards and safety devices are in place and in proper operating condition.

- Keep all equipment in safe working condition.
- Never use defective tools or equipment.
- Report any defective tools or equipment to your supervisor.
- Properly maintain and be responsible for all personal protective equipment.
- Only perform maintenance, clearing jams, and adjustments when equipment is locked/tagged out.
- Do not leave materials in aisles, walkways, stairways, roads, or other points of egress.
- Practice good housekeeping at all times [1].

Safety rules are most effective when they are written, posted, and discussed with all affected employees. Many employers emphasize the link between safety rules and the consequences of breaking them by reviewing the rules with their employees. They then ask the employees to sign a statement: "I have read and understand the safety rules, and I have received an explanation of the consequences of breaking them." The employer and the employee both keep a copy of this signed statement [2].

One of the major keys to developing safety is employee participation, as discussed in Chapter 7. When employees play a role in formulating the rules, they are more likely to understand and follow them [3].

Safety rules may also apply to specific jobs or tasks and may involve specific procedures for accomplishing a particular job or task—for example, development of job hazard analyses as described in Chapter 15. Safe work practices are generally derived from job hazard analyses. From a detailed job hazard analysis, you may determine that safe work practices must be changed or that a training program must be conducted.

Specific procedures are an important aspect of safety management and are useful in managing employee safety as well. A number of serious injuries occur while employees are performing non-routine tasks. This is something that we must understand. A concentrated effort must be made to plan and develop safe work procedures for such tasks. Remember that safe work practices should be used in conjunction with, not as a substitute for, more effective or reliable engineering controls [3].

WORK PRACTICES

Some of work practices are very general in their applicability—for example, housekeeping activities such as:

- Removal of tripping, blocking, and slipping hazards
- Removal of accumulated toxic dust on surfaces
- Wetting down surfaces to keep toxic dust out of the air [2]

Safe work practices apply to specific jobs in the workplace and involve specific procedures for accomplishing a job. To develop these procedures, you conduct a job hazard analysis. This process is clearly described with examples and illustrations in OSHA Publication 3071 (Revised 1992), "Job Hazard Analysis," and summarized in Chapter 15 [2].

In the opinion of one of the authors, the job hazard analysis should be kept separate from written procedures that employees follow to accomplish the job safely. A JHA is more detailed than a work procedure instruction (learner's checklist, etc.). Each document suffers from being combined with the other. One of the authors has experience with these methods and has spent time in a supervisory role. When you are a supervisor, what is your priority? The answer is simple: production. One suggestion is to cover the work elements first and then cover the safety hazards. This sets the tone that safety is important.

You may decide that a training program is needed. You can use the JHA as the basis for training your employees in the new procedures. A training program may be essential if your employees are working with highly toxic substances or in dangerous situations [2]. On another note: if an incident does occur, you have the JHA document to check the element that may have been missed in the initial development. This is another tool that can help you find your way to the root cause. We will discuss this in more detail in Chapter 15.

LIMITATIONS TO CONTROLLING HAZARDS WITH SAFE WORK PRACTICES

In many cases, safe work practices are a necessity to protect employees and can work, but are only as good as the management system that supports the established practices. This is because they are susceptible to human error. As we discussed, controls must be designed into the equipment before it is placed into operation. Employee training and reinforcement (consistent and reasonable) must accompany the equipment installation. Safe work practices must be used in conjunction with, and not as a substitute for, more effective or reliable engineering controls [2].

When developing a hazard control program, you should anticipate resistance when teaching new job practices and procedures to employees.

If your employees have done a job long enough without special precautions, they are likely to feel unconcerned about hazards. It is essential that they understand why special work practices are needed. Therefore, training begins with a discussion of hazards. Your employees must understand that, for an incident to occur, two things must be present: a hazard and an employee [2].

In theory, remove the hazard, and there will be no injury. Train the employee to follow proper work practices, and they can help the employee to avoid harm.

ENFORCEMENT

Under traditional safety cultures, safe work practices become a condition of employment where employees are enticed to be safe by intimidation. This is what is typically called the "bulldog" tactic. This type of method will help to reduce injuries by intimidation. Remember Theory X. This is a similar situation. However, when the person who does the "bulldogging" leaves the organization, injuries return.

When it comes to enforcement of safety polices and procedures, discipline may come into play. No matter the type of discipline system that you use, the system must be established and implemented fairly, consistently, and equally to all employees. Infractions of safety rules must be linked to the procedure for corrective action as follows:

* Taking the appropriate corrective actions on the seriousness of the infractions
* Providing employees the opportunity to correct their own at-risk behavior
* Making sure that the disciplinary system is consistent, fair, and useful

Refer to Table 11-5 for key questions concerning discipline.

The purpose of any enforcement system is not simply to punish employees; it is to control the work environment so that employees are protected and incidents are prevented.

Enforcement is like developing objectives. It should be based on making sure that employees know what is expected of them regarding published and communicated safety rules. In addition, it lets employees know how they are expected to work in relation to the goals of the organization's safety program. The best way is to use employee participation as described in Chapter 7 in the development of the enforcement system [3].

Table 11-5
Questions to Ask before You Discipline

Are you justified in disciplining the employee? Ask yourself the following:

- Have you provided the resources the employee needed?
- Have you provided the training?
- Have you provided the oversight?
- Have you provided the accountability?

OR OSHA 100o, Before You Discipline, Ask, p. 10, http://www.cbs.state.or.us/external/osha/educate/training/pages/materials.html, public domain.

WHO SHOULD INSPECT THE WORKPLACE?

Several types of individuals should be involved in inspecting the workplace. The following sections will provide a brief overview of some of the responsibilities of inspectors.

Supervisors

Supervisors should have the responsibility to inspect their own work areas at the beginning of each shift. This can be important when other shifts use the same work area and equipment or when after-hours maintenance and cleaning are routinely conducted. In addition, supervisors should review other areas for several reasons:

- Those who work in an area may not see things that they have become used to doing. It can soon become a way of life where employees become complacent. It is always good to have cross-inspections where supervisors or employees from one area look at another area. This will provide a fresh look at the equipment.
- A general site inspection will encompass areas not assigned to individual supervisors, for example, outdoor and other common areas [2].

All supervisors should have site-specific training on the hazards that they may encounter. In addition, training should be conducted on how to control these hazards. When supervisors are responsible for area inspections, they should have specific training in how to inspect [2]. For additional training needs, refer to Chapter 14.

Employees

This is one of the ways to get all employees involved in the safety process: allowing them to participate in the inspection process. Refer to Chapter 7 for more information on how to use employees in the process. One way to get employees involved is to form employee committees or joint employee–management committees to help conduct routine inspections. By implementing these methods, you can:

- Expand the number of employees doing inspections and improve hazard identification
- Increase employee awareness of hazardous situations as well as the appropriate control methods [2]

All employees should understand the potential hazards to which they might be exposed and the ways they can protect themselves and their fellow workers. Those who are involved in the inspections need training in recognizing and controlling potential hazards. They will also need written guidance, tips for inspecting, and some on-the-job training [2].

Safety Professionals

It is important for the safety professional to mentor and provide guidance on methods to conduct inspections. By using the safety professional's help in conducting inspections, you can keep the person responsible for safety in touch with the successes and/or problems in the hazard prevention and control program.

WRITTEN INSPECTION REPORTS

Written inspection reports should be designed to record hazards found during the inspection. One individual or team should be responsible for developing an action plan, and tracking hazards to completion. Refer to Appendix C for a sample report.

Written records are important because these documents will help to make sure that the following has been completed:

- Action plans are developed to track identified hazards to completion

- The assignment of responsibility for correction is accomplished
- Safety issues in the management system are identified when the same type of hazards keep appearing after correction is verified
- Safety issues in the accountability system are identified
- Hazards where no prevention or control has been planned are identified [2].

Written reports will be helpful if they are reviewed by someone knowledgeable in the inspection process. This report can provide top management with a summary of safety issues. Use in-house expertise [2].

TRACKING HAZARDS

The success or failure of inspection programs is determined by the quality of follow-up and the action planning. If corrective actions cannot be accomplished immediately after a hazard is identified, the inspection report should include interim measures that have been taken, as discussed previously, and should not be considered completed until the final action plan has been signed off. A written tracking system will improve your inspection program. Refer to Appendix C for methods of tracking hazards.

There are some important elements that you must understand about regular site inspections and the critical follow-up inspections:

- Inspections should cover all parts of the workplace as applicable.
- Inspections should be conducted at regular intervals. The frequency will depend on the size of the workplace and the nature of the hazards.
- Inspectors should be trained to recognize hazards and to bring fresh ideas.
- Information from inspections should be used to expand the hazard inventory, modify/update job hazard analyses (JHAs), and/or improve the hazard prevention and control program.
- Identified hazards must be tracked to completion [2].

An essential part of any day-to-day safety effort is to correct hazards. Developing an action plan is important in any type of environment:

- It keeps management aware of the status of long-term corrective actions

- It provides a record of what occurred, if the hazard occurs later
- It provides timely and accurate information that can be supplied to an employee who reported the hazard [2]

Many companies use forms that document hazards identified for the purpose of developing a corrective action plan. Inspection reports include discussions about hazard correction along with the information about the hazard. Forms that employees use to report hazards and incidents should provide a space for description of the action plan. Refer to Chapter 11 and Appendix B for more detail. Notice that each of these forms has a line for corrective actions [2]. Refer to Table 11-6 for a sample policy for employees to report hazards. Appendix C provides other forms for employees to report various hazards.

When documenting information about your action plan, it is important to document all interim protective measures and to include the anticipated date of completion. Otherwise, you run the risk of the action plan not being completed. This can happen when a part has to be ordered and time is needed for procurement, or when interim, less than adequate measures become substitutes for preferred but possibly more costly or time-consuming actions. On the other hand, this may not pose a problem if the hazard can be corrected in a short period of time, because someone probably will remember to see that the final completion occurs [2].

Another way to track hazards is to transfer information from the original hazard reports to a separate hazard tracking report. This system can receive information on all uncorrected hazards and not just information

Table 11-6
Sample Policy for Employees to Report Hazards

Every employee, as part of their responsibilities, is expected to report any hazards that are identified in the work area. You may make your report by contacting your supervisor or by submitting a written report. Make your report immediately or as soon as possible.

No employee, at any level, will be disciplined or harassed because of making a hazard report. Any employee found to have discriminated against another employee for this reason could be disciplined.

Remember that employee participation is needed to help keep your worksite safe.

_____

(Signature of top manager at worksite)

system establishes the actions you will take if employees do not meet your expectations [2].

A disciplinary system does not exist only to punish employees. The purpose should be to control the work environment so that employees are protected from incidents. A disciplinary system helps to make sure that safety is consistent by letting employees know what is expected of them. It provides employees with an opportunity to correct at-risk behaviors before an incident occurs [2].

A disciplinary system cannot work in a vacuum. Before you can hold any employees accountable for their actions, you must establish your safety policy and disciplinary rules. Then you need to develop safe operating procedures, train your employees on these procedures, and supervise their actions. One important thing to remember is that employees should not be disciplined for management failures. If an employee gets hurt and you have not provided the appropriate tools and materials, training, or supervision, you should never consider discipline. The management system has failed. In this case, you must determine the root cause. How many times have you investigated an incident and before even talking to the employee, someone says that the employee was at fault? This is counterproductive. I am sure that there are many different thoughts on this subject. To illustrate, in a meeting on employee discipline, I made a statement that from my many years of experience in management and safety, I had seen only 1 out of every 10 cases that warranted discipline when someone got hurt. One of the managers in the room disagreed and said that he had seen 9 out of every 10 cases that warranted discipline. Whatever are your thoughts, you have to evaluate each situation on a case-by-case condition.

On the other hand, if employees violate safety rules and procedures that they have been trained on, this is a different manner. There is also one caution here; employees sometimes violate these rules because again, they do not have the tools or equipment. One example comes to mind. One employee was cut while performing his job. This was a fairly routine job and if reasonable care were taken, there should be no issues. However, in this one particular case the manager wanted to discipline the employee. During the investigation it was discovered that the employee did violate the safety rule because he had asked several times for the proper tool and was not given the tool. He had to do his job, so he did it the only way that he could. In this case, should you discipline the employee? I will let you make your choice here. In this case, the employee was not disciplined because in this culture the manager did not provide the proper tools. Refer to Tables 11-1 through 11-7 for a sample of useful discipline systems that can be used when establishing a disciplinary program [2].

Most disciplinary systems use corrective procedures that involve three, four, or five steps. As you review each system you must remember that the use of any corrective action system will vary with the nature of the business, the issue at hand, and the frequency with which the infraction occurs.

Infractions of company rules generally are considered more serious than other employee behavior problems, but all require corrective action. Keep in mind, and tell your employees, that your primary goal is to prevent incidents by minimizing unsafe acts and conditions. Table 11-7 describes the several types of discipline systems.

Table 11-7
Sample Disciplinary Systems

Three-Step System [2]

First violation	Written warning. Usually one copy is provided to the employee and one copy is put in the employee's file.
Second violation	Written warning; suspension without pay for half or full day.
Third violation	Written report for file and immediate termination.

Four-Step System

First violation	Oral warning; notation for personnel file.
Second violation	Written warning; copy for file or Personnel Office.
Third violation	Written warning; one day's suspension without pay.
Fourth violation	Written warning and one week's suspension, or termination if warranted.

Five-Step System

First violation	Instruction/discussion concerning infractions, proper procedures, and the hazards they control; notation for supervisor's file.
Second violation	Re-instruction with notation in the employee's personnel file.
Third violation	Written warning describing the violation and actions that will be taken if it recurs.
Fourth violation	Final warning; may include suspension.
Fifth violation	Discharge.

It is useful to make a list of the kinds of infractions that you consider serious, and a second list of other types of behaviors that, although not as serious, are still not acceptable. Table 11-8 outlines some of the general workplace safety rules that can apply to a discipline system.

In another example (Table 11-9) you can link each type of infraction to a structured procedure for corrective action. Your goal is to make sure that the corrective action is appropriate to the seriousness of the infraction; that employees are given the opportunity to correct their own behavior; and that the system is realistic and used appropriately.

One key to discipline is to be fair and consistent and to make sure that a good recordkeeping system is in place. It is important to have written rules and disciplinary procedures that are communicated to and understood by all employees. It is just as important to document instances of good and/or poor safety behavior, including discussions with the employee [2].

Table 11-8
Sample Establishing General Workplace Safety Rules

Major offenses:

- Failure to follow rules regarding use of company equipment or materials
- Horseplay or creating unsafe conditions
- Tampering with machine safeguards or removing machine lockout/tagout
- Not wearing the appropriate PPE
- Provoking or engaging in an act of workplace violence against another person on company property
- Using or being under the influence of alcohol or illegal drugs
- Traffic infractions while using a company vehicle. A rating system must be developed to be consistent; one suggested method is to use the specific state driving rating system
- Other major infractions of company rules or policies

General offenses:

- Minor traffic violations while using company vehicles; refer above
- Creating unsafe or unsanitary conditions or poor housekeeping
- Threatening an act of workplace violence against another person while on company property
- Misrepresentation of facts or falsification of company records
- Unauthorized use of company property
- Other violations of company policy and rules

Table 11-9
Procedures for Corrective Action [2]

Written Warning	No safety glasses, earplugs, safety shoes
	Horseplay
	Unsafe work habits
	Infraction of other safety or health rules or requirements
Suspension (8 hours without pay). Some companies provide this 8 hours off with pay. The objective is to make the employee go and discuss with his/her family the situation and decide whether he/she wants to work for the company.	Three or more safety or health infractions of the same type
	General overall record of unsafe practices
	Refusal to follow safety and health guidelines or instructions
Termination	Excessive and repeated safety and/or health violations
	Purposely ignoring safety and/or health rules
	Unsafe actions that seriously jeopardize the safety or health of others
	General disregard for safety and health of self and others

CONTROL MEASURES

Disciplinary systems need to be proportionate to the seriousness of the safety infraction and/or the frequency of the occurrence. You will probably agree that it is inappropriate to discharge someone for occasional tardiness. What about discharging someone for one safety infraction? In this case, it is still inappropriate to issue only oral warnings to an employee who, say, repeatedly removes a machine guard. You must understand why the employee does it and why the management system failed. This is like finding the root cause for an incident. Refer to Table 11-7 for examples of a disciplinary system. Refer to Table 11-8 for a list of suggested safety infractions that can be used in general work rules.

Disciplinary procedures should not be instituted without explanation. Be sure to provide feedback to the employee on what behaviors are unacceptable, why the corrective action is necessary, and how the employee can prevent further disciplinary action. In addition, take time to recognize an employee who improves or corrects his or her behavior [2].

If your disciplinary system is to work and be accepted by all employees, you must make sure that all employees understand that the disciplinary system applies equally to everyone and is consistently applied. This includes subjecting managers and supervisors to the same type of safety rules and disciplinary procedures [2].

POLICY STATEMENT

Employees need to know where you stand on safety and what you expect of them. They need a clear understanding of the policies and procedures and the consequences of breaking the rules. This is true in all areas of work, but it is more important for employee safety. As part of the policy statement, or in an employee manual or booklet, you should have a written statement detailing your disciplinary policy [2]. Refer to Chapter 5 for a detailed discussion on policy statements.

EMPLOYEE TRAINING AND INFORMATION

It is important that employees understand why safety rules and work practices are needed. As stated, employees must understand that, for an incident to occur, two things must be present: the hazard and the employee. If you remove the hazard, there will be no injury.

An additional responsibility is to make sure that any training program has achieved its objectives: employees must understand the hazards that they work with and know how to protect themselves [3]. Just presenting training may not be sufficient.

A supervisor easily can perform informal testing to determine the results of training. This can include stopping at an employee's workstation and asking for an explanation of the hazards involved in the job and the employee's means of protection. If the training has been presented well and has been understood, each employee should be able to provide a clear, comprehensive answer [2].

It is important that all employees understand the management system and be able to answer any questions about how it works. In addition to issuing a written statement of your disciplinary policy, you should develop a list of what you consider major infractions of company policy and less serious violations. This list should specify the disciplinary actions that will be taken for first, second, or repeated offenses [2].

Training can reduce the need for disciplinary action by instructing employees in the importance of safety, the need to develop safe work habits, your operation's safe work practices and the hazards they control, and the standards of behavior that you expect. Be sure your employees understand the disciplinary system and the consequences of any deliberate, unacceptable behavior [2].

SUPERVISORS

Supervisors must be trained in developing and monitoring the corrective action system. Ongoing monitoring of safety behaviors allows the supervisor the opportunity to correct safety issues before they become serious [2].

With an effective safety culture, supervision means correcting safety issues before punishing an employee. When the relationship between employees and management is open and interactive via employee participation, safety issues are discussed and solutions are mutually agreed upon. There is buy-in into the process. This type of relationship fosters a safe work environment and changes in culture where the need for disciplinary action is not needed. When discipline is needed, the parties are more likely to perceive it as corrective rather than punitive [2].

EMPLOYEE PARTICIPATION

Employee participation plays a major role in helping to identify hazards and establishing action plans. Since employees are close to the operation, they have a unique and valuable perspective on the workplace and conditions [2]. Refer to Chapter 7.

A reliable system by which employees can report hazards is an important element of any effective safety culture and management system. A successful system will usually include the following elements:

- A written program that is consistent with other policies and procedures and encourages employees to report their concerns about safety conditions
- Timely and appropriate responses to employees concerning identified hazards with action planning where valid concern exists
- Tracking of required hazards to completion
- Protecting employees who report hazards from reprisal [2]

Once you have determined what your company policy is on how employees can report hazards, the next step is to make sure that all employees understand the policy. Employees must be reassured that the policy is genuine and that there will be no reprisal for reporting hazards.

Once the policy has been developed it should be placed on bulletin boards, distributed to all employees, and/or discussed in weekly/monthly safety meetings as applicable. You will know you have communicated the information when every employee in the organization can discuss the reporting policy. Refer to Appendix A for a sample policy statement [2].

To be effective, you should involve your employees in setting up or revising a disciplinary system. Employees who contribute their ideas to development of rules, procedures, and the disciplinary system are likely to be knowledgeable about the system. They are likely to understand that the system is designed to protect employees from the unsafe acts or at-risk behaviors of others. Note that if you have a union, you may need to involve an employee representative [2].

Employees should be encouraged to help in the enforcement of work rules and work practices. Your intent is not to turn your employees into informers, but to encourage them to keep an eye out for the safety of themselves and fellow workers. Many employers have encouraged an atmosphere where a successful safety culture exists by getting employees involved. In this type of situation, employees speak up when they see safety issues—for example, a co-worker who needs to be reminded to put on personal protective equipment [2]. In many cases, employees will discuss safety infractions with their co-workers. The key is to develop a culture where personal protective equipment is not the only safety infraction that is discussed. There are many more hazards that should be identified.

All employees deserve an opportunity to correct their own behavior. This is why an effective disciplinary system is a two-way process. Once a problem is identified, you must discuss the issues with your employee, who should be given at least one or two opportunities to change his or her behavior. If these discussions do not help employees, then discipline may be needed. Only after these discussions, and in some cases retraining, should any disciplinary action be taken [2]. Refer to Chapter 7 for a discussion of employee participation.

WORK ORDERS

When conditions are unsafe, maintenance usually will have to correct the hazard. Some companies give every employee the right to fill out work orders; others allow employees to complete work orders but require supervisory sign-off before orders are sent to the maintenance department. This system for employees to report hazards should be used only if there is a special high priority for safety work orders [2].

The work order system for reporting hazards is not sufficient if used alone. Although it can lead to hazard correction, it cannot correct unsafe practices or at-risk behaviors. In addition, the system is not useful for encouraging imaginative new approaches to improving conditions and procedures [2].

RESPONSE TIME AND ACTION PLANNING

As we discussed in Chapter 10, sometimes hazards cannot be completed in a reasonable amount of time—for example, parts or materials must be ordered, requiring a wait of several months. You need to provide your employee with the status of the report from time to time. This will send a message that the employee's concerns have not been forgotten. One important note: this program will not work if employees cannot see hazards corrected in a reasonable amount of time. Another important consideration is that when the preferred corrective activities cannot be accomplished immediately, you must provide your employees interim protection until the hazard is completed. You must take whatever steps are necessary to temporarily minimize the hazard during the abatement stage of the process [2].

Results not only must be timely, but must also address the employee's concern. As discussed, if management decides that no hazard exists, the reason behind this decision must be thoroughly discussed with all affected employees. This is where many companies fall short of their obligations to employees. You must come to an agreement that the identified hazard is or is not a safety hazard that they should or should not be concerned about. Take care to express gratitude to the employee for erring on the side of safety. It is better to have some non-hazard or condition reported than to overlook a real hazard because an employee believed that management would not respond appropriately [2]. If you are uncertain whether the reported practice or condition is hazardous, further investigation may have to be done.

Every valid hazard that is identified must have an action plan. This action plan must be tracked to completion. Hazards are sometimes quickly corrected and may not present a safety issue for employees. However, documenting corrective actions will help to determine where the management systems may have failed if the same hazard reappears. If the management system failed, different measures (activities) may have to be identified. For hazards that require complicated or time-consuming corrections, a system of tracking hazards is needed to make sure that the final completion date is not forgotten. Additionally, tracking hazards to completion enables management to keep employees informed of the action plan [2].

PROTECTING EMPLOYEES FROM REPRISAL

It is important that all employees know that reporting a hazard will not result in any reprisal from management or other co-workers. The policy on how employees report hazards should be made clear to everyone. In addition, there are several actions that you can take to help make sure that reprisal is not considered:

- Avoid performance measurements that rate supervisors negatively for employees who report hazards.
- Avoid placing policy statements dealing with hazard reporting and discipline close together on a bulletin board or in sequence in an employee handbook. A physical proximity of the two subjects can give employees the perception that they can get into trouble by reporting hazards.
- Approach all discussions and written descriptions of employee reporting of hazards as a group effort. Emphasize the positive, not the negative.
- Emphasize the responsibility that each employee has for co-workers' safety as well as their own. The safety of individual employees is everyone's responsibility.

- If you discover a case of reprisal for reporting hazards, enforce your policy clearly, consistently, and emphatically to protect the employee [2].

REPORTING SYSTEMS

The following are the most common ways in which employees can report hazards:

- Oral reports
- Suggestion programs
- A hazard card system
- Maintenance work orders
- Written forms that provide for anonymity [2]

Many companies will use a combination of some or all of the listed systems. The following will discuss each system in more detail.

Oral Reports

All employees should be able to report hazards without reprisal. When a good safety culture is in place, managers are properly trained and accept their responsibility for safety. Informal oral reporting of hazardous conditions should occur naturally. When an employee's concerns appear valid, management must have the responsibility to correct the hazard, request corrective action by maintenance or other resources, or ask for assistance from a safety professional. Most workplaces encourage this type of reporting. Used alone, it does not provide for hazard correction tracking. Nor does it enable you to look for trends and patterns.

Suggestion Programs

The most frequently used written system is where employees are encouraged to make safety suggestions. This is considered a positive approach. Not only does it provide a mechanism for reporting unsafe conditions, but it also encourages employees to develop innovative approaches to doing things safely. If a suggestion program is used for hazard reporting, management must make sure that collection points are checked several times a day and that suggestions are read at the time of collection. This will make sure that identified hazards are corrected in a timely manner. If the suggestion program is the only means of reporting hazards or the only written system, management must encourage employees to use the system for all types of hazard reporting, and not just for ideas [2].

Hazard Card System

Many companies have developed or purchased a program for employees reporting hazards. One such program includes a format for training employees in basic hazard recognition. This program uses cards where employees identify and document unsafe conditions and practices. These cards are turned in to management for reviewing and tracking of valid hazard correction [2].

Some workplaces provide recognition for the highest number of cards with valid concerns turned in over a specific period of time. Others set quotas for the number of cards turned in. The success of these special uses seems to depend upon the "culture" of the workplace.

PREVENTIVE MAINTENANCE PROGRAM

Many managers do not associate preventive maintenance with a safety program. However, good preventive maintenance plays a major role in making sure that hazard controls continue to function effectively. This will help to ensure that installed controls are still working as designed. Preventive maintenance also keeps new hazards from arising as a result of equipment malfunction [2].

The whole point of preventive maintenance is to make sure that the equipment is fixed before repairs or replacement must be done. Preventive maintenance requires reliable scheduling of maintenance activity. The scheduling depends on knowledge of what needs maintenance and how often it is used [2].

A preventive maintenance program starts with a survey of maintenance needs. It should be developed to get the work done before repair or replacement is required. Don't wait for "breakdown maintenance." As described in the process safety management (PSM) standard (refer to OSHA 29 CFR 1910.119), do not wait for something to break down until you have to fix it. A good preventive maintenance program plays a major role in making sure that equipment continues to function properly and also keeps new hazards from arising due to equipment malfunction.

Every piece of equipment or part of a system that needs maintenance, for example, oiling and/or cleaning, testing, replacement of worn parts, or evaluation, should be reviewed. You will need a complete list of all items that are to be maintained. If this list does not exist, your maintenance manager should conduct a survey and develop the list. The survey should be repeated periodically to update the list of maintenance items. Whenever new equipment is installed in the worksite, the list should be revised accordingly [2].

A good preventive maintenance program should also include a workplace survey to identify all equipment or processes that may require routine maintenance. To be effective, maintenance should be performed on a routine basis and as recommended by the manufacturer [3].

For each item considered in this inventory you must estimate the average length of time before the maintenance work becomes reactive rather than preventive or "breakdown." Plan to perform the maintenance before the average time. Review maintenance records periodically to understand how much reactive maintenance (repair or replacement of defective parts after failure) has been done. After the record review, make new estimates of average time, and adjust your maintenance timetable accordingly [2].

SUMMARY

Effective planning and design of the workplace or task can help minimize related hazards. Where eliminating hazards is not feasible, planning and design can help to minimize the exposure to unsafe conditions.

Elimination or control should be accomplished in a timely manner once a hazard or potential hazard is identified. These procedures should include the following measures:

- Using engineering controls where feasible
- Establishing safe work practices and procedures that can be understood and followed by all affected employees
- Providing PPE when engineering controls are not feasible
- Using administrative controls—for example, reducing the duration of exposure

- Maintaining the facility and equipment to minimize breakdowns
- Planning and preparing for emergencies, and conducting training and emergency drills, as required, to make sure that the proper responses to emergencies are "second nature" for all persons involved
- Establishing a medical surveillance, for example, handling first aid cases onsite as well as or using a nearby physician and emergency medical care [2]

Once a safety issue has been identified, an action plan should be developed to help resolve the issues in a timely manner. The action plan can be used to achieve compliance with applicable regulatory requirements. These action plans can also include establishing priorities, identifying deadlines, and tracking progress [2].

Even after you have conducted a comprehensive hazard survey, analyzed the possibility for changes in the workplace, routinely analyzed jobs and/or processes for hazards, and developed a program of hazard prevention and control, there still will probably be some hazards that you missed in your initial survey, or where measures have not been adequately taken to maintain prevention or control over time [2].

For the hazard reporting systems, some variations may work better for your site than others. Use the system that works best for your operation. The following are some important points to remember:

- Develop a policy that encourages employee participation in reporting hazards
- Make sure that the policy is known and understood by all employees
- Protect employees from reprisal
- Respond to employees in an appropriate and timely manner
- Track all hazards to completion
- Use the information you obtained about hazards to revise your hazard inventory and/or to improve your hazard prevention program [2]

This chapter has examined additional techniques for learning more about persistent hazards, their correction, and effective control. Regular site inspections; employee reporting hazards; incident investigations; and analyses of patterns of illness and injury, incidents, and hazards will help complete your safety program [2].

Documentation is important for a variety of reasons: It helps to track the development of a problem, corrective actions, and the impact of measures taken. It provides information to keep employees informed of issues that need correction. When you are evaluating the managerial and supervisory skills of your employees, it provides a useful record of how they handle issues [2].

If warnings, retraining, and/or corrective actions fail to achieve the desired effect, and if you decide to discharge an employee, then documentation becomes even more important. Conversely, you may want to consider an annual clearing of the personnel files of employees whose good overall safety records are marred by minor warnings [2].

Make sure that the preventive maintenance schedule is accessible. Easy availability of the schedule will help your maintenance staff plan its work. A well-communicated schedule will help to ensure the maintenance department's accountability for performing the work on time. Select a method of communication that works for your employees [2].

Preventive maintenance can be a complicated matrix of timing and activity. Keeping track of completed maintenance tasks can be as simple as adding a date and initials to the posted work schedule. Some employers use their computer systems to keep track of completed maintenance activity [2].

Records must be maintained for all preventative maintenance. Documentation will assist you in keeping up with the required maintenance and can also serve as a way of identifying and recognizing employees who have been instrumental in preventing costly repairs and accidents [3].

You should approach each category of hazard with the intention of preventing it. If total prevention is not feasible, you should control the hazard as completely as possible through equipment design. To the extent that potential exposure exists despite the designed controls, then you should use safety rules, work practices, and other administrative measures to control that exposure. Finally, you may need to use personal protective clothing and other equipment to further minimize levels of employee exposure [2].

To complement hazard controls, you also must have good management systems that involve preventive maintenance; hazard correction tracking; a fair and consistent enforcement of rules, procedures, work practices, and PPE; a solid system for responding to unexpected emergencies; and a good medical program that helps to identify hazards and minimize harm if injuries and work-related illnesses occur [2].

These are the basic components of a hazard prevention and control program. With these measures, you can provide your employees with the appropriate protection from identified hazards [2].

It can be instrumental in establishing accountability for employees [2]. Routine (weekly) safety walkthrough inspections should be conducted by top management, safety and health staff, members of safety committees, etc. These inspections are a good review of the program and can help identify areas that need to be addressed. They also keep your safety efforts visible and everyone's safety awareness up. Employees can be assigned responsibility for conducting daily inspections in specific areas.

Comprehensive surveys are not the same as inspections. Comprehensive surveys should be performed by people who can bring to your worksite a fresh vision and extensive knowledge of safety, health, or industrial hygiene, i.e., safety professionals, industrial hygienists, ergonomists, etc. Insurance carriers, corporate staffs, private consultants, and the OSHA-funded consultation projects are sources of help in this area.

A formalized mechanism should be developed that encourages the routine reporting of problems. Employees need to be made comfortable with reporting problems without fearing being labeled a whiner or complainer. Employees need to understand that not only will they not be harassed for reporting problems, but they will be rewarded for it and will be in trouble for not reporting potential problems. Managers and supervisors, on the other hand, need to be held accountable for adequately addressing concerns. When management fails to do so, appropriate disciplinary measures should be taken. When successful in establishing an effective reporting system, management will have a mechanism that will allow systematic and proactive identification of problems before accidents occur. This will enable you to be in a position to prevent losses. As the safety culture of an organization matures, reported problems will migrate from accidents and physical hazards to near-misses and at-risk behaviors [3].

Job hazard analysis breaks down a job into its component steps. This is best done by jointly analyzing each step in order of occurrence with the affected employee. Next, examine each step to determine the hazards and at-risk behaviors that exist or that might occur. Reviewing these job steps and hazards with the employee performing the job will help ensure an accurate and complete list. It also is a good way to educate employees on the risks and gets their buy-in on any needed changes. Written standard operating procedures can evolve through this process.

To determine which jobs should be analyzed first, review your job injury and illness reports. Obviously, a JHA should be conducted first for jobs with the highest rates of accidents and disabling injuries. Also, jobs where "close calls" have occurred should be given priority. Analyses of new jobs and jobs where changes have been made should follow. Eventually, a JHA should be conducted and made available to employees for all jobs in the workplace.

If an incident occurs on a specific job, the JHA should be reviewed immediately to determine whether changes are needed in the job procedure. Any time a JHA is revised, training should be provided in the new job methods or protective measures. A JHA can also be used to train new employees [3].

Change analysis should be performed whenever a significant modification or addition is made to the process. Examples include building or leasing a new building, installing new equipment, using new materials, starting up new processes, or personnel changes. An organization or process is like a web of interconnections; a change in one area throws a different part off balance. Managing these ripple effects is what makes managing change a dynamic proposition with unexpected challenges. Having a team of operators, engineers, and safety and health professionals jointly analyze potential changes or new equipment, etc., before they are put online, can identify safety and production concerns up front, ideally heading off problems before they develop. Fixing potential problems before they occur usually is less expensive than attempting to fix a problem after the fact [3].

> Trends in injuries/illness experienced should be identified and analyzed. Trends can indicate areas in need of attention. Examples of trends that might emerge are a type of injury, (e.g., eye injuries), or a number of injuries/illnesses in a specific department [3].

REFERENCES

1. Oklahoma Department of Labor, Safety and Health Management: Safety Pays, 2000, http://www.state.ok.us/~okdol/, public domain.

2. U.S. Department of Labor, Office of Cooperative Programs, Occupational Safety and Health Administration (OSHA), *Managing Worker Safety and Health*, November 1994, public domain.

3. OSHA Web Site, http://www.osha-slc.gov/SLTC/safetyhealth_ecat/mod4_factsheets.htm#, public domain.

12

Conducting Effective
Incident Investigations

INTRODUCTION

Thousands of incidents occur every day in workplaces. The failure of people, equipment, the environment, and management to behave or react as expected causes most of these incidents. Investigating incidents is important because it determines how and why failures in the management system may have occurred [6].

Investigations must be approached as a fact-finding effort to help prevent future incidents. Investigations are an analysis and account of what happened based on information. Effective investigations include the objective evaluation of all the facts, opinions, statements, related information, and the identification of the root cause(s) and actions that will be taken to prevent recurrence. It is not a repetition of the employee's explanation of the incident. Facts should be reported without regard to personalities, individual responsibilities, or actions. The key is not to blame, and fault-finding must not be a part of the investigation proceedings or results [9].

In addition, incident investigations help to identify hazards that either may have been missed earlier or may have slipped out of control during the normal process. It is useful only when conducted with the aim of identifying contributing factors to the incident, condition, and/or activity and preventing future occurrences [5].

This chapter will help you look at incidents in a different way. The bottom line: if you do not capture near misses and small incidents that do not result in an injury, you will never stop injuries.

DEFINITIONS

In industry, we use many definitions to describe an incident or an accident. Frequently both terms are used interchangeably. We use the term "loss-producing events" to describe incidents. As we stated previously, sometimes the journey is more important than the destination. The following are some of the more common definitions:

Accident. An "unexpected, undesirable event; an unforeseen incident"; "lack of intention" [8].

Causal factor chain. This is a cause and effect sequence where a specific action creates a condition that contributes to or results in an event. This creates new conditions that, in turn, result in another event, etc. Figure 12-1 summarizes a root cause analysis flow chart [2].

Chance. "An accidental or unpredictable event," "a risk or hazard" [8]. This term does not include any illnesses resulting from long-term exposure to health-related hazards.

Condition. Any as-found state, no matter the results from an event, that may have adverse safety, health, quality assurance, security, operational, or environmental implications [2].

Contributing cause. A cause that contributed to an occurrence, but by itself would not have caused the occurrence [2].

Direct cause. The cause that directly resulted in the occurrence. For example, the direct cause could have been the problem in the component or equipment that created a leak [2].

Incident. "A definite and separate occurrence; an event"; "A usual minimal event or condition subordinate to another"; "Related to or dependent on another thing" [8].

Root cause. The root cause, if corrected, would prevent recurrence of similar occurrences. The root cause does not apply to a specific occurrence only, but has generic implications to a broad group of possible occurrences. It is the most fundamental aspect of the cause that can logically be identified and corrected. There may be a series of causes that can be identified, one leading to another [2]. We must use the "5 Whys" to get to the root cause (Figure 12-2).

BENEFITS OF ROOT CAUSE ANALYSIS

Before we discuss incident investigations, you must understand that a good investigation will provide many benefits, which can include:

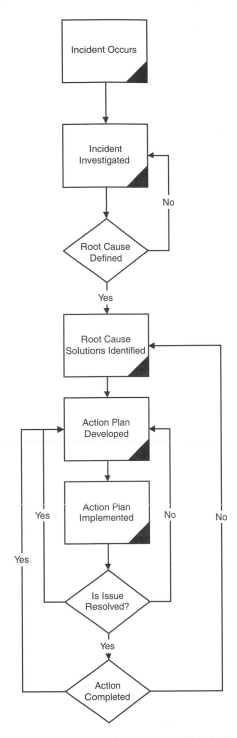

Figure 12-1 Sample root cause analysis flow chart. DOE Guideline Root Cause Analysis Guidance Document, DOE-NE-STD-1004-92, February 1992, U.S. Department of Energy, Office of Nuclear Energy, Office of Nuclear Safety Policy and Standards, Appendix E, Figure E-1, p. E-1, "Six Steps Involved in Change Analysis," modified public domain version.

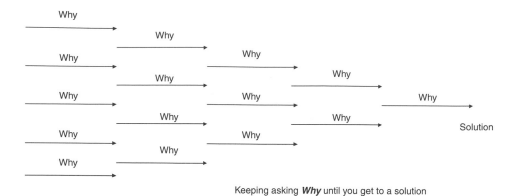

Keeping asking *Why* until you get to a solution

Figure 12-2 Root cause analysis: getting to the solution. The 5 Whys.

- Increased awareness that all incidents should be reported and investigated promptly
- The identification of the basic causes of an incident
- Prevention of recurrence of similar incidents
- Identification of incident prevention program needs and opportunities for improvement in the management system
- Reduction in employees' and families' suffering
- Reduction in workers' compensation costs due to reduced frequency of incidents
- Increased productivity [5]

An incident can be a failure in the management system because something fails. Alternatively, an incident could be a combination of work errors that have gone unnoticed or uncontrolled for a period of time [9].

INCIDENT PREVENTION

An incident may have many more events than causes. A detailed analysis of an incident will normally reveal three cause levels: basic, indirect, and direct. The direct cause is usually the result of one or more unsafe acts or unsafe conditions, or both. Unsafe acts and conditions are the indirect causes or symptoms. In turn, indirect causes are usually traceable to poor management policies and decisions, or to personal or environmental factors. These are the basic causes (Figure 12-3) [6].

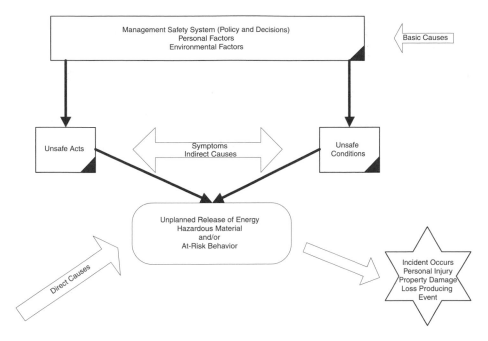

Figure 12-3 The basic causes of an incident. OSHA's Small Business Outreach Training Program Instructional Guide, Web site, Accident Investigation, http://www. osha-slc.gov/SLTC/smallbusiness/sec6.html, public domain.

Most incidents could be prevented by eliminating one or more causes. Incident investigations determine not only what happened, but also how and why. Figure 12-4 is an overview of an incident investigation flowchart. This will help you to understand the process of investigating an incident. The information gained from investigations can prevent recurrence of similar incidents. Investigators are interested in each event and the sequence of events that led to the incident [6]. Refer to Figure 12-5 for a sample incident investigation form.

BASIC CAUSATION MODELS: SEQUENCE MODELS

This section will provide an overview of an incident causation model that we believe will help you understand the investigation concepts for establishing the root cause of an incident.

In conducting hazard analysis, an incident scenario as shown in Figure 12-6 is a useful model for analyzing risk of harm due to hazards.

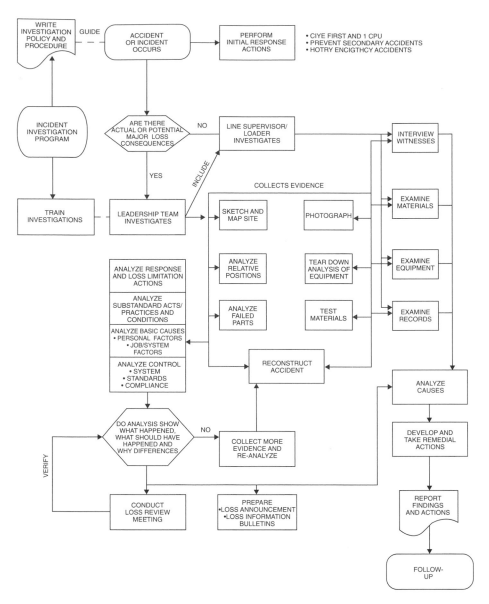

Figure 12-4 Incident investigation flowchart. Bird, Frank, George Germain, *Loss Control Management: Practical Loss Control Leadership*, revised edition, Figure 15-3, p. 346. Det Norske Veritas (U.S.A.), Inc., 1996. Reprinted with permission.

When the original models of accident causation were developed, employees were the center of the accident triangle. To some degree, employees are still being blamed for most incidents (at-risk behavior). The model depicts a sequence of events that lead to accidents. The Hein-

* Must Be Completed Within 24 Hours *			
Date of Incident			
Employee Name		Supervisor Name	
Location			
Classification			
☐ Injury	☐ Near Miss	☐ First Aid	☐ OSHA Recordable
☐ Lost Workday	☐ General Liability	☐ Vehicle	☐ Not at Fault
Description (Provide facts, describe how incident occurred, provide diagram (on back) or photos			
Analysis 1 (What unsafe acts or conditions contributed to the incident?)			
Analysis 2 (What systematic or management deficiencies contributed to incident?)			
Corrective Action(s) (List corrective action items, responsible person, scheduled completion date)			
Witnesses (Attach statements or indicate why unavailable)			
Manager			
	Print Name	Signature	Date

Figure 12-5 Sample incident investigation form.

rich Theory (1931) [4] is credited with one of the first accident models [2]. Heinrich designed a model call the "domino" theory that is based on assumptions that accident causation can be described as a chain or sequence of events leading to an accident and/or consequences [2]. Refer to Figure 12-7 for the "Domino" accident sequence model (adapted from Heinrich and other resources [1, 2, 4, 7]. In this theory events are tied together in a sequence of causes and effects. These factors are presented as a set of five dominos standing on edge in a line. As Heinrich noted, if

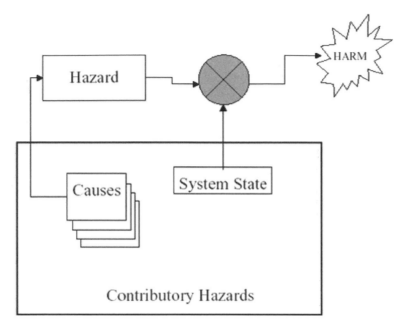

Figure 12-6 Hazard scenario model. *FAA System Safety Handbook*, Chapter 3, "Principles of System Safety," Figure 3-2, p. 3–4. December 30, 2000, public domain.

the first domino falls it knocks down the second domino, etc., until all dominos are knocked down. Removal of any domino and the causation factor will break the chain of events (causation factors) so that an injury will not occur [4]. I am sure that you have placed a set of dominos in a row to see if you could knock them down in sequence. This is an effective way of demonstrating the concept to management and employees when making a presentation.

Heinrich detailed his model using five factors that show the sequence of events: Hereditary and social environmental factors lead to a fault of the employee constituting the proximate reason for either an unsafe act or unsafe conditions (or both) that result in the accident, which leads to the injury [2].

In 1986, Frank Bird and George Germain [1] used Heinrich's [4] model to develop another accident causation model (Figure 12-8). This model used the same "domino theory" to show its key concepts of loss control.

The most obvious losses are deemed to include harm to employees, property, or processes [1]. Implied and related losses include loss-producing events—for example, business interruption and profit reductions. Bird and Germain argue that once the sequence has occurred, the

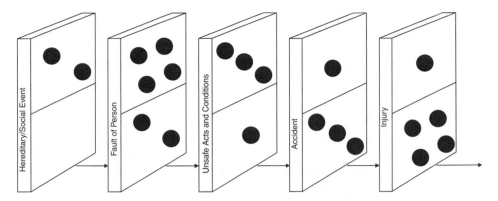

Figure 12-7 Domino accident sequence model. Cox, Sue, Tom Cox, *Safety Systems and People*, Figure 3.1 (adapted from Heinrich, 1931), p. 51, Butterworth–Heinemann, 1996. Modified with permission.

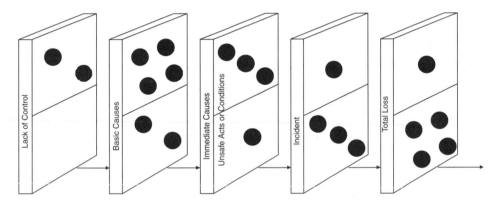

Figure 12-8 The domino theory: "loss causation" accident sequence model. Cox, Sue, Tom Cox, *Safety Systems and People*, Figure 3.2 (adapted from Bird and Loftus, 1931), p. 52, Butterworth–Heinemann, 1996. Modified with permission.

exact nature and type of loss incurred is a matter of change ranging from significant to catastrophic [1].

According to Bird and Germain, the first domino focuses on management and describes controls of the four essential management functions [planning, organizing, leadership (directing), and control] [1].

Bird and German go on to state that the lack of control (management system) is perceived to be a major factor in the incident or loss sequence and elimination of the "defect" is seen to be central and the key to accident prevention [1, 2].

Expediency, linked together with a linear causation model, thus restricts many investigations [2]. It has been further noted [7] that by narrowing down the possible combination of substandard acts and specific situations to a single stage, the identification and control of contributing causes has been severely limited [2].

Let's approach the domino theory from the format of this book. Note that management commitment and leadership and employee participation are in the middle. If the domino falls in either direction think about what will happen to your system. When these dominos fall they equate to failure in the management system. So the bottom line is that they are fragile and must be handled with care, especially in a new system. Refer to Figure 12-9.

In our opinion these approaches all have some merit, but fall short of what can be accomplished by someone looking at the incident starting at the bottom and working all the way up. Figure 12-10 shows our version of the same triangle, but as an hourglass. In our opinion, the hourglass represents this theory more realistically because employees can visualize how it works. In addition, it will help employees understand how the incident cycle works: as time running out. If you do not solve the safety issues on the top, incidents will filter to the bottom.

WHAT CAUSES INCIDENTS

There are many questions that we must ask when we are trying to determine incident causes. Some people will say that "acts of God"

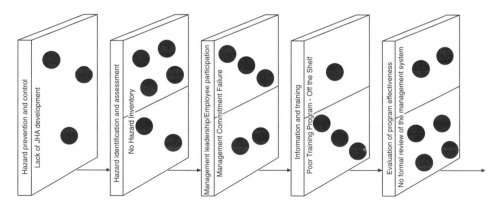

Figure 12-9 The Domino Theory: Management System Failure.

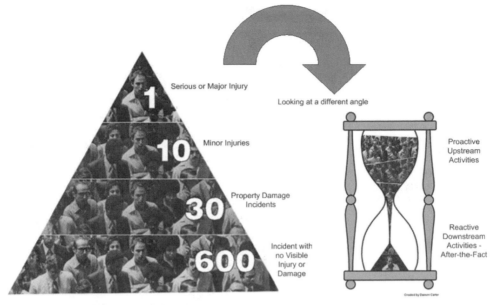

Figure 12-10 Safety accident pyramid revisited. Bird Frank E., Germain George L., *Loss Control Management: Practical Loss Control Leadership*, revised edition, Figure 1-3, p. 5, Det Norske Veritas (U.S.A.), Inc., 1996. Adapted for use—designed by Damon Carter.

account for a very small percentage of work-related incidents. The remaining 98 percent of incidents can be thought of as resulting from unsafe acts (behaviors of employees) or unsafe conditions [6]. We are led to believe that being "unsafe" includes not knowing, not caring, and not thinking about the real and potential safety hazards and exposures of a particular job task.

When an incident occurs that results in an injury or in damage to property or equipment, we often find that the employees had been doing a particular job task the same way for a long time. This time "luck ran out" [5]. Refer to Chapter 5, Figure 5-3 to understand what is considered luck and then review the safety accident pyramid.

If we work at reducing and eliminating all of the possible exposures and work errors that we face in our daily routines, we can reduce the cause of incidents. We can identify and/or anticipate possible equipment hazards [2].

One author uses a simple quiz for awareness training sessions to show that individuals are not aware of what is going on around them and that this is why incidents occur. The quiz is simple and illustrates a point for every one of us who drives a vehicle. We see this road sign every day but

Figure 12-11 Yield sign.

do not think about it. Ask yourself and others this question: Do you know the color of a yield sign? Have I got you thinking? I think you see my point. Refer to Figure 12-11 of the yield sign. In most cases, we do not know the color when asked. It sounds silly, but this simple question will show how aware you are of your surroundings. You can create awareness and also have some fun with safety. Ask different people to see how they answer the question. The key is to find new ways of presenting safety to employees to keep their interest. This is one technique. It is the same in the workplace. We are so accustomed to our jobs that sometimes we do them without thinking about the consequences. It is just natural.

ELEMENTS OF THE SAFETY SYSTEM

When we look at the reason why an incident occurs, we must look at some key elements and the interaction between these elements:

- Equipment
- Environment
- People (employees, contractors, temporary employees, visitors, etc.)
- Management [6]

Figure 12-12 depicts the 5M model of safety engineering.

Equipment. The types of equipment that an injured employee was working with, its production and maintenance requirements, its layout in the work area, and its hazards and the methods of controlling them could all be clues in the investigation—for example, guarding, noise reduction, or controls of hazardous material.

Environment. Environmental aspects may include noise; lighting; housekeeping; work inside versus outside; fumes or vapors; exhaust systems; production pressures; or stress created by the job, such as manual work versus office work, night work versus day work, weekend work, or long work days (12 hours).

People. We must explore the physical task demands of the job, such as lifting, bending, twisting; the level of training and skill of the employee; and his or her current emotional state.

Management. When we investigate an incident, the purpose is to identify the root cause of the incident as it relates to as many of the key elements as possible. We are looking at the adequacy and effectiveness of the management system.

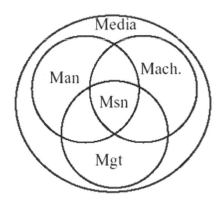

- Msn - Mission: central purpose or functions
- Man - Human element
- Mach - Machine: hardware and software
- Media - Environment: ambient and operational environment
- Mgt- Management: procedures, policies, and regulations

Figure 12-12 5M model of safety engineering. *FAA Safety Handbook*, Chapter 15, "Principles of System Safety," Figure 5-4, p. 15–11. December 30, 2000, public domain.

Table 12-1
Incident Investigation Key Points

- Set up procedures before incidents occur
- Use systematic approach
- Address both surface and root causes
- Investigate near misses
- Fact finding, not fault finding

OR OSHA 100o, Incident Investigation Key Points, http://www.cbs.state.or.us/external/osha/educate/training/pages/materials.html, public domain.

Investigations are a barometer that can be used to measure the management system and the safety climate (culture) of an organization. If we investigate only incidents that result in an injury, illness, or property damage, and ignore near misses and minor first aid incidents, then we are missing many valuable opportunities [7].

Note that it can be costly to react to incidents. The key is to make sure that you are proactive in your management system: if there is a series of incidents, you must review them. In most cases, if your management system is working, you should stay the course. In the worst case, you may need to change your management system slightly. Most important is to stay the course and keep your safety process focused and moving in the right direction. Table 12-1 summarizes some key points in investigating an incident.

UNDERSTANDING THE AUDIT TRAIL

First, if you identify safety glasses and ear plugs as your main safety program, then you do not have a successful program. One question comes to mind, how many eye injuries have occurred in your facility? You need to look deeper and evaluate the hidden hazards. The question is, "Why do we look at PPE?" The simple answer is because it is easy to find. It takes no effort. Some implement a safety glass policy in areas where there have been no eye injuries. Again, this is used to make it easy to administer the program. I remember one example where a customer had one eye injury; a small chip of metal came over the safety glass, and got into the employee's eye. The solution to the problem was to immediately replace all safety glasses with ones that had a shield over the top of the glass to prevent anything from getting into the eye. This was a case where there was no root cause of the injury determined. It was reactionary to the injury without much thought. Take the time to understand the issues and act appropriately.

Let's take an example of two eye injuries. The first step is to review the incident report by evaluating the background information to understand why the injury occurred. In this case, the incident report listed the cause of the injury as employee failure to wear eye protection [9].

When the employees were interviewed, one employee felt that their goggles did not fit. The other employee complained that they fogged up and gave that as the reason for not wearing the goggles. Interviews with the manager revealed that the manager was reluctant to initiate the company disciplinary policy for the two employees, who were excellent workers. Typical of most companies, the incident report blamed the employee. The company had a written safety program, performed record-keeping accurately, and had trained their employees. What is missing from this type of safety program [9]?

If this company had applied basic management principles, they would have encouraged evaluation of the employees' reasons for not wearing the goggles, rather than turning to discipline. If management had questioned all of the employees who performed the job that resulted in two eye injuries, they would have found that 75 percent of the employees indicated that they did not wear their goggles because they forgot them, 10 percent felt that the goggles did not fit or slipped off, and 15 percent felt that goggle fogging was a big problem. By addressing each of these issues separately, non-conformance could be significantly reduced. One solution: the manager could have purchased retainer clips that attached the goggles directly to the hard hats. Changing the style of goggles could have eliminated the problem of fogging. The manager also could have purchased an extra supply so that any goggle damaged by chemicals could be immediately changed.

If only the eye injury cases had been evaluated, a significant reason for not using the goggles (representing significant risk) would not have been addressed. In other words, the 75 percent of the employees who did not wear their goggles because they had left them in their lockers would not have been addressed [9].

If the employees had participated in the incident investigation, additional input might have been solicited at the time of the first incident, thereby preventing the second incident. In addition, there was no discussion of the injuries or near misses during the monthly safety meeting, indicating a failure to use the safety committee to address real, practical, and soluble issues. If the safety committee had addressed the root cause, they might have been able to address all of the instances of non-conformance before the second injury [9].

DOCUMENTING STEPS

After the first injury, the following sequence of events should have taken place:

- Perform incident analysis and appropriate recordkeeping
- Review incident reports during the safety committee meetings for further input and develop a corrective action plan
- Determine if any near misses had occurred for this job, and if so, why they were not reported
- Review the JHA for the task and revise as necessary
- Reevaluate disciplinary/incentive programs and how they might affect employee participation and near-miss reporting; modify the programs if they are disincentives
- Retrain the employees and managers accordingly
- Perform an analysis of conformance with identified critical behaviors to gauge the effectiveness of corrective action [9]

If these actions had been completed, a number of positive results might have occurred. First, by addressing the root causes, management could have demonstrated their commitment to preventing injuries. Employee participation in the safety committee, incident analysis, JHA, and retraining of employees could have heightened employee awareness and sensitivity to this behavior, and increased employee morale and feeling of being part of a team. As a result of goggle use becoming routine, eye injuries might have been reduced or eliminated entirely [9].

Although reviewing the OSHA log is a good tool for auditing a management system, it is only one of many tools that can be used. Such an approach takes only a proactive look at the management system once a safety issue has occurred. Ideally, a company should try to be proactive in their approach [9].

What this audit approach does is to see if the management system is working, with linkages that are integrated with other parts of the process. If an incident or a near miss occurs, this process can be followed to see if there was the appropriate "ripple" effect. Would the near miss be reported? Once reported, would the JHA be reviewed, or would employees be retrained? Making an incident report, because it is expected, but not seeing evidence of the "ripple" effect, such as changing a JHA or retraining employees or purchasing new equipment, may be evidence that a safety program is a "paper program" and not a vital working document [9].

One other word of caution. If you insist that incidents be reported, you must investigate all cases that are reported and not select the easy reports. If you do not investigate all of the cases and treat them equally, what do you think will happen? The following are some potential issues: Employees will find no reason to report incidents because nothing is getting done and/or employees will continue to get hurt.

WHAT SHOULD BE INVESTIGATED?

Promptness is the most important element when investigating incidents. Delays will usually result in a failure to collect the facts.

Table 12-2 provides an overview of investigation procedures. We will continue to discuss these techniques later in this chapter.

Incidents should be investigated as soon as possible for a number of reasons. First, if the cause of the incident is still present, we want to try to prevent other employees from being injured. Second, the scene can change after an incident. Valuable clues may be lost if the area is cleaned up. Changes in operations, shift and employee changes, weather, and lighting are some of the factors that could affect the quality of the investigation and reconstruction of the events [5].

The third reason is to make sure that you can gather the facts while the information and impressions are fresh in any witness's mind. As time passes, what originally may have been a critical piece of information becomes obscured. Employees begin to form their own opinions or can be influenced by the opinions of others, turning some facts into fiction. The employees involved in the incident may change their stories after having time to think about what occurred, fearing that if they contributed to the incident they might be blamed and/or disciplined. In other cases, they may want to protect themselves or co-workers [5].

The scene of the incident should be evaluated promptly before any changes that may interfere with fact finding. In the event of a serious incident, photographs should be taken before the area is returned to normal use.

Table 12-2
Investigation Procedures

The procedure used in an investigation depends on the nature and results of the incident. The following steps can be used to get to the root cause:

1. Define the scope of the investigation
2. Select trained investigators
3. Present a preliminary briefing to the investigating team
   - Description of the incident and damage estimates
   - Normal operating procedures
   - Maps (local and general)
   - Location of the incident site
   - List of witnesses
   - Events that preceded the accident
4. Visit the incident site to get updated information
5. Inspect the incident site
   - Secure the area
   - Prepare the sketches and photographs. Label each and keep accurate records
6. Interview each employee and witness. Also, interview those who were present before the incident and those who arrived at the site after the incident. Keep accurate records of each interview
7. Determine the following
   - What was not normal before the incident
   - Where the abnormality occurred
   - When it was first noted
   - How it occurred
8. Analyze the data obtained in step 7. Repeat any of the prior steps, as necessary
9. Determine:
   - Why the incident occurred
   - A likely sequence of events and probable causes (direct, indirect, basic)
   - Alternative sequences
10. Check each sequence against the data from step 7
11. Determine the most likely sequence of events and most probable causes
12. Conduct a post-investigation briefing
13. Prepare a summary report including the recommended actions to prevent a recurrence.

An investigation is not complete until all data are analyzed and a final report is completed. In practice, the investigative work, data analysis, and report preparation proceed simultaneously over much of the time spent on the investigation.

OSHA's Small Business Outreach Training Program, Investigation Procedures, http://www.osha-slc.gov/SLTC/smallbusiness/sec6.html, public domain.

Something to consider: a "near miss" is an incident where there was a possibility of an injury but no property was damaged and no personal injury sustained—but where, given a slight shift in time or position, damage and/or injury easily could have occurred [5].

WHO SHOULD INVESTIGATE AN INCIDENT?

The success or failure of the incident investigation processes can frequently be traced directly to the actions of management or failure of the management system.

The responsibility for investigation of incidents could be assigned to any level of management. Workplace rules should be established and communicated requiring employees to report all incidents immediately to supervision. Once an incident is reported, the immediate supervisor should initiate the investigation as soon as possible, by the end of the shift when the incident occurred or no later than 24 hours after the incident was reported. Written statements should be obtained from the injured employee and witness(es) as promptly as possible following the incident. This will be discussed in more detail later in this chapter.

Responsibilities of management, at every level, should include the following:

- Prompt review and analysis of all reports
- Prompt and positive assistance to make sure that there are proper corrective actions
- Review of all major injuries or property damage incidents by meeting with affected supervision and the injured employee
- Constructive, purposeful, and timely criticism of incident reports
- Active participation to demonstrate interest and support [5]

In some cases, injuries should be immediately reported to the individual who files the workers' compensation claims. Most of the time, results of the investigation are critical in determining compensability of the injury [5].

Follow-up investigations may be requested either by the affected supervisors or by the safety professional to verify facts and to obtain more information about an incident. In either case, the immediate supervisor should lead the initial investigation and take full responsibility for making all necessary arrangements.

Table 12-3
Be Ready If An Incident Happens

- Write a clear policy statement
- Identify those authorized to notify outside agencies (fire, police, etc.)
- Designate those responsible to investigate accidents
- Train all accident investigators
- Establish timetables for conducting the investigation and taking corrective action
- Identify those who will receive the report and take corrective action

OR OSHA 100o, "Be Ready When Incidents Happen," http://www.cbs.state.or.us/external/osha/educate/training/pages/materials.html, public domain.

Many companies use a team to investigate incidents involving serious injuries or extensive property damage. This may supplement the supervisor's investigation or may serve as a second-level investigation and peer review. When a team investigates, the team leader must have enough authority and status in the organization to do whatever is necessary to correct the condition [5].

Those individuals who investigate incidents (supervisors, employees, teams, etc.) should be held accountable for describing causes carefully and clearly. When reviewing investigation reports, the safety professional should review the report for catch phrases—for example, "Employee did not plan the job properly"; "The employee was not trained properly." Although such statements may suggest an underlying problem with this employee, they are not conducive to identifying all possible causes, prevention, and controls [5]. Table 12-3 summarizes what you need to do to be ready if an injury occurs.

ANALYSIS OF PATTERNS

A review of the OSHA log is the most common method to determine pattern analysis. However, you must remember that these are only recordable injuries and do no reflect first aid cases, near misses, etc. These logs are not the only useful source of information. Records of hazard analysis can be analyzed for patterns—for example, inspection records and employee hazard reporting records [5].

Records being analyzed must contain enough information to allow for a specific pattern. A workplace with few employees or very little hazardous work may require a review of 3 to 5 years of information. Because

Table 12-4
Fact-Finding

Gather evidence from many sources during an investigation. Get information from witnesses and reports and observation. Get copies of all reports (documents containing normal operating procedures, flow diagrams, etc., maintenance charts, or reports of difficulties or abnormalities). Keep complete and accurate notes. Record pre-incident conditions, the incident sequence, and post-incident conditions. In addition, document the location of employee, witnesses, equipment, energy sources, and hazardous materials.

In some investigations, a particular physical or chemical exposure, principle, or property may explain the sequence of events.

OSHA's Small Business Outreach Training Program, Fact-Finding, http://www.osha-slc.gov/SLTC/smallbusiness/sec6.html, public domain.

a site is small or has little hazardous work activity does not mean that pattern analysis is useless. Even if an office area has only one or two injuries each year, a 5-year review may indicate uncontrolled cumulative trauma disorder or lack of attention to tripping hazards. Larger sites will find useful information in monthly, quarterly, or yearly reviews [5]. Table 12-4 gives an overview of fact-finding.

Repetitions of the same type of injury or illness indicate that hazard controls are not working properly. Injuries do not have to be identical to constitute a repetition—for example, they could occur to the same part of the body. Any clue that suggests a previously unnoticed connection between several injuries is worth further investigation [5].

INTERVIEWING INJURED EMPLOYEES

Before interviewing the injured employee, the supervisor should explain briefly that the purpose of the interview is to learn what happened and how it happened so that recurrence of similar incidents can be prevented. Reassurance should be given that the purpose of the interview is not to blame the injured employee. The following questions should be asked to establish some facts:

- What was the employee doing when the injury occurred?
- How was the employee performing the task?
- What happened?

Such questions as to why the employee did what he/she did should be deferred until after agreement on what happened is established. Do not phrase questions so they are antagonistic and do not try to corner the employee, even though the employee's version may conflict with itself. If this occurs, explain the point of conflict tactfully.

Injured employees should be asked to complete an incident investigation report describing what happened, why it happened, and what could have been done to prevent occurrence of the incident.

To bring an interview to a close, discuss how to prevent recurrence of the incident. Get the injured employee's ideas and discuss them. Emphasis should be placed on the precautions that will prevent recurrence, and the interview should close on a friendly note.

INTERVIEWING WITNESSES

Witnesses are an important source of information. There are several categories of witnesses who could have information helpful in determining the causes of an incident:

- Eyewitnesses, those who saw the incident happen and/or were involved in the incident
- Those who came on the scene immediately following the incident
- Those who saw events leading up to the incident

One important element is to establish if the witness saw the incident. Be sure to gather facts, as distinguished from speculation and conjecture on what might have happened. Circumstantial evidence is also important. If a witness saw the incident site immediately before the incident, this evidence may be of value.

Interview witnesses as promptly as possible after the incident. As time passes, witnesses could forget important details or change their stories after having had time to talk with other employees. Never interview witnesses in a group. Witnesses should be interviewed in private. Explain how knowing the full story may prevent a serious injury to another employee, and explain the purpose of the interview. Make sure that the witnesses know that they are doing a service by giving a frank and honest account of what they saw and what they know.

Use an incident witness statement to collect statements. Be sure to gather the witnesses' input regarding possible at-risk behaviors they know of or had seen that may have contributed to the incident, as well as information regarding other possible causes.

No questions should be asked that imply answers wanted or not wanted. Refrain from asking questions or saying anything that blames or threatens the employee who had the incident. Do not badger witnesses or give them a bad time, and do not resort to sarcasm, open skepticism, or accusation. Handle all discrepancies with tact and let the witnesses feel that they are partners in the investigation [9].

The following are some key points to remember during the interview process:

- Use an informal setting for the interview. If you sit across a desk from the employees being interviewed, they will be more intimidated than if you sit beside them.
- Explain to each witness that the investigation is being conducted to try to understand the cause of the incident so that it does not happen again. Let them know that the information they provide will aid in helping to understand what happened.
- Use open-ended questions. Ask questions that cannot be answered by a "yes" or "no"—for example, "Tell me what you saw" is much more likely to uncover valuable details that the witness might have temporarily forgotten or might think is not important. Do not try to finish their sentences or interrupt them during the interview. Wait until they are finished before asking any questions. Let the witnesses proceed at their own pace without interfering in the conversation. Tell them you are taking notes to document specific details. Keep the witness talking about the sequence of events and do not allow him or her to begin drawing conclusions. Allow the witness to read your notes to see if they are accurate. This will show that you are interested in reducing incidents [5, 10].
- Conclude the interview with a "thank you." Encourage the employees to call you if they remember any other pertinent information later. Give them some feedback on specifics of the information that they provided that you found to be particularly valuable.

Table 12-5 is a reference on interviewing techniques.

RECREATING THE INCIDENT

When additional information is required or when some facts must be verified, you may need to re-create the incident. One caution: if this re-creation takes place, you must make sure that the investigation techniques do not result in another incident.

Table 12-5
Interview Techniques

Experienced individuals should conduct interviews. In some cases, the team assigned should include an individual from legal or human resources. The following are recommended steps in interviewing:

- Appoint a speaker for the group
- Get preliminary statements as soon as possible from all witnesses
- Locate the position of each witness on a master chart (including the direction of view)
- Arrange for a convenient time and place to talk to each witness
- Explain the purpose of the investigation (incident prevention) and put each witness at ease
- Let each witness speak freely. Listen to the employee. Be courteous and considerate
- Take notes without distracting the witness
- Use sketches and diagrams to help the witness remember any important facts
- Emphasize areas of direct observation. Label hearsay accordingly (stick to the facts)
- Be sincere and do not argue with the witness
- Record the exact words used by the witness to describe each observation. Do not put words into a witness's mouth
- Word each question carefully and make sure that the witness understands what is being asked
- Identify the qualifications of each witness
- Supply each witness with a copy of his/her statements. Signed statements are desirable

After interviewing all witnesses, the team should analyze each statement. In some cases, the team may wish to re-interview one or more witnesses to confirm or verify key points. While there may be inconsistencies in statements, investigators should assemble the available testimony into a logical order. Analyze this information along with data from the incident site.

OSHA's Small Business Outreach Training Program, "Interview Techniques," http://www. osha-slc.gov/SLTC/smallbusiness/sec6.html, public domain.

DETERMINING CAUSE

The emphasis is on correcting the problem so that it will not be repeated. The following is a list of criteria for making sure that corrective actions are viable:

- Will these corrective actions prevent recurrence of the condition?
- Is the corrective action within the capability of the organization to implement?
- Have assumed risks been clearly stated?

The corrective actions should address the specific circumstances of the event that occurred and management system improvements aimed at the root cause. They should address options for reducing the frequency, minimizing the personnel exposures, and/or lessening the consequences of one or more of the root causes [5].

The following steps outline the fact analysis process:

- Identify the facts and list them in order of occurrence. One method is to use a storyboard to outline the facts.
- For each fact, develop a list of questions that you would like to have answered. Use questions that ask Who, What, Where, When, Why, and How. Keep in mind the five elements of the safety system: People, Machine, Materials, Method, and/or Environment.
- List the answers to each question. If you have done a thorough job of interviewing and gathering information, you should have many of the answers to your questions. Other unanswered questions may require that you go back and dig for some more information.
- Identify those answers to your fact-based questions that you feel are some of the root causes behind the sequence [5].

Once all of the possible causes of the incident have been identified, it is time to develop the recommended solutions that will help to prevent a recurrence of similar incidents.

The facts identified during the investigation should lead to conclusions. These conclusions should lead to recommendations. Recommendations lead to action planning—it is important that the corrective action plan be based on the conclusions from the investigation analysis. Action planning leads to prevention. Any facts that do not add to the sequence of events or that do not support a conclusion should not be included in the incident investigation report [5].

CORRECTIVE ACTION PLANS

When you develop and implement a corrective action plan, consider the following questions:

- Is there at least one corrective action plan associated with each root cause?
- What are the consequences of implementing or not implementing the corrective action plan?
- What is the cost of implementing the corrective action plan?
- Will training be required as part of implementing the corrective action plan?
- What period is required for the corrective action plan to be implemented?
- What resources are required for implementation and continued effectiveness of the corrective action plan?
- What impact will the development and implementation of the corrective action plan have on employee and/or other work areas?
- Can implementation of the corrective action plan be measured [5]?

In some cases, implementation of the recommendation may be outside of the supervisor's responsibility. Top management and the safety professional would have the responsibility to locate the resource and provide the funding. This is another reason why top management should review the investigation reports. They have the authority to assign the necessary resources to achieve some of the more complex corrective actions.

We must make sure that the solutions are implemented and are effective. Following through on corrective actions is an "opportunity for improvement." In this way, we can continuously improve our management system. Holding employees responsible and accountable for completing their assigned tasks is essential. Close the loop on the continuous improvement process by monitoring the effectiveness of the solutions after they have been implemented. What may have seemed like the best solution at the time of the investigation may not be the ultimate best solution.

PROBLEM SOLVING TECHNIQUES

Incidents represent problems that must be solved through investigations. Several formal procedures solve problems of any degree of complexity. This section discusses two of the most common procedures: change analysis and job hazard analysis.

Change Analysis

As the name implies, this technique emphasizes change. Consider all problems to result from some unanticipated change. Make an analysis of the change to determine its causes. Use the following steps:

- Define the problem. What happened?
- Establish the norm. What should have happened?
- Identify, locate, and describe the change (what, where, when, and to what extent).
- Specify what was and what was not affected.
- Identify the distinctive features of the change.
- List the possible causes.
- Select the most likely causes.

Any time something new is brought into the workplace, no matter if it is a piece of equipment, different materials, a new process, or a new building, new hazards may unintentionally be introduced into the environment. Before considering a change for a worksite, analyze it thoroughly beforehand. Change analysis helps in heading off a problem before it develops. Figure 12-13 gives an overview of change analysis. You may also find change analysis useful when staffing changes occur [9].

To solve a problem, an investigator must look for deviations from the norm (i.e., what changed?) [6].

Job Hazard Analysis

Refer to Chapter 15 for a complete explanation of how to conduct a JHA.

REPORT OF INVESTIGATION

Results of the investigation, including the identification of the root causes and the preventive/corrective actions, should be shared with all employees in the workplace who are affected by the findings.

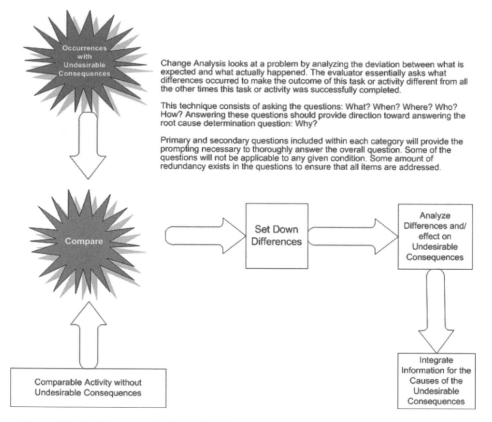

Change Analysis looks at a problem by analyzing the deviation between what is expected and what actually happened. The evaluator essentially asks what differences occurred to make the outcome of this task or activity different from all the other times this task or activity was successfully completed.

This technique consists of asking the questions: What? When? Where? Who? How? Answering these questions should provide direction toward answering the root cause determination question: Why?

Primary and secondary questions included within each category will provide the prompting necessary to thoroughly answer the overall question. Some of the questions will not be applicable to any given condition. Some amount of redundancy exists in the questions to ensure that all items are addressed.

Figure 12-13 Change analysis. DOE Guideline Root Cause Analysis Guidance Document, DOE-NE-STD-1004-92, February 1992, U.S. Department of Energy Office of Nuclear Energy, Office of Nuclear Safety Policy and Standards, Appendix E, Figure E-1, p. E-1. Modified public domain version.

The following outline can be useful in developing the information to be included in the incident report:

- Background information
  Where and when the accident occurred
  Who and what were involved
  Operating personnel and other witnesses
- Account of the accident (What happened?)
  Sequence of events
  Extent of damage
  Accident type
  Agency or source (of energy or hazardous material)
- Discussion (Analysis of the accident—how; why)
  Direct causes (energy sources; hazardous materials)

Indirect causes (unsafe acts and conditions)

Basic causes (management policies; personal or environmental factors)

- Recommendations (to prevent a recurrence) for immediate and long-range action to remedy:

Basic causes

Indirect causes

Direct causes (such as reduced quantities or protective equipment or structures) [3]

SUMMARY

Investigations are often thought of as a burdensome process that requires a lot of time and effort. Generally, the time and effort does not seem to produce the expected results of incident reduction. This can be the case if the investigator and top management do not fully appreciate the value and benefits of a good, thorough investigation process [6].

A successful incident investigation not only determines what happened, but also finds how and why the incident occurred. Incident investigation is another tool for identifying hazards that were missed earlier or that slipped by the planned controls. However, it's only useful when the process is positive and focuses on finding the root cause, not someone to blame!

All incidents should be investigated. "Near-misses" are considered incidents, because, given a slight change in time or position, injury or damage could have occurred. This allows you to identify and control hazards before they cause a more serious incident.

Remember the six key questions that should be answered in the incident investigation and report: who, what, when, where, why, and how. Thorough interviews with everyone involved are necessary.

The primary purpose of the incident investigation is to prevent future occurrences. Therefore, the results of the investigation should be used to initiate corrective action.

Most incident investigations follow formal procedures. This chapter discussed two of the most common procedures: change analysis and job hazard analysis. An investigation is not complete until a final report is issued. Responsible individuals can use the information and recommendations to prevent future incidents [6].

When the benefits are recognized and management's commitment is visibly demonstrated, the time spent will pay for itself many times over.

The investigator for all incidents should be the supervisor in charge of the involved area and/or activity. Incident investigations represent a good way to involve employees in the safety process. Employee participation not only will give you additional expertise and insight, but in the eyes of the workers, will lend credibility to the results. Employee participation also benefits the involved employees by educating them on potential hazards, and the experience usually makes them believers in the importance of safety, thus strengthening the safety culture of the organization. The safety professional should participate in the investigation or review the investigative findings and recommendations. Many companies use a team or a subcommittee or the joint employee–management committee to investigate incidents involving serious injury or extensive property damage.

All supervisors and others who investigate incidents should be held accountable for describing causes carefully and clearly. When reviewing incident investigation reports, the safety professional should be on the lookout for catch phrases—for example, "Employee did not plan job properly." Although such a statement may suggest an underlying problem with this employee, it is not conducive to identifying all possible causes, preventions, and controls. It is too late to plan a job when the employee is about to do it. Further, it is unlikely that safe work will always result when each employee is expected to plan procedures alone.

Recommended preventive actions should make it very difficult, if not impossible, for the incident to recur. The investigative report should list all the ways to "foolproof" the condition or activity. Considerations of cost or engineering should not enter in at this stage. The primary purpose of incident investigations is to prevent future occurrences. Beyond this immediate purpose, the information obtained through the investigation should be used to update and revise the inventory of hazards, and/or the program for hazard prevention and control. For example, the Job Safety Analysis should be revised and employees retrained to the extent that it fully reflects the recommendations made by an incident report. Implications from the root causes of the accident need to be analyzed for their impact on all other operations and procedures [6].

REFERENCES

1. Bird, Frank, George Germain, *Loss Control Management: Practical Loss Control Leadership*, revised ed., Det Norske Veritas (U.S.A.) Inc., 1996.

2. Cox, Sue, Tom Cox, *Safety Systems and People*, Butterworth–Heinemann, 1996. Modified with permission.

3. DOE Guideline Root Cause Analysis Guidance Document, DOE-NE-STD-1004-92, February 1992, U.S. Department of Energy Office of Nuclear Energy, Office of Nuclear Safety Policy and Standards, public domain.

4. Heinrich, H. W., *Industrial Accident Prevention*, 4th ed., with Dan Petersen and N. Ross. McGraw-Hill, New York, 1969.

5. U.S. Department of Labor, Office of Cooperative Programs, Occupational Safety and Health Administration (OSHA), *Managing Worker Safety and Health*, November 1994, public domain.

6. OSHA's Small Business Outreach Training Program, http://www.osha-slc.gov/SLTC/smallbusiness/sec6.html, public domain.

7. Petersen, Dan, *Techniques of Safety Management*, McGraw-Hill, New York, 1971.

8. *The American Heritage College Dictionary*, 3rd ed., Houghton Mifflin, Boston, 1997.

9. OSHA Web site, http://www.osha-slc.gov/SLTC/safetyhealth_ecat/comp1_review_program.htm#, public domain.

13

Developing and Administering a Medical Surveillance Program

INTRODUCTION

A medical surveillance program is a system that is put in place to make sure that the level of health expertise identified in the safety program is sufficient. Having a medical surveillance program does not mean that you have to hire a physician to work at your facility. There are many ways you can find and use medical health expertise [2].

Establishing a comprehensive medical surveillance program is more than an after-the-fact response to work-related injury or illness. It also includes activities that identify safety hazards and will help you to develop an action plan for prevention or control of identified hazards. You may find it more difficult to establish the goals and objectives for your medical provider than for the other parts of your safety program. Illnesses may not appear obvious at first. For example, an employee who is experiencing hand pain and who gradually is developing a cumulative trauma disorder (CTD) or MSD may seem to have a less serious problem than the employee who has a severe cut or broken bone [2]. This chapter will provide some guidance to help you decide what will work best for your operation.

WHY DO YOU NEED A MEDICAL SURVEILLANCE PROGRAM?

In an effort to be proactive, you identify hazards and involve employees in the identification of hazards to help prevent injuries. We must accept the reality that despite your best planning and prevention efforts, injuries can occur. It is how you manage the injury, if it happens, that matters. In such a case, you must be prepared to deal with medical emergencies. If you do not have a nurse on staff, then you have to be prepared to make other arrangements. First aid and cardiopulmonary resuscitation (CPR) assistance should be available on every shift. This is the second option under the OSHA standards (29 CFR 1910.151(b)) for workplaces that are not close to medical facilities; OSHA has interpreted this rule to mean within a five-minute response time [1].

Medical programs can provide health care for an organization, both onsite and at a nearby provider. A medical program is another name for the system that can be put in place to make sure that health expertise is included in the safety program.

A medical surveillance program may include activities that cover safety hazards in your workplace and that will help you to formulate a plan to minimize and/or control related hazards.

BENEFITS OF A MEDICAL PROGRAM

A medical program can consist of elements ranging from basic first aid and CPR to sophisticated approaches for the diagnosis and resolution of various safety and health related issues, including ergonomic issues. The nature and extent of your medical program will depend on a number of factors. If the use of a nearby medical provider appears to be the best solution for your needs, make sure that you meet with representatives of the facility to discuss your medical needs. Have the medical provider visit your facility to get familiar with your process. The medical provider should also be asked to conduct periodic visits and walkthroughs of your facility to maintain familiarity with the job tasks being performed and participate in a safety analysis of each job [1].

An effective medical surveillance program will help to reduce all types of safety hazards and the related injuries. The positive results from such a program can be measured by a decrease in lost workdays and workers' compensation costs. You can also expect the medical program to help increase worker productivity and morale [1].

WHO SHOULD MANAGE THE MEDICAL PROVIDER?

In some cases, you may find that your medical surveillance program works best when managed by a health professional. A physician or a registered nurse with specialized training, experience, and knowledge in occupational health works with you, but not as your employee. This arrangement works best because safety professionals, industrial hygienists, physicians, and nurses all have their own areas of specialized knowledge. You cannot expect to get all the information and service your safety program needs with one type of specialist. If you tried, you might overlook or misidentify a dangerous hazard in your business [1].

Refer to Appendix D for a description of the different ways in which physicians and registered nurses receive specialty training in medicine and the different services that they can provide you [2].

No matter what medical surveillance program you decide to use, it is important to use a medical specialist with occupational health/medical training. Not every medical provider is trained to understand the relationships among the workplace, the work performed, and specific medical symptoms. The size and complexity of your medical surveillance program will depend on the size of your workplace, its location in relation to your medical provider, and the nature of the hazards that exist.

The medical surveillance program works best when managed by a health professional. As we have discussed, this does not mean that you have to hire a health professional as a full-time employee.

PREVENTING HAZARDS

Make sure that the safety policy shows that the organization is as concerned about its employees' health as it is about safety. Use a medical provider knowledgeable in medical case management when developing and prescribing health training and other preventive activities [1].

WHAT SERVICES DO YOU NEED FROM A MEDICAL PROVIDER?

There is no such thing as a standard medical provider. There is no substitute for examining your business's special characteristics and finding a medical provider who is right for you. These special characteristics include:

- The actual processes in which your employees are engaged
- The type of materials handled by your employees
- The type of facilities where your employees are working
- The number of employees at each worksite
- The characteristics of your work force—for example, age, gender, ethnic group, and education level
- The location of each operation and the distance from health care facilities [1, 2]

As you look at the characteristics of your employees and workplace, you should be asking yourself questions such as:

- Are there hazards in the process, materials used, or facilities that make it likely that employees will get sick or hurt or will suffer abnormal health effects from their work environments?
- Do the numbers of employees at your facility require occupational health resources on site or is it less practical than offsite contract services?
- Are there so many employees that time and funds will be saved by implementing onsite resources?
- Are there any special characteristics of the employees that make them more vulnerable to illness or injury or less likely to understand the safety hazards of the worksite [1]?
- Are there any specific hazards or conditions in the workplace that would warrant having a medical provider near the facility?

Getting the answers to these questions will put you in a better position to decide what type of medical provider services you need. The range of services are described in the next section. Refer to Appendix D for some examples of how different companies have tailored specific medical provider activities [2].

One thing to consider as you set up your program is the Americans with Disabilities Act (ADA). Under this act employers may require employees to submit to medical examination only when justified by business necessity. The results of any medical examination are subject to certain disclosure and record retention requirements (29 CFR 1910.20), but also are subject to confidentiality requirements of the ADA. The ADA's employment-related provisions are enforced primarily by the U.S. Equal Employment Opportunity Commission [1, 2].

THE RANGE OF MEDICAL PROVIDER FUNCTIONS

There are three types of basic medical provider activities:

- Provide recommendations for prevention of hazards that could cause illnesses
- Provide early recognition and treatment of work-related illnesses and injuries
- Limit the severity of work-related illness and injury by working with local doctors and hospitals [2]

Providing Recommendations for Preventing Hazards

- Make sure that your safety policy outlines how concerned you are about your employees' safety (refer to Chapter 5). Enlist a qualified medical professional to help you identify the hazards and potential hazards of your workplace (Chapter 10).
- Use your medical provider in the development and presentation of health training and other activities that include various measures, such as OSHA's Bloodborne Pathogens standard. Refer to Chapter 14 and OSHA Publication 2254 (revised 1992), "Training Requirements in OSHA Standards and Training Guidelines."
- No matter which service you choose, it is still your responsibility to determine if you have employees who fall in the scope of the specific requirements and to make arrangements for the appropriate training for all affected employees.
- Provide professional occupational health expertise as a resource to your safety and health committee, and be sure to include your medical provider in your annual self-evaluation. (Refer to Chapter 6.)

Early Recognition and Treatment

- Use the medical provider to help you decide when you may need to conduct baseline and periodic testing of your employees and new hires for evidence of potential exposure. This will depend on your workplace and the associated hazards—for example if you do welding. This is called "medical surveillance" and is part of some specific OSHA requirements for specific types of exposures. Use your medical provider to do the testing needed to accomplish medical surveillance.
- Make sure records are kept of all visits for first cases. Have your medical provider review the symptoms reported and the diagnoses to see if there is a pattern of any health-related issue.
- Provide first aid and CPR assistance through properly trained employees on every shift. Make sure that these employees keep their certifications current and that they receive adequate training in specific hazards specifically associated with the worksite.
- Involve the medical provider in alcohol and drug abuse interventions, smoking cessation programs, and any other company programs geared toward helping employees to recognize and obtain treatment for any substance abuse problem.
- Make sure that the medical provider has current credentials, has had recent continuing education, and understands the hazards of your worksite. These standards will help to ensure his or her ability to recognize early symptoms of health issues and begin prompt and appropriate treatment to prevent disability.
- Make sure that standardized procedures (protocols) are used throughout the medical surveillance program, particularly if you are using more than one contractor for health services. (Refer to Appendix D.)
- Have one of your medical providers keep your employee injury and illness records, where feasible. Make sure your recordkeeping system effectively protects the confidentiality of individual employee medical records [2].

The following are some methods on how to develop and deliver health care in accordance with federal and state regulations:

- Coordinate emergency response at your facility and of all external emergency organizations, such as the fire department, any contractual organization, or nearby community hospital. Everyone needs to know exactly what to do and what to expect from others [1, 2].

- Maintain contact through your medical provider to discuss transitional work programs and job assignments with any employee who is off work due to an injury. Keep in contact with the medical provider who provides treatment and care to make sure that the treatment is appropriate and that the employee is responding as expected.
- Use a registered nurse or physician to help advise an employee who is off work for an extended period about workers' compensation rights and benefits and ongoing care.
- Use these medical providers to provide evaluations aimed at determining if an employee can resume full duty after missing work or if work duties need to be modified.
- Consult your physician or registered nurse for help with the development of a transitional duty position, to make sure that the employee can perform the work and benefit from feeling productive again [2].
- Develop and deliver health care in accordance with federal and state regulations, for example, applicable OSHA standards, workers' compensation laws, public health regulations, and/or company policies.

SUMMARY

Your medical surveillance program is an important part of your safety program. It can help to deliver services aimed at evaluating hazards that can cause injuries, recognizing and treating injuries, and limiting their severity.

To determine the appropriate services, you need to consider your site-specific needs such as the type of process and materials your employees work with and the resulting or potential hazards, the type of facilities where employees work, the number of workers at each site, and the characteristics of the work force, such as age, gender, cultural background, and educational level. In addition, the location of each operation and its proximity to a medical provider are important [1].

Whether you hire a full-time professional or contract with a medical provider, make sure that they have specialized training, experience, and up-to-date credentials. The medical provider should be trained, experienced, and/or certified in the identification, treatment, and rehabilitation of injuries [1, 2].

In addition, the medical provider must be familiar with applicable OSHA regulations and recordkeeping requirements.

An early-return-to-work program should be in place at the facility. The job descriptions must be complete for all tasks that include the physical requirements of the job. Transitional duties must be identified that are productive, creative, and not demanding to the employee.

The employer should follow a prescribed protocol for early contact and close communication with the injured employees and medical providers, who will help to facilitate return to regular or modified work at the earliest possible date [1].

Finally, all employees must be aware of, and fully support, the early-return-to-work program.

REFERENCES

1. Oklahoma Department of Labor, Safety and Health Management: Safety Pays, 2000, http://www.state.ok.us/~okdol/, Chapter 12, pp. 73–80, public domain.

2. U.S. Department of Labor, Office of Cooperative Programs, Occupational Safety and Health Administration (OSHA), *Managing Worker Safety and Health*, November 1994, public domain.

14

Defining Safety and Health Training Needs

INTRODUCTION

After establishing organizational policies, goals, objectives, and activity-based indicators and communicating those issues to all employees, the next step is to provide training for managers, supervisors, and employees. This training will assist employees in meeting and exceeding objectives [1].

One common method of responding to employee performance deficiencies is to provide training. If it is improved performance we seek, then training must be based on specific, measurable, performance-based objectives (not what the students will know, but what the student will be able to do upon successful completion of any training course). The appropriate training must be designed with specific guidelines and must supplement and enhance other educational and training objectives [6].

When you understand the many ingredients that make up a good training program, the task of blending the required elements can be challenging. What you can say and the way you say it can go a long way toward effective communication with the trainee. For most presenters, the bottom-line objective of training is to change or improve someone's performance or behavior and increase the trainee's knowledge.

Some of us who become instructors tend to be content with the way we present information or are technical experts who have a tendency to focus on delivering content and performance. This does not always create a good learning environment. It is time to get serious about assessing students' knowledge of a subject before attempting to train. We must validate post-training to determine retention at various times after training [5]. Refer to Table 14-1 for an overview of why training is not always successful. Note the distractions that are presented. These distractions are

Table 14-1
Reducing Distractions to Enhance Learning

Distraction	Solution
• Past trainings produced little or no change in safety culture.	Emphasize the current commitment (if genuine) to safety, as well as its personal value. Reinforce management's commitment to making safety a top priority. Productivity gains should net.
• Productivity will slip during the training.	
• Employees feel they've heard it before!	Listen to frustrations from the past. Emphasize to employees the current value and benefit of the skills/information to their safety today.
• Employees are asked to attend trainings before or after a shift, or on their day off, and resent being there.	Allow employees to express their discontent before the beginning of the training. Acknowledge them for being there. Encourage participants to make the time worthwhile.
• Downsizing has lowered morale.	Communicate to employees that a safe workplace and open dialogue with management will improve morale. Underscore the special value of this training, especially with a smaller workforce.
• Emphasis on the numbers, not on personal well-being.	
• Tension or poor communication exists between management and line employees.	Help employees believe their health and well-being is a high priority to their employer. Involve all levels of employees and management in pre-training planning/issue sessions.

Topf, Michael D., *The SAFOR Report*, A Publication from the Topf Organization, pp. 3, Winter 1998. Modified with permission.

real. One distraction is worth noting here. Employees who are asked to come in before their shift, stay after their shift, or come in on their day off are not inclined to participate in the training session. You are sending a message that "we cannot afford to support safety."

DEFINING A TRAINING PROGRAM

To be effective, a good training program must be results oriented. Knowledge of many activities and processes involved in a given task will help in designing a productive and beneficial learning experience. The objective of an effective training program is to provide two-way communication and help the student (employee) to learn the required techniques. By changing the way we communicate, we become a more powerful and effective trainer [5].

The following questions should be asked to satisfy yourself that you have a successful training program: How effective is my safety training? Can my employees explain existing and potential hazards that they are exposed to? Do employees know how to protect themselves and their co-workers from these hazards? Can they tell me what they must do in the event of a fire or other emergency [6]?

Training needs will vary based on the established goals and objectives. No matter the operation, new employees, existing employees, and contractors must be trained to do the job and to recognize and avoid potential hazards in the workplace [1].

Training can help employees develop the knowledge and skills they need to understand workplace hazards and protect themselves. OSHA considers safety training vital to every workplace. This is the fourth major element in OSHA's Safety and Health Program Management Guidelines [3]. Refer to Figures 14-1 through 14-3 for an overview of designing a training program. National Institute of Occupational Safety and Health (NIOSH) conducts training research using flow diagrams. The flow diagrams have been modified to fit the needs of this book.

The installation of new equipment or process changes may create a need for training that either increases existing hazards or creates new hazards. Finally, all employees may require "refresher" training to keep them prepared for potential problems or emergencies.

Safety education is most effective when integrated into the overall company's training in performance requirements and job practices. It can range from the simple precautionary warnings given to new employees when they are first shown the job to more elaborate, formalized instruction [1, 3].

Instructing your supervisors and employees does not have to be complex or lengthy. This is one of the mistakes that supervisors make: they think that training classes must be long to be effective. In some cases, supervisors think that if a class is intended to be one hour long, then they must do everything to keep it one hour long. This is always the case, depending on the nature of the meeting. Sometimes, a safety meeting can last for 15 minutes and the message can be conveyed. Do not get caught in this trap.

This chapter will help you to design (identify needs), revise, implement, and evaluate your safety training. Refer to Table 14-2 for an overview of site-specific training.

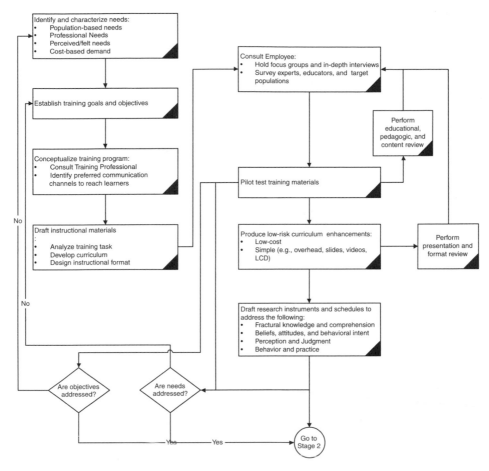

Figure 14-1 Stage 1: Needs assessment. Adapted from NIOSH Publication No. 99-142, A Model for Research on Training Effectiveness, www.cdc.gov/niosh, October 1999, p. 11, Figure 3, public domain.

MANAGEMENT COMMITMENT AND EMPLOYEE PARTICIPATION

Before training begins, make sure that your company policy clearly states the company's commitment to safety and to the training program. This commitment must include, for example, paid work time for training (a major consideration) and training in the specific language that the employee understands. Involve both management and employees in developing the program [2]. Refer to Chapter 6, "Management Leader-

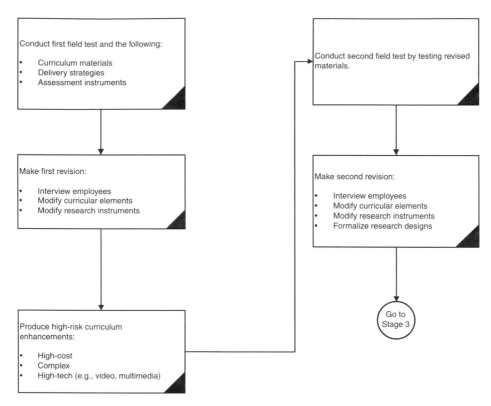

Figure 14-2 Stage 2: Conducting field test. Adapted from NIOSH Publication No. 99-142, A Model for Research on Training Effectiveness, www.cdc.gov/niosh, October 1999, p. 12, Figure 4, public domain.

ship," and Chapter 7, "Employee Participation," for a detailed approach to safety management.

The success of training depends on the level of commitment from management, supervisors, and employees. Some questions that should be considered before designing and implementing a training program include the following:

- Does the safety policy clearly address a commitment to training?
- Did employees participate in policy development?
- Has the policy been clearly communicated to everyone?
- Does the safety commitment include paid work time for training?
- What methods can be used to get management and employees involved in developing the training program?
- What languages are best understood by employees?
- What resources are available to assist in developing a training program [1, 2]?

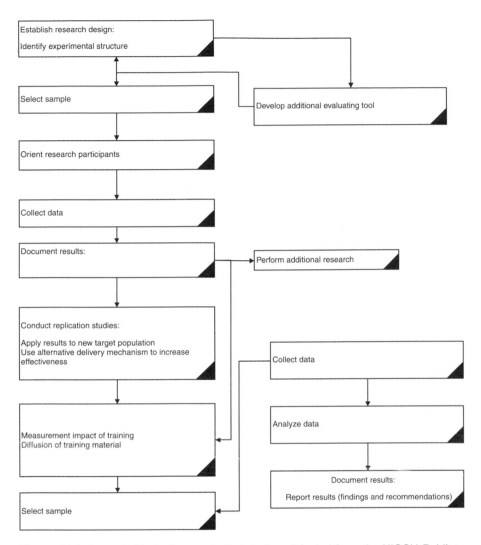

Figure 14-3 Stage 3: Evaluating controlled studies. Adapted from the NIOSH Publication No. 99-142, A Model for Research on Training Effectiveness, www.cdc.gov/niosh, October 1999, pp. 13–14, Figure 5 and 6, public domain.

You must make sure that all employees understand the hazards to which they may be exposed and how to prevent harm to themselves and others, so that employees accept and follow established safety rules. So that managers and supervisors will carry out their safety responsibilities effectively, make sure that they understand those responsibilities and the reasons for them. Refer to Chapter 8 for assigning responsibilities.

Remember that the primary focus must be on safety concerns that can best be addressed by training [1].

Table 14-2
Safety and Health Training Overview

How do we know what the hazards are?

For example, does everyone in the workplace know:

• The workplace plan in case of a fire or other emergency?
• When and where PPE is required?
• The types of chemicals used in the workplace?
• The precautions when handling them?

Training can help to develop the knowledge and skills needed to understand workplace hazards and safe procedures.

It is most effective when integrated into a company's overall training in performance requirements and job practices. Refer to Table 14-1 for an overview of reducing distraction to enhance learning. The content of a training program and the methods of presentation should reflect the needs and characteristics of the particular workforce. Therefore, identification of needs is an important early step in training design. Involving everyone in this process and in the subsequent teaching can be highly effective.

The following five principles of teaching and learning should be followed to maximize program effectiveness:

• Trainees should understand the purpose of the training.
• Information should be organized to maximize effectiveness.
• People learn best when they can immediately practice and apply newly acquired knowledge and skills.
• As trainees practice, they should receive feedback.
• People learn in different ways, so an effective program should incorporate a variety of training methods.

The following are some types of safety training needed:

• Orientation training for site workers and contracts
• JHAs, Standard Operating Practices (SOPs), and other hazard recognition training
• Training required by OSHA standards, including the Process Safety Management standard
• Training for emergency response individual and emergency drills
• Accident investigation training

Who Needs Training?

Training should target new hires, contract employees, employees who wear PPE, and employees in high-risk areas. Managers and supervisors should also

Table 14-2
Continued

be included in the training plan. Training for managers should emphasize the importance of their role in visibly supporting the safety program and setting a good example. Supervisors should receive instruction in company policies and procedures, as well as hazard detection and control, incident investigation, handling of emergencies, and how to train and reinforce training.

Long-term employees whose jobs change as a result of new processes or materials should not be overlooked. The entire workforce needs periodic refresher training in responding to emergencies.

Plan how to evaluate the training program when you are initially designing it. If the evaluation is done properly, it can identify your program's strengths and opportunities for improvement, and provide a basis for future program changes.

Keeping training records will help make sure that everyone who should get training does. A simple form can document the training record for each employee.

OSHA Web site, http://www.osha-slc.gov/SLTC/safetyhealth_ecat/comp4.htm, public domain.

IDENTIFYING TRAINING NEEDS

You will want your training program to focus on safety concerns that are most appropriately addressed by training. As discussed in Chapter 11, it is important to determine the best way to deal with a particular hazard. When a hazard is identified, try to remove it. If it is not feasible to remove the hazard, you must train employees to protect themselves against the hazard until such time as it can be corrected. Once you have decided that a safety problem can be addressed by training (or by another method combined with training), be sure to follow up by developing specific training goals based on your particular needs [2].

NEEDS ANALYSIS

The workplace analysis and hazard prevention and control program is a good source of information to identify training needs. Another source is to use your incident reports. New work practices, processes, equipment, and materials (including hazardous materials) may introduce new hazards into the workplace. Your training must address these changes.

Conducting a needs analysis is one of the most important points when developing a successful training program. Too often trainers pull programs "off of the shelf" and try to adapt them to their specific application. This sometimes falls short because of the generic application for which the program was developed.

One must remember that training is not the answer to every issue. When a safety issue does arise, it must be thoroughly evaluated before training is chosen as the best alternative. Determining training needs can be time-consuming, but with the proper investment, it will pay off because of better planning. Even when employees cause the problem, you must analyze the problem to determine what kind of problem you are dealing with.

You may find that the problem is a skill deficiency issue that must be dealt with. Often you will find a problem in performance that seems highly conducive to high productivity. The problem can be related to trainee attitudes. He or she may be skilled but for a variety of reasons does not care about learning [5].

Understanding these areas will help to identify the problem and determine the appropriate training solution. By analyzing the problem you can determine the role, if any, that training will play in resolving the training need.

The assessment of training needs requires a three-part analysis. Organization analysis involves a study of the entire organization—determining where the organization training emphasis can and should be placed. The first step is to get a clear understanding of short- and long-term goals. What is the company trying to achieve in safety, in general, and specifically by department? The second step is an inventory of the company's attempts to meet goals through its human and physical resources. The final step is an analysis of the climate of the organization [4].

All of the required data must be collected and analyzed to find relevant work situations. However, in the evaluations you can gather job descriptions, task analyses, and other information relevant to the problem when training is being proposed [5].

New employees need to be trained to perform their jobs safely, to recognize, understand, and avoid potential hazards to themselves and others. If new employees are likely to be exposed to hazardous substances, they must be trained before they begin to work [3].

In addition, contractors (temporary employees, subcontractor employees, visitors, etc.) may need specific training to recognize workplace hazards. One often-forgotten individual is the experienced employee who will need training if the installation of new equipment creates changes to the job, or if process changes create new hazards or increase existing hazards. All employees may need refresher training to keep them prepared for emergencies and alert to ongoing housekeeping problems [3].

NEW EMPLOYEE ORIENTATION

One thing that we often forget is new employee orientation. It is critical to get new employees off on the right path. No matter the type of position, an orientation must be conducted. There is one thing for sure, benefits will be discussed with all new employees—why not safety? The following is a list of those who are often overlooked:

- Training of contractor and/or temporary employees
- Training employees who work in high hazard and/or special hazard areas
- Training in the use and maintenance of personal protective equipment
- Attitudes and perceptions of managers, supervisors, and employees toward safety and health

This needs assessment will help you understand where there are opportunities for improvements that can be incorporated into your management system. It will help you design a program that is geared toward reinforcing the positive aspects of safety and changing the negative [1].

Specific hazards that employees need to know about should be identified through site safety surveys, change analysis, and job hazard analysis as discussed in Chapter 15. Incident records may reveal additional hazards and the need for training. In addition, near-miss reports, maintenance requests, and employee suggestions may uncover still other hazards requiring employee training.

In larger organizations that have extensive training needs, more formalized training may be required. In either case, five basic principles should guide your training program:

- *Communicate the perceived purpose.* The employee must understand the purpose of the instruction. The beginning of any training program should focus on why this instruction will be useful.
- *Organize the order of presentation.* Information should be organized to maximize understanding. For example, if you are instructing

employees on the proper way to use a respirator, the order in which you present the material should match the steps the employee must follow to fit, wear, and maintain the respirator.

- *Provide appropriate work practices.* We learn best when we can immediately practice and apply newly acquired skills and knowledge. Instruction is best provided where demonstration, practice, and application can be immediate. When onsite training is not feasible, arrange for your employees to practice and apply the new skills and knowledge as soon as possible.
- *Provide immediate feedback (knowledge of results).* As we practice, we need to know as soon as possible if we are correct. Practicing a task incorrectly can lead to harmful patterns. Praising the employees for correct actions enhances motivation and encourages formation of desirable habits.
- *Account for the individual differences.* We are individuals. We learn in different ways. A successful safety training program incorporates a variety of learning techniques—for example, written instruction, audiovisual instruction, lectures, and hands-on coaching. In addition, we learn at different speeds. The pace of the training should recognize these differences. One effective way to learn is by instructing others. Therefore, after the initial instruction and some practice, it can be helpful to divide the group into teacher/learner teams, sometimes pairing a rapid learner with a slower one, but also giving the slower learner a chance to teach [1].

DEVELOPING LEARNING ACTIVITIES

Develop learning activities to meet identified training needs. Be imaginative in your choice of methods and materials, and make sure that you use your resources. One way to get ideas is by looking at the training programs of other organizations in your industry [4].

ESTABLISHING LEARNING OBJECTIVES

A learning objective is distinct from a learning goal. The objective defines a level of performance of quality and the specific level of quality of the performance. A learning goal is defined as a general skill or behavioral change that will occur as the result of training [5].

Unless we know what we want our training program to do, how can we measure its success? We must document the desired behavior and performance standards that the program is to achieve.

COURSE CONTENT DEVELOPMENT

When the specific objectives are identified, you can select the training methods that will be effective in achieving those objectives. A number of training methods can be used, such as role playing, case studies, icebreakers, games (puzzles, Jeopardy, etc.), and/or group discussions and exercises. Each method must be evaluated in relation to the stated objective. The objective is to make the training fun and interesting so the trainee can learn. Try new things to increase awareness.

The task is to identify which method will meet your objectives. The trainees' willingness and desire to learn will affect the selected training techniques. If trainees are motivated when they arrive at the training session, you will have no problems getting the message across. If they don't want to be in the course or if they feel they can't learn anything, then you must use training techniques that will build interest and a willingness to learn [5].

After you have completed the training program design, conduct a test run. Conduct the program just as you would have for a group of trainees. This test run will help to work out the kinks and allow you to make the necessary changes to make it flow properly [5]. Any problems that don't show up as you go through the training portion of the program will show up when you begin evaluating the effectiveness of the program [5].

CONDUCTING THE TRAINING

A good learning climate maximizes learning by offering the trainees a reason to learn, preparing them for the stated objectives of the training, and developing their trust in the instructor.

After a brief introduction of the training topics and how you are going to conduct the training, ask the trainees to think of what benefits they want to get from the training. At this point, the trainees can see and establish in their own minds the value they will get from the time in the classroom. Since the group developed the list, rather than the instructor it lends more validity to the training. The trainees will be more motivated to receive the training so they may realize the benefits [5].

Using a discussion period serves two purposes: it keeps the trainees interested and involved in the learning process. It also provides them an opportunity to fulfill their personal needs for expression. During the discussion period, trainees can socialize and relieve any tension that may have built up during the time they were listening to the instructor.

Design activities that promote participation. This will help to make sure that everyone in the process is involved. Make sure that participation is used to provide freedom to make mistakes. Most people do not participate in group activities or are reluctant because they fear that they will be ridiculed for making mistakes. You can reduce this feeling by letting employees know up front that making mistakes is OK and that it is a good way of learning. Once the trainees accept this, they will participate [5].

A good management system provides employee safety training in those specific elements of their jobs. In an effective management system employees will feel that they have received adequate safety training and that they understand how to work safely [4].

Employee performance depends on their ability and motivation to work safely. Training helps employees to perform their jobs safely. Therefore, in many cases training is essential.

We hire all types of individuals. Some are susceptible to getting hurt while others are not. Perhaps some are actually "accident-prone." If you believe in this concept, you are a traditional manager: you think that employees are hurt by chance. In this case, the safety knowledge may not be up to your requirements. Some employees are eager to work safely, while others are indifferent to safety. It leaves us with a group of employees who may or may not have to be trained on safety. Where do we go from here [4]?

The logical process would be to attempt to get the answers to the following questions:

- Where are we going?
- How will we get there?
- What do we do when we get there [4]?

Management tells us, through its policy statements, where we want to be and what each individual's function is once we get there [4]. Remember when we discussed the idea "Begin with the end in mind"? We must then find out how able our employees are to fulfill these duties and responsibilities. "Can they do it?" Defining training needs is the key point in the safety process, but not always the complete answer. It provides the objective and sets the criteria for performance measurement. We would expect that any function this vital to the success of training would be widely used in industry [3].

The climate of an organization is, in essence, a reflection of its employees' attitudes toward various aspects of work, management, company policies and procedures, goals and objectives, and membership in the organization. These attitudes are learned and are a product of the employee's experiences both inside and outside the work environment. A training program may be designed to effect certain changes in the organizational climate. For example, our safety training hopes to influence the employees' attitudes (awareness) toward safety [4].

Job analysis for training purposes involves a careful study of job tasks. This is a further effort to define the specific content of training—determining what should be the contents of training in terms of what an employee must do to perform a task, job, or assignment in an effective way [4]. It requires an orderly, systematic collection of data (task analysis) about the job. We are familiar with this through our job Hazard Analysis procedures. Refer to Chapter 15 for an overview of Job Hazard Analysis. The following methods are also available for job analysis:

- *Observations*. Is there obvious evidence of unsafe acts (at-risk behavior)/conditions or poor work methods? Are there instances on the part of individuals or groups that reveal poor personnel relationships, emotionally charged attitudes, frustrations, lack of understanding, or personal limitations? Do these situations imply training need?
- *Interviews* with supervisors, top management, and employees to understand safety issues.
- *Group conferences* with interdepartmental groups and safety committees to discuss organizational objectives, major operational issues, action plans for meeting objectives, and areas where training would be a value
- *Comparative studies* of safe versus at-risk employees to underline the bases for differentiating successful from unsuccessful performance
- *Surveys*
- *Tests or examinations* of safety knowledge; analyses of safety sampling

- *Supervisors' reports* on the safety performance of employees
- *Incident records*
- Actually performing the job/task [4]

Employee analysis focuses on the individual and job performance as it relates to safety—determining what skills, knowledge, or attitudes an indi-

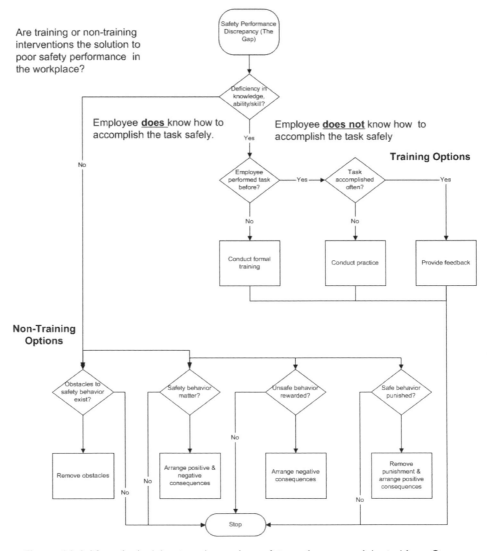

Figure 14-4 Mayer's decision tree: improving safety performance. Adapted from Oregon State OSHA 115, "Training the Safety Trainer II. Presenting On-the-Job Training," March 2000, pp.4. Adapted from Mayer, Robert, Peter Pipe, *Analyzing Performance Problems*, p. 3.

vidual employee must develop if he or she is to perform the tasks that constitute the employee's job in the organization. Refer to Figure 14-4 to help with this analysis [4].

If employees are to learn and improve their knowledge about safety they must feel motivated and comfortable in the learning environment. The following are some suggestions for enhancing the success of your training program:

- Prepare employees for training by putting them at ease
- Recruit employees who show signs of being good trainers. Prepare them to conduct peer training (train-the-trainer)
- Explain the job or training topic. Determine how much your employees already know about the subject
- Boost your employees' interest in training by helping them understand the benefits.
- Pace the instruction at the trainees' learning speed. Present the material clearly and patiently
- Present only as much information in one session as your employees can handle
- Have your employees perform each step of the task and repeat your instructions and explanations. Have them repeat a task until you are satisfied they know how to do it
- Encourage employees to help each other by dividing into groups
- Check frequently for correct performance during the initial practice period
- Reduce surveillance as the trainees become more proficient
- Encourage your employees to build the new skill into the way they work best, but caution them not to change the newly learned procedure without first checking with you or their supervisor [4]

HOW TO TRAIN

There is more empirical data here than perhaps anywhere else in psychology. Here are some of the "knowns" [4].

Motivation and Learning

According to Petersen, learning theorists generally agree that an individual will learn most efficiently when motivated toward some goal that is attainable by learning the subject matter presented. It is necessary that the goal be desired, and the learning behavior must appear directly related to achieving that goal. If the training doesn't seem to directly relate to the goal, other kinds of behavior that, to the learner, appear to be relevant to the goal will be tried [4].

When a safety training course is conducted, some employees may feel that they have more important production problems to worry about and will spend time thinking about these issues. They will usually complain about being taken away from the job to learn a lot of nonsense (useless information). The behavior of the employee is oriented toward relevant goals, no matter whether these goals are safety, increased recognition, production, or socialization. Employees attempt to achieve those goals that are salient at the moment, regardless of the trainer's intent. Therefore, making sure that employees understand the direct relationship between the training and the goal is a critical aspect of training [4].

Reinforcement and Learning

Positive recognition for specific behaviors increases the probability that employee behaviors will occur again. Negative rewards decrease the probability [4].

Whether or not an event is reinforcing will depend on the perceptions of the individual who is learning. What one employee regards as a rewarding experience, another may regard as neutral or non-rewarding, or even punishing. In general, there are various classes of reinforcers: food, status, recognition, money, and companionship. These are reinforcing to almost everyone at one time or another [4].

Practice and Learning

An individual learns through practice. In practicing a skill, those behaviors that are performed and that are reinforced will be acquired and maintained. Without practice, learned skills are quickly lost. In training, the follow-up and practice are as important as the initial training. Also, spaced practice (a little at a time) seems to be more effective than massed practice. This seems to be true both for learning rates and for retention. It is better to have a number of short sessions than one long session. Employees seem to learn faster [4]. Figure 14-5 depicts the average retention rates of employees with various learning methods.

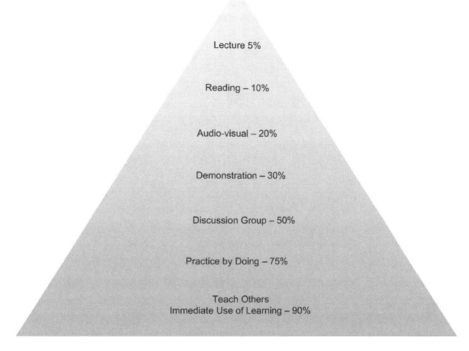

Figure 14-5 Average retention rates.

Feedback and Learning

Most experts agree that providing feedback is essential to good training. It is difficult for the employee to improve performance unless you provide specific feedback and knowledge about his or her performance [3].

An employee who is not performing correctly needs to know what is the nature of the errors. How can they be corrected? This is essential. Some theorists strongly emphasize immediate reinforcement for each bit learned [3].

Meaningfulness

Meaningful material is learned and remembered better than material that is not meaningful [3]. Figure 14-6 diagrams a useful performance model.

The concept of meaningfulness has implications for the way that the material is presented to the employee. The trainer must try to think in the employees' terms and get down to their level, to put the material across

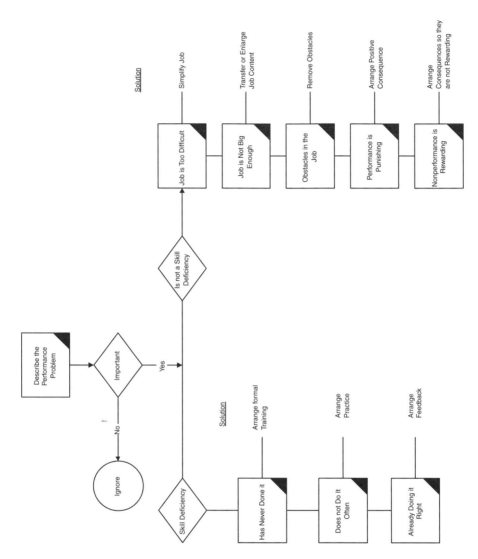

Figure 14-6 Performance model. Peterson, Dan, *The Challenge of Change: Creating a New Safety Culture, Resource Manual*, p. 75, CoreMedia Development, Inc., 1993. Reproduced with permission.

with examples and plain language familiar to them. It is important to supply as many word associations for new ideas and concepts as possible so that they become more meaningful [4].

Climate and Learning

The classroom environment makes a difference. Research has concluded that to encourage high rates of achievement in highly technical subjects, the environment must be challenging. To encourage achievement in non-technical areas, classes should be socially cohesive and satisfying [4].

MANAGEMENT TRAINING

A good safety training program is impossible if you do not have top management support and understanding. Training managers in their roles and responsibilities is necessary to make sure of their continuing support and understanding. In many cases formal classroom training may not be necessary. The subject can be covered periodically as a part of regularly scheduled management meetings [4].

It is import to realize that it is management's responsibility to communicate program goals and objectives to all employees. Management's role includes making clear assignments of safety responsibilities, providing authority and resources to carry out assigned tasks, and holding managers and supervisors accountable for their actions [4].

Training should emphasize the importance of managers' visibly demonstrating their support for the management system. Managers should be expected to set a good example by following all established safety policies and procedures, and should actively encourage employee participation in safety problem identification and resolution. Refer to Chapter 8, "Assigning Safety Responsibilities" [4].

SUPERVISOR TRAINING

Line employees and production leaders are often promoted to leadership positions without adequate knowledge of how to train other employees. It is not unusual for supervisors to lack knowledge of required policies and procedures. Since this is the case, they may need additional training in hazard identification and control and in incident investigation [4].

One important question to ask is: Are supervisors perceived to be well trained and able to handle safety-related issues? On the other hand, are supervisors skilled in performing training and communicating the correct message to employees [4]?

The first step in measuring supervisory performance is to define their roles and responsibilities by defining the tasks and/or activities that supervisors need to accomplish. This can be done authoritatively or participatively. Once roles and responsibilities are defined, the supervisor must know how to accomplish the assigned tasks. This section will help you to determine if the supervisors have been given sufficient training to carry out their assigned roles and responsibilities. The other essentials to supervisory training performance are measurement and recognition [3].

When training is done systematically the following process applies:

- It determines where employees are—their current knowledge and skill level
- It determines where employees should be—the behaviors required to perform the job safely
- It identifies a systematic way to provide the difference (looking for opportunities for improvement, thereby closing the gap) [4]

These three steps concentrate on defining training needs. Historically in safety training, the concentration has been on other areas such as choosing the training method or on specific content. If we do a good job of defining the need, everything else falls into place. The assessment of training needs is detailed in a three-part analysis:

- An organization analysis should identify the company resources and objectives as they relate to site-specific training.
- Conducting a job analysis will help to define jobs-specific and associated tasks.
- A manpower analysis will explore the human dimensions of attitudes, skills, and knowledge as they relate to the company and the employee's job [3].

The analysis involves a study of the following: the objectives, resources, allocation of resources in meeting its objectives, and the total environment of the organization. These things determine the training philosophy [4].

Job analysis for training purposes involves a careful and detailed review of jobs or tasks in an organization. Refer to Chapter 15 on JHAs. This is a great tool to identify specific job-related tasks for training. This review will help to define a specific content for training. It will require an orderly, systematic collection of data about the job, similar to job hazard analysis procedures. The analysis might include the following questions and observations:

- Is there obvious evidence of unsafe acts (at-risk behavior) or poor procedures?
- Are there incidents on the part of individuals or groups that reveal poor employee relationships, emotionally charged attitudes, frustrations, lack of understanding or personal limitations?
- Do these situations imply training needs?
- Is there a management request for training of employees?
- Do interviews with supervisors, top management, and employees reveal information about safety issues?
- Do group conferences (with interdepartmental groups and safety committees), where organizational objectives, major operational problems, and plans for meeting objectives are discussed, reveal areas where training could be of value?
- Have comparative studies of safe versus at-risk behaviors been conducted to determine the basis for differentiating successful from unsuccessful performance?
- Are supervisors' reports on the safe performance of employees reviewed to determine training needs?
- Are incident records reviewed to determine the need for retraining?
- Is actual job performance observed [4]?

JOB ORIENTATION

The format and extent of orientation training will depend on the complexity of hazards and on work practices. For small businesses, job orientation may consist of a quick review of site-specific safety rules. A safety professional and/or the new employee's supervisor usually present this training [5].

For larger businesses with more complex hazards and work practices to control them, orientation should be structured carefully. You want to make sure that all new employees start the job with clear expectations and understanding of the hazards and how to protect themselves and others. Employers frequently provide a combination of classroom and on-the-job training [5].

Many employers have found it useful to have some employees trained to provide peer training. Others have followed up supervisory training with a buddy system, where an employee with long experience is assigned to watch over and coach a new employee, either for a set period of time or until it is determined that training is complete [3].

No matter if the orientation is brief or lengthy, the supervisor should make sure that before new employees begin the job, they receive instruction in how to respond to emergencies [3].

VEHICULAR SAFETY

One thing that is commonly overlooked in a safety program is vehicle safety. Over-the-road motor vehicle incidents in some cases are the leading cause of work-related fatalities. Given the reality of this hazard, all employees operating a motor vehicle on the job should be trained in its safe operation [5, 6]

Do not overlook the training of on-premises vehicle drivers. These drivers can be exposed to such hazards as vehicle imbalance, loads tipping while the vehicle is cornering, and dangers related to battery charging [5].

PERSONAL PROTECTIVE EQUIPMENT (PPE)

Supervisors and employees must be taught the proper selection, use, and maintenance of PPE. Since PPE sometimes can be cumbersome, employees may need to be motivated to wear it in every situation where protection is necessary. Training should begin with a clear explanation of why PPE is necessary, how its use will benefit the user, and the limitations of the equipment. Remind your employees of your desire to protect them and of your efforts, not only to eliminate and reduce the hazards, but also to provide suitable PPE where needed. Explain how essential it is that they do their part to protect their health and safety [6].

EMERGENCY RESPONSE

Train your employees to respond to emergency situations. Every employee needs to understand the following [1]:

- Emergency telephone numbers and who may use them
- Emergency exits and how they are marked

- Evacuation routes
- Signals that alert employees to the need to evacuate

Refer to Chapter 8 [5].

PERIODIC TRAINING

At some worksites, complex work practices are necessary to control hazards. At such sites, it is important that employees receive periodic safety training to refresh their memories and to instruct them on new methods of control. New training may also be necessary if OSHA requirements change or new standards are issued. It is important to keep these sessions interesting. Some companies have found it effective to give employees the responsibility to plan and present periodic safety training. The success of this method depends upon employees' being provided adequate training resources and support to develop their presentations [5].

Many companies use monthly safety meetings for this training. In construction and other high-hazard industries where the work situation changes rapidly, weekly meetings often are needed. These meetings serve to remind employees of the upcoming work tasks, the environmental changes that may affect them, and the procedures they may need to follow to protect themselves and others [5].

One-on-one training is the most effective training method. The supervisor periodically spends some time watching an individual employee work. Then he/she meets with the employee to discuss safe work practices, provide positive feedback, and/or provide additional instruction to counteract any observed unsafe practices. One-on-one training is most effective when applied to all employees and not just when there appears to be a problem. This is because the positive feedback provided for safe work practices is this method's most powerful tool. It helps employees establish new safe behavior patterns. It also recognizes and thereby reinforces the desired behavior [5].

EVALUATING THE PROGRAM

What is usually missing from the training design is evaluation. The evaluation of training is not simple. We try to determine what changes in skill, knowledge, and attitudes have taken place as a result of training [4].

Too often in safety training there simply is no measurement of any kind to evaluate whether or not the training has accomplished its objectives, or if it has accomplished anything. Employees in the training process, regardless of organizational level, should have a measurable improvement in some skill or behavior. Management must insist that the training be evaluated in terms of improvement of the skill or in terms of measurement of behavior change. Refer to Appendix E [4].

You must evaluate each training program to determine if it is effective—whether the training you have provided achieved the intended goal of improving safety and performance. When carefully developed and carried out, the evaluation will highlight the training program's strengths and identify areas of challenges that need change or improvement [4].

EVALUATING TRAINING

You should design a plan to evaluate the training sessions as needs are being identified and training content developed. This important part of your training effort should not be put off until training is completed. The following are some ways you can evaluate your training program:

- Before training begins, determine what areas need improvement by observing employees and soliciting their input. When training is completed, test for improvement by observing employees performing their job. Ask them to explain their jobs' hazards, protective measures, and newly learned skills and knowledge.
- Keep track of employee attendance at training sessions. Training will not work for an employee who does not show up. Absenteeism can signal a problem with the employee, but it also can indicate a weakness in training content and presentation. Refer to Appendix E.
- At the end of training, ask participants to rate the course and the trainer. This can be done in informal discussion, or confidentiality can be assured by a written unsigned questionnaire. Refer to Appendix E.
- Compare pre- and post-training injury rates. The periods of time being compared must be long enough to allow significant differences to emerge [3].

It often is easier to conduct an activity than to judge it. However, do not ignore this evaluation phase. It will allow you to calculate your training program's bottom line profitability and the effectiveness of training programs. Once you have made the effort to provide employee safety training, you must answer these questions:

- Have the goals of training been achieved?
- Do the results justify offering the training program again?
- Can the training program be improved [1]?

Evaluation of any training program is an ongoing process. End-of-course formal evaluation must be part of every program. Rating sheets that are completed by trainees are helpful if they are properly constructed. Be cautious—don't take any evaluations to heart. You have to understand that all trainees will not like what you say or do and will sometimes be critical. Look at it this way; evaluations are a feedback system that measures the effectiveness of the trainer's actions. They can be used to learn and correct errors in the training program [1]. Such records will help make sure that everyone who needs training receives it, that refresher courses are provided at regular intervals, and that documentation is available (when needed) to show that appropriate training was conducted.

The next training session will always be better if you learn to accept constructive suggestions from the trainees, your peers, management, and other trainees. If the goal was not attained, major revisions may be in order. If goals were reached and the evaluation response and results were favorable, it is still good advice to review and revise bits and pieces [5].

RECORDKEEPING

A recordkeeping system for training programs offers specific benefits to the organization. It is important to keep training records to verify that training was conducted. A simple form is all you need: one that identifies the trainee, the topic or job, and the training date, with space for a brief evaluation of the employee's participation and success. These records will help you make sure that everyone who needs training receives it, that refresher courses are provided at regular intervals, and that documentation is available, when needed, to show that appropriate training was provided. Refer to Appendix E for one example of an easy-to-maintain training record [2].

SUMMARY

The content of your training program and the methods of presentation should reflect your company's training needs and the particular characteristics of employees. Identification of training needs is an important early step in training design. Involving employees in this process and in the subsequent training can be effective [6].

Training needs analysis must be researched before a training program can be developed. You can begin to realize that an effective training program doesn't "just happen." A well-designed training program is an integration of knowledge, skills, and attitudes, all blended to mold specific programs, the desired learning, under specified conditions.

Communicate the purpose of training. Present information in a clear, understandable manner and a logical order. Provide trainees the opportunity to practice the skills being taught. Let employees know if they are performing a new skill incorrectly, and provide positive feedback when employees are performing correctly. Recognize that we are all individuals, and that we learn in different ways. Provide a variety of different learning opportunities, and pace your instruction and practice period, so that all trainees, slow and fast learners, have the time they need to absorb the new skills and knowledge.

Your program should be geared toward employees' recognizing hazards and learning ways to protect themselves and their co-workers. You should target new hires, contractors, employees who need to wear personal protective equipment, and employees in high-risk areas. Do not overlook the seasoned employee whose job changes as a result of new processes or materials. In addition, the entire population needs periodic refresher training in responding to emergencies [6].

Plan from the initial design stage to evaluate your training program. An effective evaluation will identify your program's strengths and opportunities for improvements, determine whether training goals are being met, and provide a basis for future program changes. Recordkeeping will help make sure that all who need training receive it. A simple form can document both your efforts to teach and your employees' success at learning hazard recognition and protection.

As we consider particular training requirements, we might first evaluate the need by asking if training is what is required. Experienced training professionals often find that misinformed people assume that all problems are training problems and that developing a training program can solve them.

Training is more complicated than telling or showing someone how to perform a task. Training is the transfer of knowledge to trainees in such a way that the trainees accept and use the knowledge in the performance of their jobs. The knowledge should be specific, and the training should be directed at identified behavior of the trainee. The trainer should learn specific skills or techniques that can be demonstrated and observed on the job [5].

Can all employees explain every existing and potential hazard to which they are exposed? Do they know how to protect themselves and their co-workers from these hazards? Can they explain precisely what they must do in the event of a fire or other emergency?

Training can help employees develop the knowledge and skills they need to understand workplace hazards. OSHA considers safety and health training vital to every workplace.

REFERENCES

1. Oklahoma Department of Labor, Safety and Health Management: Safety Pays, 2000, http://www.state.ok.us/~okdol/, Chapter 13, pp. 81–90, public domain.

2. OSHA Publication 2254 (Revised 1992), "Training Requirements in OSHA Standards and Training Guidelines," public domain.

3. U.S. Department of Labor, Office of Cooperative Programs, Occupational Safety and Health Administration (OSHA), *Managing Worker Safety and Health*, November 1994, public domain.

4. Peterson, Dan, *The Challenge of Change, Creating a New Safety Culture, Implementation Guide*, CoreMedia Development, Inc., 1993, Resource Manual, Category 16, Supervisor Training, pp. 73–75. Modified with permission.

5. Roughton, James, Nancy Whiting, *Safety Training Basics, A Handbook for Safety Training Program Development*, Government Institute, 2000.

6. Roughton, J., J. Lyons, April 1999, "Training Program Design, Delivery, and Evaluating Effectiveness: An Overview," American Industrial Association, *The Synergist*, pp. 31–33.

15

Understanding Job Hazard Analysis

INTRODUCTION

What is a job hazard analysis (JHA)? Can it work for you? You may have heard other terms, for example, job safety analysis (JSA) or task analysis. No matter what you call the process, a JHA can be used to help you to develop safe work practices or procedures.

A JHA is a written procedure that you can use to review job methods and identify hazards that may have been overlooked during initial task design, process changes, and the like. A JHA is a systematic method of identifying jobs and tasks to help pinpoint hazards associated with the task and developing procedures that will help reduce or eliminate identified risks. You can also use JHAs to document changes in a workplace and provide consistent training.

Some hazards are obvious and can be identified through safety reviews. Others are less obvious and can only be identified by conducting a systematic analysis of each job task to identify potential hazards.

BENEFITS OF JHAs

The main reason for identifying the benefits in the beginning stages of the JHA process is to help management understand what they can expect from the process. This is an opportunity to outline the positive impact that developing JHAs will have on the organization. As the process moves forward, these benefits will become more evident because hazards will be identified that are usually overlooked in the normal course of inspections [4].

One thing to remember is that the benefits of a JHA go beyond safety. As suggested in OSHA's training model, you should conduct a job analysis to understand what training is needed. The JHA provides step-by-step safety procedures for performing each task. Figure 15-1 gives a sample job hazard analysis form for cleaning the inside surface of a chemical tank [1].

Another benefit of developing a JHA is that it provides a consistent message to each new employee on a specific task or for seasoned employees who need safety awareness training or review of their specific or nonroutine task [5].

In addition, a properly designed JHA is a good learning tool that you can use to evaluate incidents. Incidents often occur because employees are not trained in the proper job procedures. One way to reduce these incidents is to develop safer and more efficient work methods or procedures and train all employees in them.

The JHA allows you to identify opportunities for improvement in the system. Once you discover the opportunity, you can update the JHA to reflect the needed changes.

Establishing clear job procedures is one of the benefits of conducting a JHA—carefully reviewing and recording each step of a job or related tasks that make up the job, identifying existing or potential job safety hazards, and determining the best method to perform the job or to minimize or eliminate the associated hazards. Table 15-1 lists reasons why JHAs are important.

However, with all of their benefits, there is one drawback. A JHA program takes time both to document and to implement effectively and is a continuous improvement process through a "living document." As you go through the process, you will see that the process is time consuming, but the end results will outweigh this drawback.

ASSIGNING RESPONSIBILITY

Assigning the responsibilities for JHAs to a specific individual or group of individuals reinforces management's commitment to safety. In addition, it provides a point person or group of employees that will serve as a clearinghouse for development of all JHAs. The individual or group who is selected should be respected by their peers, understand the concept and value of developing JHAs, and understand their relationship to incidents. They should also be able to help overcome barriers that may occur during

Job Title		
Date of Analysis		
Job Location		
Step	**Hazard**	**New Procedure or Protection**
1. Determine the contents of the tank, what process is going on in the tank, and what hazards this can pose.	Explosive gas. Improper oxygen level. Chemical exposure Gas, dust, vapor: Irritant Toxic Liquid: Irritant Toxic Corrosive Heated Solid: Irritant Corrosive Moving blades/equipment.	• Establish confined space entry procedures (29 CFR 1910.146). • Obtain work permit signed by safety, maintenance, and supervisors. • Test air by qualified person. • Ventilate to 19.5%–23.5% oxygen and <10% LEL of any flammable gas. Steaming inside of tank, flushing and draining, then ventilating, as previously described, may be required. • Provide appropriate respiratory equipment SCBA or airline respirator. • Provide PPE for head, eyes, body, and feet. • Provide harness and lifeline. (Reference: OSHA standards: 1910.106, 1910.146, 1926.100, 1926.21(b)(6); and NIOSH Doc #80–406). • Tanks should be cleaned from outside, if possible.
2. Select and train operators.	Operator with respiratory or heart problem; other physical limitation. Untrained operator-failure to perform task.	• Examination by industrial physician for suitability to work. • Train operators. • Dry run. (Reference: National Institute for Occupational Safety and Health (NIOSH) Doc. #80–406)
3. Set up equipment	Hoses, cord, equipment-tripping hazards. Electrical-voltage too high, exposed conductors. Motors not locked out and tagged.	• Arrange hoses, cords, lines, and equipment in orderly fashion, with room to maneuver safely. • Use ground-fault circuit interrupter. • Lockout and tag mixing motor, if present.
4. Install ladder in tank	Ladder slipping.	• Secure to manhole top or rigid structure.
5. Prepare to enter tank	Gas or liquid in tank.	• Empty tank through existing piping. • Review emergency procedures. • Open tank. • Check of jobsite by safety professional or industrial hygienist. • Install blanks in flanges in piping to tank (isolate tank). • Test atmosphere in tank by qualified person (long probe)
6. Place equipment at tank-entry position	Trip or fall.	• Use mechanical-handling equipment. • Provide guardrails around work positions at tank top.
7. Enter tank.	Ladder-tripping hazard. Exposure to hazardous atmosphere	• Provide personal protective equipment for conditions found. (Reference: NIOSH Doc. #80–406; 29 CFR 1910.134). • Provide outside helper to watch, instruct, and guide operator-entering tank, with capability to lift operator from tank in emergency.
8. Cleaning tank	Reaction to chemicals, causing mist or expulsion of air contaminant.	• Provide PPE and equipment for all operators and helpers. • Provide lighting for tank (Class I, Div. 1). • Provide exhaust ventilation. • Provide air supply to interior of tank. • Frequent monitoring of air in tank. • Replace operator or provide rest periods. • Provide means of communication to get help, if needed. • Provide tow-man standby for any emergency.
9. Cleaning up.	Handling of equipment, causing injury.	• Dry run. • Use material-handling equipment.

Table 15-1
Why JHAs Are Important

- Help detect hazards
- Help to develop training plan
- Help supervisors understand what the employee knows
- Help recognize changes in equipment or procedures
- Improve employee involvement

OR OSHA 103o, Why JHAs Are Important, p. 5, http://www.cbs.state.or.us/external/osha/educate/training/pages/materials.html, public domain.

implementation. Someone in a management role should be part of the group [2]. This should help to facilitate the JHA process.

Before training the individual or group, top management must define their expectations and show their commitment. This should address such issues as:

- How much time will be allowed each day for developing JHAs?
- What is expected in terms of document quality?
- What resources will be provided in training and skill development?
- What are the long-term objectives and time requirements for completing selected JHAs? [4]

These expectations must be communicated to other managers and the JHA developers.

CONDUCTING A JHA

The process for conducting a JHA is simple. Before beginning the JHA, one objective is to observe the general work area. Since each job involves a different sequence of activity or task, you should observe how the job is performed. It is important to list each task that is being performed before going any further. Refer to Table 15-2 for some sample questions you might ask when you are conducting a JHA.

Figure 15-1 Sample job hazard analysis form for cleaning inside surface of chemical tank. Adapted from United States Department of Labor, OSHA: Job Hazard Analysis, OSHA 3071, 1998 (revised), public domain.

Table 15-2
Sample Questions You Might Ask When Conducting a JHA

Yes	No	
		Are there materials on the floor that could cause a tripping hazard?
		Is there adequate lighting?
		Are there any live electrical hazards?
		Are there any chemical, physical, biological, or radiation hazards associated with the job? Are any of these hazards likely to develop?
		Are tools—for example, hand tools, machines, and equipment—in need of repair?
		Is there excessive noise that may hinder communication or is likely to cause hearing loss?
		Are job procedures understood and followed and/or modified as applicable?
		Are emergency exits clearly marked?
		Are industrial trucks or motorized vehicles properly equipped with brakes, overhead guards, backup signals, horns, steering gear, seat belts, etc.? Are they properly maintained?
		Are all employees who operate vehicles and equipment authorized and properly trained?
		Are employees wearing proper personal protective equipment (PPE)?
		Have any employees complained of headaches, breathing problems, dizziness, or strong odors?
		Have tests been made for oxygen deficiency, toxic vapors, or flammable materials in confined spaces before entry? Is ventilation adequate, especially in confined or enclosed spaces?
		Are workstations and tools designed to prevent twisting motions?
		Are employees trained in the event of a fire, explosion, or toxic gas release?

U.S. Department of Labor, Occupational Safety and Health Administration (OSHA) 3071 Job Hazard Analysis, 1998 (revised), public domain.

The list in Table 15-2 is only a sample of some of the hazards that you may note when conducting a JHA. The list is by no means complete. Each workplace has its own unique requirements and conditions. You should add your own questions to the list.

It is important that management understand that they need to look at the overall objective of the JHA and develop a strategy to integrate these JHAs into their process.

BREAKING DOWN THE JOB

The essence of a JHA is to break down every job into basic tasks or steps. You can do this by listing each step in the order of occurrence as you watch the employee perform the job. No basic step should be omitted. Make sure you record enough information to describe each action. When you are finished, review each step with the employee. Figure 15-2 gives a sample JHA worksheet that should be used to list the task in a logical order.

Brief Job Description		
Task Description	Existing and Potential Hazards	Recommended Corrective Actions
Task Description	Existing and Potential Hazards	Recommended Corrective Actions
Task Description	Existing and Potential Hazards	Recommended Corrective Actions
Task Description	Existing and Potential Hazards	Recommended Corrective Actions
Task Description	Existing and Potential Hazards	Recommended Corrective Actions
Supervisor:		Date:
Analysis Conducted by:		Date:
Other:		Date:
Other:		Date:

Figure 15-2 Sample job hazard analysis worksheet.

One key point is to be careful and not to omit any steps. However, also take care not to make the job hazards too detailed; too much detail will make a JHA ineffective and unenforceable. One common mistake is to mix work elements with job hazards. A JHA is not intended to document work process instructions, although some people believe that they should be included [2].

Talk to as many employees as possible: new, experienced, transfer and/or temporary employees, managers, maintenance personnel, safety professionals, etc. Common problems will soon become apparent. Not only will you base your decision on better information, but also employees will be pleased at being consulted. Discuss potential solutions with employees and other technical specialists. The flow diagram in Figure 15-3 gives an overview of the process of conducting a JHA.

JHAs in some form may already exist for some or all jobs. If so, you need to determine if the JHAs are effective. For example, is the facility experiencing incidents that result in recordable cases or workers' compensation losses, quality or productivity problems, etc.? If incidents do exist, work procedures or processes may not be effective or the management system may not provide the necessary support. Refer to Figure 15-4 for a comparative analysis on the successful completion of the JHA.

After you record the steps of the job, review each step to determine the hazards that exist or that might occur. There are several ways to identify job hazards: Evaluate the ways human error (at-risk behavior) might contribute to the hazard. Record the types of potential incidents and the physical agents involved. Make sure that the procedure is clearly written.

Evaluate each step as often as possible to identify all real hazards. Both physical and mechanical hazards should be considered. Review the actions and positions of the employees. Refer to Table 15-3 for a sample checklist.

Note equipment that is difficult to operate and could be used incorrectly. Make sure that all equipment is in proper working condition. Determine what stress level the employee is experiencing [5].

DEVELOPING A PRIORITY LIST OF JOBS

Developing specific priorities will help you to understand the tasks that have the highest level of hazards. There are several advantages in having a group establish these priorities. The first advantage is that knowing the

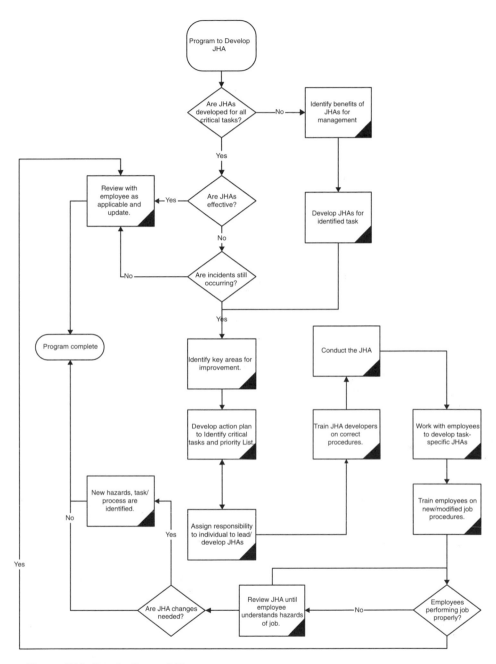

Figure 15-3 Conducting a JHA.

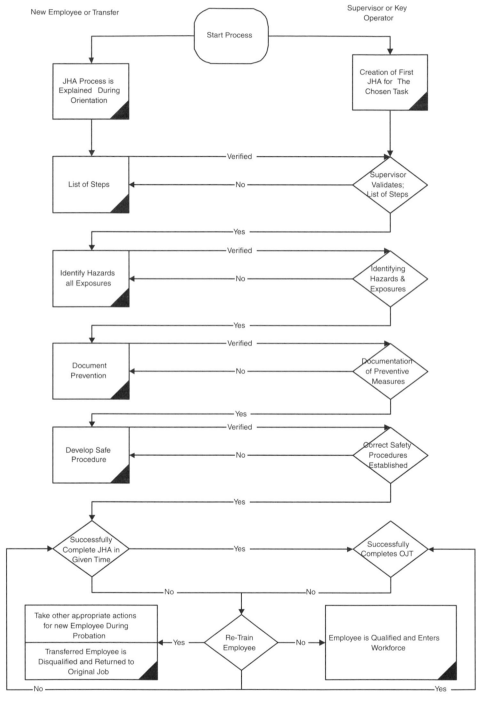

Figure 15-4 Comparative analysis of training. OR OSHA 103, Comparative Analysis, p. 14, http://www.cbs.state.or.us/external/osha/educate/training/pages/materials.html, public domain.

Table 15-3
Sample Checklist for Evaluating Each Job Step

Yes   No   Is the employee wearing PPE?

Are work positions, machinery, pits or holes, and/or hazardous operations adequately guarded?

Are lockout procedures used for machinery deactivation during maintenance?

Are there fixed objects that may cause injury, such as sharp edges on equipment?

Is the flow of work properly organized, for example, is the employee required to make movements that are rapid?

Can reaching over moving machinery parts or materials injure the employee?

Is the employee at any time in an off-balance position?

Is the employee positioned at a machine in a way that is potentially dangerous?

Is the employee required to make movements that could lead to or cause hand or foot injuries, or strain from lifting, the hazards of repetitive motions?

Do environmental hazards such as: dust, chemicals, radiation, welding rays, heat, or excessive noise result from the performance of the job?

Is there danger of striking against, being struck by, or contacting a harmful object?

Can employees be injured if they are forcefully struck by an object or contact a harmful material?

Can employees be caught in, on, by, or between objects? Can employees be injured if their body or part of their clothing or equipment is caught on an object that is either stationary or moving? Can they be pinched, crushed, or caught between either a moving object and a stationary object, or two moving objects?

Is there a potential for a slip, trip, or fall? Can employees fall from the same level or a different level?

Can employees strain themselves by pushing, pulling, lifting, bending, or twisting?

Can employees overextend or strain themselves while doing a task? Can they strain their backs by twisting and bending?

What other hazards not discussed have the potential to cause an incident? Repeat the job observations as often as necessary until all hazards have been identified.

U.S. Department of Labor, Occupational Safety and Health Administration (OSHA) 3071 Job Hazard Analysis, 1998 (revised), public domain.

higher priority task will help to target specific management responsibilities. The second advantage is that it will reduce the amount of time
needed to train the JHA developers. If the developers are expected to
select the task, they will have to be trained in incident data analysis and
trending techniques [3]. Refer to Chapter 12, "Conducting Effective Incident Investigations," for a description of these techniques. Table 15-4 lists
typical types of jobs that are considered high priority.

At the higher management level, there will probably be more information available on accident statistics, quality and productivity statistics,
employee turnover, and strategic long-term prospects for the tasks in
question.

To evaluate a job effectively, you should have some experience and be
trained on the intended purpose of the JHA, have an open mind, and have
examples of correct methods. To determine which jobs should be analyzed
first, review injury and illness reports such as the OSHA log, medical case
histories, first aid cases, and workers' compensation claims. First, you
should conduct a JHA for jobs with the highest rates of disabling injuries
and illnesses. Do not forget jobs where you have had "close calls" or "near
misses." You should give these incidents a high priority. Analyses of new
jobs and jobs where changes have been made in processes and procedures
should be the next priority [4].

When selecting the job for analysis, the following points can be useful
in establishing priorities:

Table 15-4
High Priority Jobs

If a task includes any two or more of the following elements, consider the
application of a JHA:

- High frequency
- High duration
- High force
- Posture (The way employees are positioned at the workstation. For example, ergonomic-related hazard.)
- Point of operation (Where machine parts come together to create pinch point where employees can get caught in the equipment.)
- Mechanical pressure
- Vibration
- Environmental exposure (To toxic materials, heat and cold, etc.)

OR OSHA 103o, "What Jobs Need JHAs?," p. 6, http://www.cbs.state.or.us/external/osha/
educate/training/pages/materials.html, public domain.

- Injury and occupational illness severity—those jobs that have involved serious incidents. There may be a basic problem in the work environment or in the job performance itself.
- Accident frequency. The higher the frequency of incidents, the greater the reason for developing and implementing a JHA.
- Task potential for illness or injury, even if no such incident has occurred.
- A new job or task where there is no history or statistical information about its potential for incidents. Many incidents occur in a job or task where the employee is not accustomed to the job.

To be effective, the creation of a task or modification of a task through the introduction of new processes, equipment, etc., should automatically require you to develop a new or revised JHA. Jobs with many steps are usually good candidates. As stated before, you should assign each job selected a priority based on the injury potential and the severity of potential hazards.

After you have generated a list of hazards or potential hazards and have reviewed them with the employee, determine if the employee can perform the job another way to eliminate the hazards, such as combining steps or changing the sequence. You should be aware if safety equipment and precautions are needed to control the hazards.

If safer and better job methods can be used, list each new step, such as describing a new method for disposing of materials. As in establishing training objectives, list exactly what the employee needs to know to perform the job using a new method. Do not make general statements about the procedure, such as "Be careful." Be as specific as you can in your recommendations.

USING EMPLOYEE PARTICIPATION TO DEVELOP TASK-SPECIFIC JHAs

One of the key concepts that are usually overlooked is to make sure that the specific employee who performs the task participates in the discussion of the JHA. The worst thing to do is to develop the JHA, make it final, and *then* communicate it to the operator.

JHAs allow managers and employees to identify risks together. The manager works with the employee to record each step of the job as it is performed; consulting with the employee to identify any hazards involved in each step, and, finally, enlisting the employee's help on how to eliminate any hazards noted. When you develop a JHA collectively, you create a sense of ownership that encourages teamwork between the manager and the employee. This systematic gathering of information and teamwork is essential to avoid snap judgments [4].

Once you have selected the job for analysis, discuss the procedure with the employee who performs the job and explain the intended purpose. Point out that you are studying the job itself and not checking on the employee's job performance. Involve the employee in all phases of the analysis, from reviewing the job steps and procedures to discussing potential hazards and recommended solutions. You should also talk to other employees who have performed the same job in the past.

Employees are the best source for identifying job hazards, and they appreciate you consulting with them on matters that affect them. Employees become more receptive to changes in their job procedures when you give them an opportunity to help develop the change.

Review the JHA Until Employees Understand the Hazards of Each Task

With the completion of JHA documents, the employees on those jobs may need to be trained in the new procedure. The extent of this training may vary depending on the complexity of the job. In some cases, it may be more of an informal communication effort with a work group; in others, it could be formal on-the-job training. It would depend on how different the JHA procedure is from what the employees were doing before.

When you have completed your training, each employee should sign an acknowledgement stating that they understand the identified hazards. This form should be signed before the employee performs any nonroutine task or any new task. Figure 15-5 shows a sample sign-off form that can be used to document this review.

By signing this form the employee is verifying that he or she has reviewed the job hazard analysis for the specific task that they are performing or about to perform and agree with the JHA as it relates to the job steps (task), the hazard associated with the task, and the recommended corrective measures. At this point, they can also bring up any questions concerning the hazards of the job. You may ask, what have I gained from this action? Think about it—you have now placed some ownership on the employee to help the employee understand the hazards of the job.

Employee Name	Date Reviewed

This form should be signed off prior to performing any non-routine task or any new task. In addition, I understand that I must review this JHA with my supervisor periodically.

By signing this form, I am verifying that I have reviewed this Job Hazard Analysis for the specific task that I am performing or about to perform. I agree with the JHA as it relates to the job steps (task), the hazard associated with the task, and the recommended corrective measures. I also understand that if I have any questions concerning the hazards of this task that I can stop the job and notify my supervisor for resolution.

Figure 15-5 Sample JHA review and acknowledgement form.

DEVELOPING AN ACTION PLAN TO IDENTIFY INCIDENTS

If hazards are still present, try to reduce the necessity for performing the job or the frequency of performing it. Go over the recommendations with all employees performing the job. Their ideas about the hazards and proposed recommendations are valuable. Be sure that employees understand what they are required to do and the reasons for changes in job procedures.

SUMMARY

The JHA is one component of an overall strategy. The facility needs to understand other components of a safety management system that can affect the quality of the job procedures. A program should be developed to evaluate whether employees are consistently following documented procedures. The action planning process is the preferred method to help accomplish this goal.

A JHA documents procedures that can be used to review job methods and identify hazards that may exist in the workplace. JHAs can also be used to document changes in work tasks. Some solutions to potential hazards may be physical changes or a modified job procedure that eliminates or controls the hazard.

All employees should be trained in how to use the JHA. Managers are in the best position to do the training by observing the job as it is being performed to determine if the employee is doing the job in accordance with the job procedures.

A JHA should be monitored to determine its effectiveness in reducing or eliminating hazards. You should also find out if the employee is following the JHA when performing the job. If so, evaluate the effectiveness. If not, try to find out the reason.

It is important to assign both authority and specific responsibility to implement each protective measure. A safety professional may need to provide the training; the manager should provide safe tools and equipment; and the employees should inspect their tools to ensure that they are in safe condition.

Everyone has seen the demonstration where you start off by telling a story to the first person in a group. The story is then passed on to the next person, etc. By the time the story gets back to the original storyteller, the message has changed. If they had a written script similar to a JHA, however, the story would have been the same message around the room.

We need to remember that JHAs should be easily readable and the hazards need to be easily understood. For readability, JHAs need to be typed. They should be placed at the workstation. It is important to highlight the most critical hazards for special attention. The objective is to make a JHA a user-friendly document that everyone can read to understand the hazards of the identified task.

The bottom line: a JHA is not a mandatory requirement or a standard and you are not required to use the recommended methods. It is considered a management tool and a best management practice (BMP), going beyond the OSHA standard.

This chapter was written as a guide only and is based on the U.S. Department of Labor, Occupational Safety and Health Administration OSHA 3071 Job Hazard Analysis, 1998 (Revised) [1] found on the OSHA Web site, www.OSHA.gov under "Publications."

REFERENCES

1. U.S. Department of Labor, Occupational Safety and Health Administration (OSHA) 3071 Job Hazard Analysis, 1998 (revised), public domain.

2. Roughton, J., C. Florczak, "Job Safety Analysis: A Better Method," *Safety and Health*, January 1999, National Safety Council, pp. 72–75.

3. Roughton, J., "Job Hazard Analysis: A Critical Part of your Job as Supervisor Is Evaluating and Controlling Workplace Hazards," *Occupational Health and Safety*, Canada, January/February 1996, pp. 41–44.

4. Roughton, J., "Job Hazard Analysis: An Essential Safety Tool," J. J. Keller's OSHA Safety Training Newsletter, October 1995, pp. 2–3.

5. Roughton, J., April 1995, "How to Develop a Written Job Hazard Analysis," Presentation, National Environmental Training Association Conference, Professional Development Course (PDC) Orlando, Florida.

Making Sense of
the Behavior-Based
Safety Process

INTRODUCTION

Pick up almost any safety literature today and you will see an article discussing behavior-based safety. Each author has a different opinion on the use of behavior-based safety and how to implement the process. Nowhere will you find a discussion of integrating the process into the existing management system. In many cases, as the behavior-based process is presented, the process stands out like a "sore thumb." Since there is such an emphasis on behavior-based safety, many managers think that this will solve all safety issues. It is important to remember that no matter how it is presented, the behavior-based process is only one element of your management system and should not stand alone. For example, there are still hazard recognition techniques, employee participation, and other types of efforts that must be used to identify hazards. You must remember that behavior-based safety does not catch all of the hazards since it focuses only on employees' habits (at-risk behaviors). The JHA is the one method that can be used to help find and correct other hazards outside of behavior. Refer to Chapter 15.

WHAT CONTRIBUTES TO AT-RISK BEHAVIORS?

Employees have a tendency to take on the attitudes of management and co-workers and repeat those behaviors that the culture reinforces through a direct or an implied recognition and/or punishment system. Therefore, we can say that in the work environment human behavior is a function of the management system and the perceived safety culture. Figure 16-1 highlights cultural influences on safety behavior; Figure 16-2 summarizes positive attitudes toward safety that go beyond safe behavior at the workstation. The low probability that any one at-risk behavior will result in an employee being injured continues to reinforce employees' perceptions that they can take risks and not get hurt.

To reduce on-the-job incidents, management and employees must all be actively involved in identifying and encouraging the practice of safe behavior and identifying hazards. Chapter 11 spells out the steps in hazard recognition. The safety culture in your organization must allow active

Figure 16-1 Cultural influences. Eckhardt, Robert, "Practitioner's Influence on Safety Culture," *Professional Safety*, July 1996, p. 25, Figure 2. Reproduced with permission.

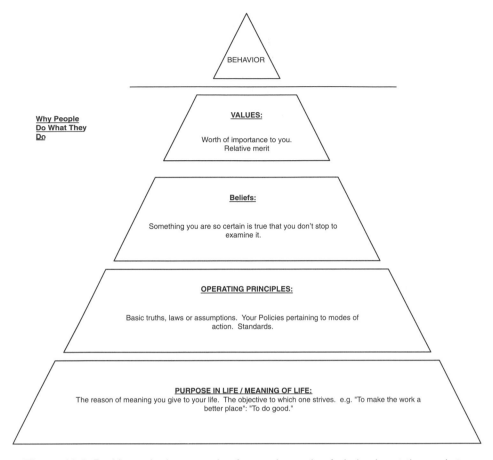

Figure 16-2 Positive attitudes toward safety go beyond safe behavior at the workstation. Topf, Michael D., Richard A. Petrino, "The Topf Organization," *Professional Safety*, December 1995, p. 26. Reproduced with permission.

employee participation. Chapter 7 discusses how you can use employee participation as part of the process. Behavior-based safety observations can help an organization build a safety culture that fosters proactive safe behaviors. Again, remember this is not the only approach to developing an effective safety culture.

Behavior-based safety can be summarized as several fundamental principles:

- Losses are the downstream measures of at-risk behaviors. At-risk behavior may or may not be under the direct control of the employee. In some cases, employees may have no alternative but to

Table 16-1
Activators and Consequences Take Many Forms

Activators: Safety meetings, goal setting, rules and regulations, pledge signing, policies and procedures, incentives/disincentives, instructions, signs, training, modeling, saying please.

Behaviors: putting on PPE, locking out power, using equipment guards, giving a safety talk, cleaning up spills, coaching others about their behaviors.

Consequences: Self-approval, reprimand, peer approval, penalty, feedback, injury, prize or trinket, inconvenience, thank-you, time savings, comfort.

Figure 16-3 shows an ABC model that explains why we do what we do. Figure 16-4 gives an indication that directions alone may not be sufficient to maintain behavior.

Adapted from French, Anne R., "Achieving a Total Safety Culture: Integrating Behavioral Safety into the Construction Environment," Safety Performance Solutions, Presentation June 30, 1999, Atlanta. Reproduced with permission.

work under "at-risk" conditions because of production pressures from supervision, procedural deficiencies, resource issues, etc. Refer to Table 16-1.

- Recognize employees. This is similar to an activity-based safety system. This system encourages employees to continue to use safe operating practices.
- Implement a management system that not only encompasses at-risk behaviors, but integrates this process into the management system so that it does not stand alone.

A common management system with a behavior component can consist of the following:

- Identifying and understanding safe behaviors as well as at-risk behaviors
- Conducting observations in areas where at-risk behaviors have occurred or are likely to occur
- Providing feedback on observations
- Collecting observation data
- Integrating behavior-based safety into the management system to identify and recommend solutions for employees at risk
- Identify management behaviors that may be systemic in nature
- Recognizing successes and achievement of the management system

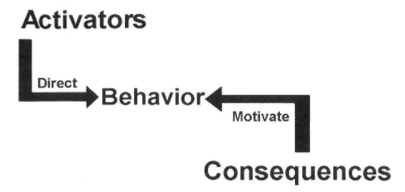

Figure 16-3 The ABC Model explains why we do what we do. French, Anne R., "Achieving a Total Safety Culture: Integrating Behavioral Safety into the Construction Environment," Safety Performance Solutions, Presentation June 30, 1999, Atlanta. Reproduced with permission.

Figure 16-4 Directions alone may not be sufficient to maintain behavior. French, Anne R., "Achieving a Total Safety Culture: Integrating Behavioral Safety into the Construction Environment," Safety Performance Solutions, Presentation June 30, 1999, Atlanta. Reproduced with permission.

UNDERSTANDING WHY EMPLOYEES PUT THEMSELVES AT RISK

Understanding at-risk behaviors helps in identifying and understanding what safe behavior looks like. What would you observe if an employee was doing a specific task safely? What would you observe if an employee was doing a task at-risk? Think about the effects of the pyramid that we discussed in Chapter 12. Figure 16-5 is presented here as a reminder why we must look at at-risk behaviors and near-misses to help reduce the probability of an incident. Without this information, when we develop safety rules and operating procedures, investigate incidents, conduct safety meetings, and train employees, we may not identify some at-risk behavior. This is true when we develop job hazard analyses. Safety issues occur when job descriptions or learners' checklists are vague and unclear, are not communicated in terms of what to do, and are not tied to the job hazard analysis.

We think that you will agree that everyone engages in thousands of behaviors per day. Think about what you do on a daily basis. Focusing on behaviors can reduce injuries (Figure 16-6). Consider the amount of risk that you create for yourself and others, for example, driving on a major highway moving from one lane to another in heavy traffic. It is important that you understand behavior categories that account for incidents (near misses or loss-producing events). Once you understand these at-risk behaviors, you can create awareness that, if done correctly, will help to minimize the chance of incidents. Figure 16-7 gives an overview of unsafe behaviors as the results of system influences.

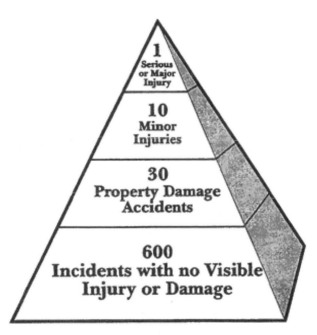

Figure 16-5 The 1969 U.S. Ratio Study. Bird, Frank E., George L. Germain, *Loss Control Management: Practical Loss Control Leadership*, revised ed., p. 5, Figure 1-3, Det Norske Veritas (U.S.A.), Inc., 1996. Reprinted with permission.

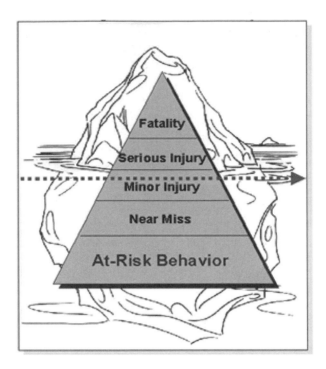

Figure 16-6 Focusing on behaviors can reduce injuries. French, Anne R., "Achieving a Total Safety Culture: Integrating Behavioral Safety into the Construction Environment," Safety Performance Solutions, Presentation June 30, 1999, Atlanta. Reproduced with permission.

THE COMPONENTS OF A BEHAVIOR-BASED SYSTEM

One key component of a behavior-based system approach is observing employees while they are performing their normal tasks. This process allows observers to quantify the frequency of safe behavior, recognize employees who are following safe work practices or at-risk behaviors, and provide feedback. In addition, you can identify at-risk behaviors where employees have no controls over their task, where they cannot do the job safely because of the way it was designed or using the tools and equipment provided. If you put behavioral objectives in easy-to-understand terms, you can approach employees and begin by discussing what they are doing safely. Once you have discussed the positive safety aspects, then you can discuss at-risk behaviors. At this point, you and the employee should agree or disagree on the observed risk. In most cases checklists are developed to help observers collect information on a co-worker's behavior. This checklist should be simple, easy to use, and give

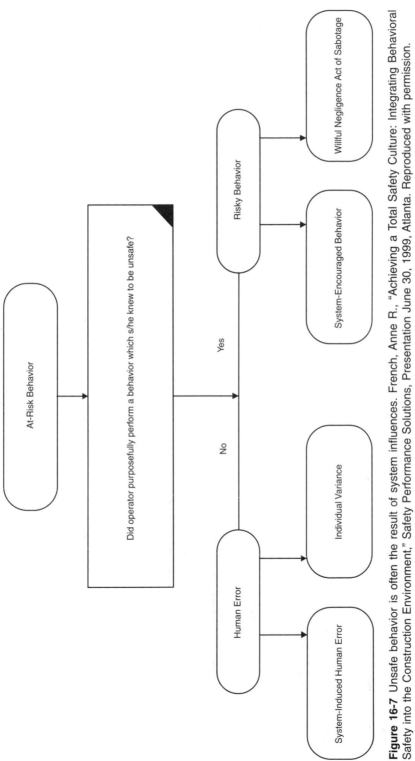

Figure 16-7 Unsafe behavior is often the result of system influences. French, Anne R., "Achieving a Total Safety Culture: Integrating Behavioral Safety into the Construction Environment," Safety Performance Solutions, Presentation June 30, 1999, Atlanta. Reproduced with permission.

measurable results. Using a site-specific checklist will provide consistent observations and a higher degree of success.

Success of any behavior-based safety process relies on trust among the employee (observers), management, and the employees being observed. Observers are trained to ask the employee for permission to make an observation where feedback is immediate. When this occurs, this is a win–win situation for everyone. One key to remember is when providing employees feedback; you need to consider how you would like to be approached. The way you say the words and how you act toward the employee are keys to successful two-way communication. Observers need specific training in how to observe and provide feedback in a positive, constructive manner.

Employees use "at-risk" behaviors for a number of reasons—for example, because of the way the job is designed, because of a lack of training, or because they are recognized for taking risks based on the safety culture and the management system. The management team members need to ask themselves the following questions:

- What safety management system improvements are needed?
- What changes can be made in the activators or consequences to increase the frequency of the desired safe behavior? Figure 16-8 reviews consequences that motivate behavior.

Through the behavior-based process, employees can be trained in the application of an effective management system. Figure 16-9 lists some principles for managing motivation.

Safety improvement teams and action planning are required elements of the behavior-based process to continually address at-risk behaviors. Team members should periodically review the percentage of safe versus the percentage of at-risk behaviors and other factors to understand where action planning is needed. In many cases, some of the at-risk behaviors can be shaped into safe behaviors through a process of regular observation and feedback.

MANAGEMENT SYSTEMS

According to Geller, in a total safety culture, everyone feels responsible for safety and pursues it on a daily basis. At work, employees will go beyond "the call of duty" to identify hazards and at-risk behaviors. Then

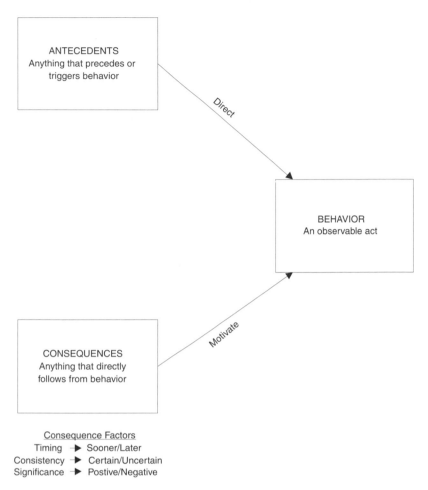

Figure 16-8 Consequences motivate behavior. Veazie, Bob, "Foundation Principles: Keys to Success of Behavior-Based Safety Initiatives," *Professional Safety*, April 1999, p. 26, Figure 4. Reproduced with permission.

they intervene to correct related hazards. According to Geller, a total culture requires continual attention to the three domains outlined in Figure 16-10. Safe work practices are supported with proper recognition procedures. As we have discussed, safety is not a priority to be shifted according to situational demands. Rather, safety is a value linked to all situational priorities. Geller refers to helping people as "actively caring." Figure 16-11 shows that actively caring behaviors can focus on environmental, personal, and behavioral aspects of safety. Geller describes in Figure 16-12 how the actively caring behavior model proposes that feelings of empowerment are enhanced by the perception of personal control [2].

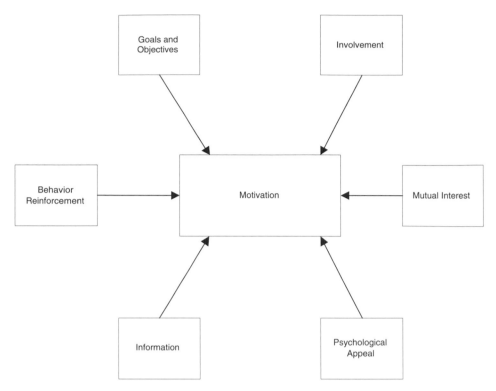

The Goals and Objectives Principle - Motivation to accomplish results tends to increase when people have meaningful goals toward which to work.

The Principle of involvement - Meaningful involvement increases motivation and support.

The Principle of Mutual Interest - Programs, projects, and ideas are best sold when they bridge the wants and desires of both parties.

The Psychological Appeal Principle - Communication that appeals to feelings and attitudes tends to be more motivational than that which appeals only to reason.

The Information Principle - Effective communication increases motivation.

The Principle of Behavior Reinforcement - Behavior with negative effects tends to decrease or stop.

Figure 16-9 Practical principles for managing motivation. Germain, George L., *FEBCO, Safety and the Bottom Line*, Chapter 7, "Beyond Behaviorism to Holistic Motivation," p. 231, Figure 7-2. Reproduced with permission.

It is important to develop a set of comprehensive principles on which to base safety procedures and policies. In the end, it is more useful to teach comprehensive principles. As the old saying goes, "Give a man a fish and you feed him for a day; teach him how to fish and you feed him for life" [2].

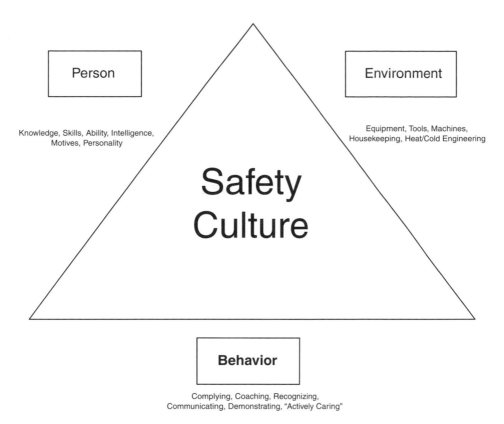

Figure 16-10 A Total Safety Culture requires continual attention to three domains. Adapted from E. Scott Geller, Ph.D., The Psychology of Safety Professional Development Conference and Exposition, San Diego, 1996, p. 10, and *Professional Safety*, September 1994, Figure 1, p. 19. Reproduced with permission.

SAFETY CULTURE CONCEPTS

Readers familiar with the writings of W. Edwards Deming and Steven R. Covey will recognize the similarity of each author's theory to safety [7]. If you have ever studied the successes of Dr. Deming, you will find that some of his philosophies can be adapted to safety quite well. According to Dr. Deming, 94 percent of the problems in business are due to lack of management commitment. In his book *Out of the Crisis*, Deming addresses 14 key points that all companies must work to enhance. These points can be aligned to a safety culture change for any organization. If you study his teachings in detail and apply some of his 14 points to safety,

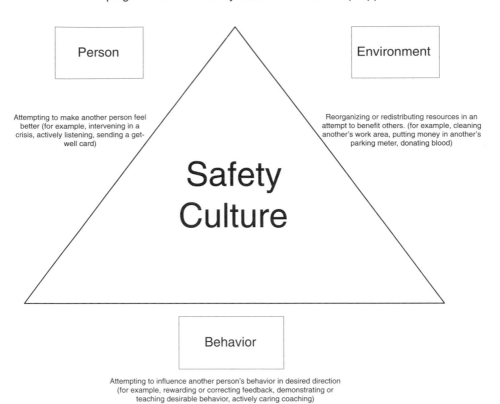

Figure 16-11 Actively caring behaviors. Adapted with permission from E. Scott Geller, Ph.D., *Professional Safety*, September 1994, Figure 2, p. 21. Reproduced with permission.

you will start to understand some basic concepts that can be utilized on a daily basis [7].

It is important to note that not all 14 points may fit the safety culture or philosophy of your organization. One must adopt a safety culture that fits an organization's specific needs. Figure 16-13 compares Dr. Deming's 14 points with a safety culture change. One of the problems is that people forget his "seven deadly sins" [7].

The 14 points constitute a theory of management. Applied properly, they will transform your style of management. Unfortunately, deadly diseases stand in the way of transformation. We try here to understand their deadly effects. Alas, cure of some of the diseases requires complete shakeup of management style (fear of takeover, for example, and short-term profit) [7].

There are diseases and there are obstacles. The distinction is intended to be partly in terms of difficulty of eradication and partly in terms of severity of the injury inflicted [7].

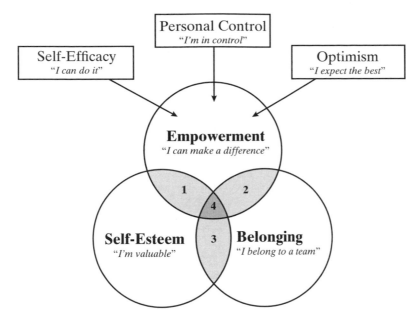

1. I can make valuable differences.
2. We can make a difference.
3. I'm a valuable team member.
4. We can make valuable differences.

Figure 16-12 Five person factors. Adapted from E. Scott Geller, Ph.D., The Psychology of Safety Professional Development Conference and Exposition, San Diego, 1996, p. 10, and *Professional Safety*, September 1994, Figure 3, p. 23. Reproduced with permission.

DISEASES AND OBSTACLES

Deadly diseases afflict most organizations. Briefly described, the deadly diseases are:

- Lack of constancy of purpose to plan products and services that will have a market, keep the company in business, and provide jobs
- Emphasis on short-term profits: short-term thinking (just the opposite of constancy of purpose to stay in business), fed by fear of unfriendly takeover, and by pushes from bankers and owners for dividends
- Evaluation of performance, merit rating, or annual review
- Mobility of management; job hopping
- Management by use only of visible figures, with little or no consideration of figures that are unknown or unknowable

#	Deming Quality Points	Safety Culture
1	Create constancy of purpose for improvement of product and service.	Create a new culture where top management is committed to zero incidents. Provide for continuous improvement in the safety system to minimize cost from workers' compensation and other related incident expenses.
2	Adopt the new philosophy.	Adopt the new philosophy that all lost time incidents are preventable. Zero incidents is the only target or goal to achieve! Do not accept anything less. Convey this message to all employees.
3	Cease dependence on inspection to achieve quality.	Eliminate facility inspections to determine compliance to standards. Utilize demonstrated management commitment and audit the management system. If management is committed, safety compliance will be automatic. It is management's job to continually review the system.
4	End the practice of awarding business on the basis of price tag alone. Instead, minimize total cost by working with a single supplier.	Do not use contractors or subcontractors that have poor safety records. Develop and implement a safety prequalification system to select the right contractor. Do not reward or recognize managers that have completed a job on time and under budget when the job had an incident as it related to safety.
5	Improve constantly and forever, every process for planning, production, and service.	Seek continuous improvement in policies and procedures utilizing employees as well as management as peer review team members. Identify ways to reduce costs and maintain appropriate health and safety measures.
6	Institute training on the job.	Provide required safety training for the job. Commit the necessary time, funding, and resources. Invest in training for ergonomics for office workers and hazard recognition training for the rank and file.
7	Adopt and institute leadership.	Ensure that there is executive level leadership for safety. Once this is in place, upper level managers must then coach and guide employees on what is the proper attitude toward safety.
8	Drive out fear.	Eliminate the circle of mistrust. Employees must be encouraged to report all incidents, near misses, and close calls, no matter how minor.
9	Break down barriers between staff areas.	Ensure that there is both employee participation and demonstrated management commitment. All departments must work together for a single purpose and build upon this relationship as a partnership for safety excellence.
10	Eliminate slogans, exhortations, and targets for the work force.	Do not rely on safety signs or contests. Recognize that incentives are short-term improvements and tend to breed competition among peer groups. At the same time, recognize that signs and contests can be productive and useful in a mature safety culture.
11	Eliminate numerical quotas for the work force and numerical goals for management.	Eliminate posting goals for the employees. Keep this as an internal measurement exclusively for management. Review trends regularly.
12	Remove barriers that rob people of pride of workmanship. Eliminate the annual rating or merit system.	Provide a rating system that promotes continual positive reinforcement. Do not wait for the year-end to discuss safety performance. It must be an ongoing process.
13	Institute a vigorous program of education and self-improvement for everyone.	Plan for safety education and cultural growth of the zero incident concepts at all levels of the organization.
14	Put everybody in the company to work to accomplish the transformation.	Put everyone in the company to work to achieve a change in the safety mind set. Create safe work practices or safety procedures for the entire operation. Make things consistent. Work toward a common vision of zero incidents. Promote a positive safety culture change. (J. Roughton, J. Mercurio)

Figure 16-13 Deming's quality versus culture change.

#	Covey's Seven Habits	Safety Culture
1	Be Positive	Becoming involved in the safety culture breeds safety awareness. Employees must be encouraged to report unsafe acts and conditions and encourage correcting these conditions immediately. Be proactive, not reactive.
2	Begin with an End in Mind	Create a new culture where top management is committed to zero incidents. Ensure continuous improvement to help minimize cost of workers' compensation and other related incident expenses. Adopt the new philosophy of zero incidents. Do not accept anything less.
3	Put First Things First	Using the Dupont theories of making Safety equal to Quality, Cost, Productivity, and Morale, you have to put all of the key items in first place.
4	Think Win-Win	Every time we prevent an incident, everyone wins. No one benefits from an injury or even worse, a lost workday case.
5	Seek First to Understand then be Understood	The way most of us operate is the way that we know best, feel comfortable with, and is the easiest way for us. We all have perceived notions and must adapt to change. We must listen to what the employee is saying and make our minds up to what is right.
6	Synergize	Create policies and procedures for the entire company. Make the programs consistent where everyone gets the same safety message. Work toward a common vision, one that involves the zero incident concepts.
7	Sharpen the Saw	The sharper the saw blade, the easier the cut. The more action and involvement by management, the easier the transformation to zero incidents. This is vital in the safety and quality culture. We must continually reinforce the importance of demonstrating management commitment by visual actions.

Figure 16-14 Covey's seven habits versus safety culture [6]. Sitek, Greg, "Seven Habits of Safety," *Equipment World*, August 1997, used with permission.

- Excessive medical costs
- Excessive costs of liability, swelled by lawyers who work on contingency fees [7]

The bottom line is that if you watch out for the deadly diseases, your management system will not fail.

On the other hand, according to Steven Covey, "Seek first to understand, before being understood" [6]. This is his fifth habit of highly effective people. You can review the seven habits of highly effective people and blend and align the concepts with the zero incident concept. Figure 16-14 compares Dr. Covey's seven habits and a safety culture [6].

RISKY BEHAVIOR

Rewards for risky behavior mean you are likely to take more chances. As you gain work experience, you often master dangerous shortcuts from other seasoned employees. Since these at-risk behaviors are not followed by a near miss or injury, they remain unpunished, and the at-risk behavior persists until an injury occurs [2].

One method to instill a safety culture into a management is to start group meetings, by asking employees to discuss something that they have done for safety since the last meeting (a short safety topic at each meeting). As we have discussed, this will help you to get buy-in into the process. Because of the shared safety knowledge (awareness), safety is given a special status and integrated into the overall business agenda [2].

Figure 16-15 compares Maslow's motivation theories to safety. Maslow's hierarchy of needs as illustrated (the order of needs that people want) shows how people feel. Note that there is a comparison to the self-actualization safety approach [2, 8].

The challenge in achieving a total safety culture is to convince everyone to have the responsibility to intervene for safety. A social norm or expectancy must be established in everyone to share equally in the responsibility to keep everyone safe. Through training, employees learn how to translate principles and rules into specific behaviors. If people are motivated to maximize positive consequences and minimize negative consequences, actively caring behavior will only occur if perceived rewards outweigh perceived costs [2].

PRINCIPLES OF BEHAVIOR-BASED PSYCHOLOGY

A prime principle of behavior-based psychology is that it is easier, especially for large-scale cultures, to "act a person into the safe thinking" than it is to address attitudes and values directly in an attempt to "think a person into safe acting." Another key principle of behavior-based psychology is that the consequences of our behavior influence how we feel about the behavior. Generally, positive consequences lead to good feelings or attitudes and negative consequences lead to bad feelings or attitudes [2]. Refer to Figure16-16 for some principles of behavioral-based safety.

Supporting the actively caring model as detailed in Figure 16-11, employees with higher self-esteem help others more frequently than those scoring low on a self-esteem scale. Long-term behavioral changes require employees to change "inside" and "outside" [2].

Self-actualization needs, for development of capability to the fullest potential.	First Level	**Self-actualization safety approach** Creative safety Participative (by employees) safety Highest order achievement
Esteem needs, for strength, achievement, adequacy, confidence, independence and for reputation, recognition, attention, importance, etc.	Second Level	**Esteem safety approach** Achievement orientated safety Participative (by employees) safety Safety suggestion schemes - recognition for contributors Safety committees (employee-orientated) Safety surveys (point out good and some poor points)
Belongingness needs, receiving and giving love and affection	Third Level	**Belongingness and love safety approach** Soft sell safety (for example, identify safety with family and love) Safety committees Safety suggestion scheme Recognition of individual safety awareness Recognition of group safety awards Positive safety
Safety and security needs, in the sense of stability, dependency, protection, freedom from fear, anxiety, etc.	Fourth Level	**Safety and security safety approach** Safety incentive plans (for example, money) Hard sell safety (for example, blood and gore advertisements) Negative safety audits (find something wrong) Physical guarding rules and regulations Scare procedures
Physiological needs (drives) - for example, hunger, thirst	Fifth Level	**Physiological safety approach** Threat of termination for violation of safety rules Hard sell safety (for example, blood and gore advertisements) Scare procedures

Figure 16-15 Comparison of Maslow's motivation theories to safety. Cox, S, Tom Cox, *Safety Systems and People*, Table 7.2, p. 166, 1996. Modified with permission.

As Geller states, in a total safety culture, employees actively care on a continuing basis for safety. To achieve a total safety culture, employees must believe they have personal control over the safety of their organization [2].

One key is to reduce the probability that the learner will make an error and feel lowered effectiveness or self-efficacy. Celebrating accomplishments in a small way builds self-efficacy and enables support that averts a self-fulfilling prophecy [2].

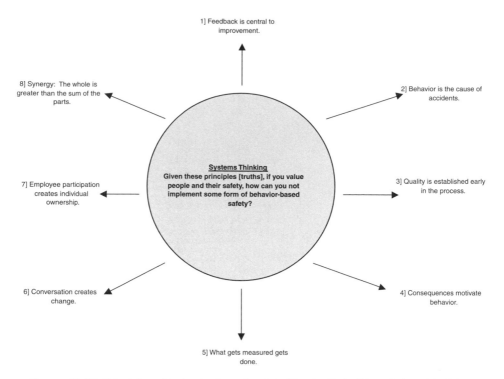

Figure 16-16 Principles of behavior-based safety. Veazie, Bob, "Foundation Principles: Keys to Success of a Behavior-Based Safety Initiative," *Professional Safety*, April 1999, p. 26, Figure 5. Reproduced with permission.

When we see a consequence consistent with decisions we have made from among our choice opportunities, we increase our trust in the people who gave us the power to choose. We have gained confidence and ability to take personal control of the situation. A kind (positive) word can go a long way [2].

As we saw in Figure 16-14, Dr. Covey reminds us that we all come into the world dependent on others to survive. As we mature and learn, we reach a level of independence where we strive to achieve success or avoid failure. However, higher levels of achievement and quality of life are usually reached after we develop a perspective of interdependence and act accordingly. Figure 16-17 shows that a total safety culture requires a shift from dependence to interdependence. According to Dr. Covey, action to achieve independence means we actively listen (habit 5: "Seek first to understand, then be understood") and develop relationships or contingencies with people that reflect positive outcomes for everyone involved (habit 4: "Think win/win"). Consistently practicing these habits leads to synergy (habit 6) [2, 6].

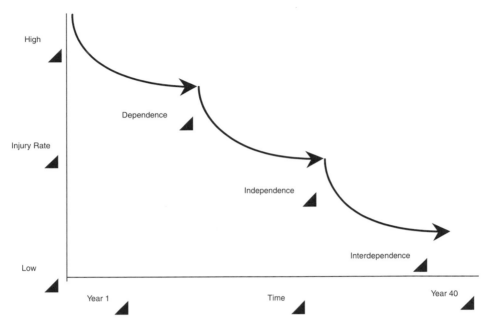

Figure 16-17 A total safety culture requires a shift from dependence to interdependence. French, Anne R. "Achieving a Total Safety Culture: Integrating Behavioral Safety into the Construction Environment," Safety Performance Solutions, Presentation June 30, 1999, Atlanta. Reproduced with permission.

EMPLOYEE ACTIVITIES

If employees can link their daily activities to safety results, then celebrating a reduction in injury rate can be useful, even motivating. However, it is critical to recognize behaviors, procedures, and processes that lead to fewer injuries or lower workers' compensation costs [2]. Refer to Figure 16-18 for an overview of incident causes.

As we have discussed, changes in an environment and behaviors can be assessed directly through systematic observations; changes in knowledge, perceptions, police, attitudes, and intentions are only assessable indirectly through survey techniques, usually questionnaires [2].

AWARENESS APPROACH TO BEHAVIOR MANAGEMENT

Although the goal of many safety programs is to change behavior, it should be emphasized that unwanted behavior will not change on a

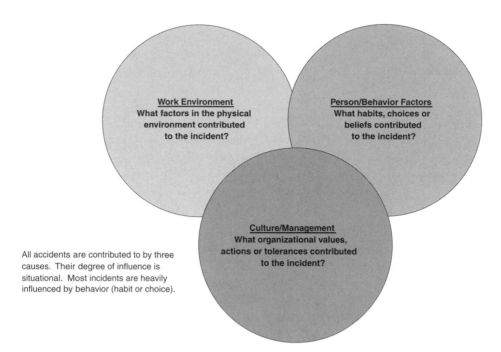

All accidents are contributed to by three causes. Their degree of influence is situational. Most incidents are heavily influenced by behavior (habit or choice).

Figure 16-18 Incident causes. Source: Veazie, Bob, "Foundation Principles: Keys to Success of a Behavior-Based Safety Initiative," *Professional Safety*, April 1999, p. 26, Figure 3. Reproduced with permission.

lasting basis unless the underlying beliefs, attitudes, and values that cause and support that behavior in the first place are addressed (see Figure 16-14). A multidimensional behavior modification approach is required for lasting improvements. This approach includes three primary types of behavior modification strategies that interact to uncover the beliefs, attitudes, and values that drive unsafe behavior, while simultaneously increasing individual responsibility: self-mastery; self-management of safety behavior; and self-observation skills, which are essential to monitor the unconscious human mechanisms, such as daydreaming and distractibility, that can cause incidents even when employees are wearing PPE or following safety procedures. Awareness and self-observation are also essential for an individual to discover self-defeating, unsafe beliefs, attitudes, and behaviors. Topf calls this the awareness approach to behavior management, and includes among others, three key behavioral components:

- *Cognitive behavior management.* Understanding how beliefs, attitudes, values, thoughts, and other aspects of people's thinking processes contribute to both "automatic" and calculated behaviors that are unsafe. This includes the way people listen to the instructions of others and how they interpret that listening. Attentional skills are essential for people to focus and stay present. For example, as we discussed, recall the common exercise where you are in a group and the instructor starts on one side of the room and provides a written statement to the first person. This first person reads the statement and puts it into memory. He/she then tells the next person, who tells the next person, etc. What happens when it gets back to you? The message is usually so off base that it is not funny. This is a good example to show how important communication is in the safety arena.

- *Reality behavior management.* Dealing with current behaviors and their consequences to achieve a shift to personal responsibility and self-management that leads to realistic, corrective action plans.
- *Management by commitment.* Once individuals realize that they are inherently committed to their own safety, and that management shares and supports that commitment (demonstration), they are motivated by the mutual benefit to themselves and the company to act in a safe manner and in greater cooperation with safety requirements. They are also motivated to be safe in off-the-job situations [3]. Refer to Chapter 1 for the cost of an incident.

Programs that focus on employee awareness, attitudes, and thinking are effective tools for significantly reducing incidents on and off the job. These programs provide a substantial return on investment in terms of employee productivity and, ultimately, improve an organization's competitiveness [3].

ASSESSING THE ORGANIZATIONAL CULTURE

Human beings are motivated by positive and negative forces; by rewards and punishments. When it comes to safety there is no compromise. Although employees still need to be motivated, the motivation for safe behavior ultimately needs to be self-generated, arising from the commitment of each employee to remain safe. This commitment extends to co-workers, families, and the community [3].

Employees need to be invited and encouraged to examine their own thinking and their own commitments about safety. They need to focus on issues of responsibility and accountability and need to understand that they have the power and ability to cause their own safety, rather than being victimized by a lack of personal attention to safety procedures and practices [3].

The process to assess the safety culture is to determine the collective attitudes (perception) about safety. (Safety perception surveys are discussed in Chapter 17.) Confidential interviews are conducted with all levels of employees and management (see the discussion of independent reviews in Chapter 17). A strategic action plan model is developed where a new course is charted for instilling safety into the organization. The point is to design a new culture that is consistent with management and employees' perception of the safety objectives [3].

In most organizations today, safety is the responsibility of the safety director. In these cases, the safety professional is sometimes known as the "safety cop," because of lack of management commitment. In most current safety cultures, the safety director must compete with an existing, yet unspoken, agreement among employees that states, "If you don't call me on my mistake, I won't call you on yours," or, "It's not my responsibility, it's someone else's." Even in organizations with larger safety staffs, the job is difficult, sometimes because of lack of management commitment. In the past decade, the culture of many organizations has promoted maximum production through increased efficiency, often at the expense of safety: employees are injured trying to maintain production levels [3].

To increase efficiency, many companies adopted the learning theory's stimulus/response model of behavior modification, where the employees who worked the fastest and produced the most results were rewarded with raises, promotions, or other positive reinforcers [3]. This relates to our discussion on rewards and doing a job faster to keep being rewarded.

As we discussed, in this model, employees learned that working quickly was the preferred method for increasing efficiency, even though they may have started taking safety shortcuts, often putting their lives, or the lives of others, at risk. In effect, they were conditioned to believe that shutting down equipment, even in the name of safety, could be cause for being fired or reprimanded [3].

According to Topf, safety is instilled in the culture of an organization as a key value and a key requirement. The objective is to alter attitudes and behaviors about safety and to cause a long-lasting cultural shift in the organization. In this model, everyone in the organization is a supporter of safety from top management to the employee who sweeps the floors. It is total employee participation in the process that, when a shift of the culture occurs, causes the conversation among employees to shift to, "I observe and support you in being safe; you observe and support me." A team effort is established with employee participation [3].

What are the advantages of this method? Increased efficiency and safety occur in the same model, for example, increased efficiency is achieved not only by working faster, but also by practicing stringent safety methods and working smarter. Employees begin to think about safety in such a way that they foster an attitude change about the value and importance of safety on and off the job. Through an established safety program that is supported by all levels of management and employees, employees develop a sense of personal responsibility for safety performance and begin to associate safety with increased efficiency and production [3].

Steps that must be taken to make sure of safe actions—human nature often causes people to take risks or act without regard for their safety. The key steps to safe actions and behaviors are shown in Figure 16-19.

According to Topf, going away, even if just for a moment, means inattention, daydreaming, not listening to instructions, and losing focus while working on repetitive or familiar activities. Lack of awareness means not remaining alert to hazardous situation at all times. Lack of focus means difficulty concentrating on tasks that increases the probability of incidents. Failure to distinguish means the inability to see, hear, and smell potential safety hazards as a result of irrational thinking, conscious or unconscious attitudes, and behaviors, resulting in decreased effectiveness

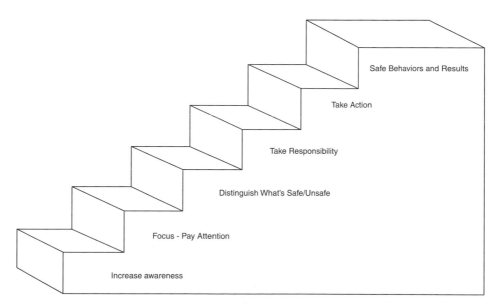

Figure 16-19 Key steps to safe actions and behaviors. Topf, Michael D., *Safety and Employee Involvement, Making Your Company SAFOR*, 1998, p. 60, Figure 2. Reproduced with permission.

in taking necessary actions to remain safe. Lack of responsibility means not responding appropriately when safety demands immediate action, because of a lack of understanding or acceptance of how to manage personal attitudes that may support or undermine the appropriate response for safety. Resistance to changing behavior means resisting certain safety behaviors and feedback from others because of experiences, beliefs, and attitudes regarding safety. Habits are not changed as needed for safety because of a lack of awareness of what is really at stake. Ignoring early warning signs (the mental alarm system or "inner voice") or other signals that may be warning you of impending danger and the necessary actions to take to avoid incidents and near misses will cause a system to fail [3].

A characteristic of any concept is its emphasis on raising employees' awareness as a means to reduce incidents caused by a lapse in judgment or "going away" [4]. As discussed, training focuses on both the emotional and intellectual aspects of paying attention.

According to Theune, one method states that despite its roots in cognitive behavior methodology, the process is a proven and pragmatic approach that incorporates a number of theories, including behavior-based change. Companies have recognized that incremental and short-term results are no longer enough. They seek change on a large and lasting scale change that occurs on an individual, group, and cultural level [4].

CORE OF THE PROCESS

At the core of this process are many essential steps:

- *Assess*. Survey the culture to determine prevailing attitudes, beliefs, and behaviors related to safety and the environment by way of confidential questionnaires (perception survey), interviews (independent review process), focus group meetings, and visits to assess workplace conditions and possible contributions to incidents [4]. We will cover this in more detail in Chapter 17.
- *Train all employees*. Training focuses on what causes people to become distracted, to take risks, or to act unsafely, even when they know what they should do. It also addresses the underlying "human mechanisms" that cause people to place themselves at risk. Leaders are trained in communication, empowerment coaching, behavior management, and leadership skills [4].

- *Involve employees (employee participation)*. Refer to Chapter 7, "Employee Participation." Teamwork, acceptance, participation, positive buy-in, and problem solving skills are the desired results [4]. Refer to Chapter 14 for training requirements.
- *Reinforce*. Ongoing reinforcement for all employees keeps key concepts fresh and alive [4].
- Offer ongoing support: As we have discussed, management leaders must make the transition from "safety cops" to "safety coaches," mentors. They must support employees' safe behaviors and bring the process back into alignment if it veers off course [4].
- *Customize observation and review*. As we have discussed, management, supervisors, and employees must work together to develop meaningful processes for observation and feedback, support, and empowerment, to include activity-based measurements [4].

The visual image that best illustrates the value of the process is a bridge under construction. Figure 16-20 shows a bridge to a successful safety culture. Theune states that the land forms to be connected are knowing (the safe processes and procedures that employees have been taught) and doing (the execution of that knowledge at appropriate times in appropriate circumstances). We bridge the gap by creating strong, self-sustaining pilings we identify as awareness, skills, structure, and support. The pilings are spanned by improved attitudes, trust, communication, and participation [4].

If we take another look at another bridge from the loss control side, you will see similar elements. Figure 16-21 depicts a bridge that will take you from the traditional method of injury management (injury-oriented safety program) to the proactive state (accident-oriented safety program). Many managers have bridged the gap—they have put to practical use the

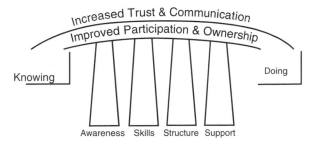

Figure 16-20 A culture change. Topf, Michael D., *Making Your Company SAFOR*, The TOPF Organization, King of Prussia, Pennsylvania, 1998, p. 4. Reprinted with permission.

Figure 16-21 Damage control bridges the gap. Bird, Frank, George Germain, *Loss Control Management: Practical Loss Control Leadership*, revised ed., Det Norske Veritas (U.S.A.), Inc., 1996, Figure 15-3, p. 346. Reprinted with permission.

definition of accident we saw in Chapter 12 and have developed safety programs that support the management system.

SUSTAINING THE CHANGE

Management appreciates built-in reinforcements to sustain growth and change, transforming short-term benefits into long-term cultural changes. Another benefit is consistency. Yet another means of sustaining the success is the success itself. Significant reductions in OSHA recordable incidents, lost time rates, near misses, and workers' compensation costs create a sense of shared purpose [4]. Refer to Figure 16-22 to understand how injuries occur.

To put into the proper perspective, the following was summarized and modified to fit this text from an article written by Donald H. Theune [9]. Note that the following is a modified quote from this approved source.

A new driver starts to drive a car. He/she drives with great care (cautious and alert) for a couple of weeks. One day this person glides through a yellow light, feeling slightly guilty, but relieved that the incident did not cause harm to anyone. The second time this person runs the yellow light with the same consequence, they feel much differently. However, over time, this habit becomes a common practice and, eventually, the urge to hit the accelerator as the light turns from green to yellow becomes a conditioned response. "I beat the system." "I took a risk and didn't get hurt." These are the thoughts running through his or her mind. As we described before, the driver sees a bad accident, but remains convinced, "Accidents

- Peer behavior will be characterized by employees looking out for one another's safety and working together as a team.
- Leaders will be eager to coach, counsel, and inspire employees to act and work safely. The environment they create will be one where employees actively participate in the safety process and consider the process their own.
- Managers will participate proactively and empower teams to identify and solve issues before they become safety problems. They will involve all levels of employees in a positive safety culture that emphasizes the "we" over the "us vs. them" [9].

According to Theune, former National Safety Council President and OSHA administrator Jerry Scannell put it well when he observed: "It is difficult for a company to be healthy and productive if its workers are not." When we reach employees at the deepest level, they "feel" the value and benefit of working safely, not because they seek to adhere to company policy, not because they seek to avoid citations for their employer, but because they recognize that they are not immune from accidents [9].

The myth: "Accidents don't happen to me" has been replaced by the belief that individuals are responsible for their own well-being. Underlying it all is a closely held value that says working and living safely is worthy of everyone's attention [9].

Resulting cultural changes also help carry the process forward and create other lasting and positive changes. Some of the additional benefits can include improved attitudes, increased trust, better communication, and increased participation that boost quality, productivity, and morale. Employees perceive, rightly, that they are respected, that their input carries weight, and that they have more control over their own destiny. They have more trust in management and fewer issues [4].

According to Theune, there are three other built-in systems to help make sure that there is continued employee participation and ongoing improvement:

- The Agreements Process addresses rules on an ongoing basis that are not clear, are not being followed, or require updating
- The Breakdown Process addresses system and policy issues that stand in the way of achieving safety excellence
- The Customized Observation and Review Process provides measurement and feedback on the organization's progress in supporting safe attitudes and behaviors, in safety problem identification and resolution throughout the process period [4]

SUMMARY

From where we sit at the beginning of the 21st century, it is axiomatic that the "problem" we face in safety is not wayward employees who spitefully place their hands in unguarded machinery or purposefully pollute the environment. What we face is a set of larger, harder to fix woes that stem from a misalignment of attitudes and behaviors [4].

Past articles written by notable psychologists have promoted behavior modification as a technique to achieve success in safety. This approach follows the belief that by changing employees you will eliminate injuries.

The behavior modification technique is a valuable method to use for industries with well-defined organizations (such as a plant or manufacturing setting). For those companies with a diversified and constantly changing workforce, the idea of instituting a zero-incidents concept may be more readily absorbed. For industries faced with turnover and cultural norms that interfere with safety performance, a safety movement that follows the zero-incidents concept may be right for you.

Aside from behavior-based safety, in a zero-incidents culture, one focuses on real-time issues. Nobody should ever think that it is OK to suffer a disabling injury while at work or home. It is up to management to convince the skeptics that zero is a reachable vision, a reality.

This goes to show that everyone has a different way of solving problems. We all learn by trial and error. Sometimes we get things to work right the first time; other times we do not. We must always strive for the best, always look for proven methods, and avoid "reinventing the wheel." The management system must always be adaptable to enable continuous improvement. Any company that institutes a cultural change toward the zero-incidents concept is bound to see safety improvements that the entire workforce can be proud of.

One element of safety management is to look at the behavior of employees and the organizational culture. Everyone has a responsibility for safety and should participate in management system efforts. Modern organization safety has progressed from "safety by compliance" to a more appropriate concept of "prevention by planning." Reliance on compliance could translate to after-the-fact hazard detection that does not identify organizational errors that are often the contributors to incidents [5].

Behavior modification is not new, although some would lead you to believe differently. The basic process involves systematically reinforcing positive behavior while at the same time ignoring or exercising negative reinforcements to eliminate unwanted behavior. There are two primary approaches used in behavior modification programs. One is an attempt to eliminate unwanted behavior that detracts from attaining an organizational goal, and the other is the learning of new responses. In safety, the objective is to eliminate unsafe acts (at-risk behaviors) [1]. The second major goal of behavior modification is to create acceptable new responses to an environmental stimulus [1]. Figure 16-23 gives an overview of the route to total commitment and the reinforcement cycle that maintains it.

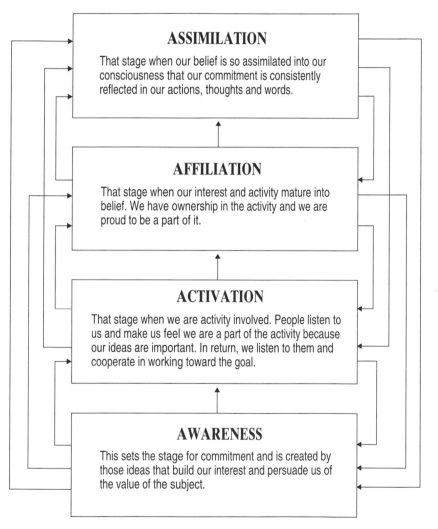

ASSIMILATION

That stage when our belief is so assimilated into our consciousness that our commitment is consistently reflected in our actions, thoughts and words.

AFFILIATION

That stage when our interest and activity mature into belief. We have ownership in the activity and we are proud to be a part of it.

ACTIVATION

That stage when we are activity involved. People listen to us and make us feel we are a part of the activity because our ideas are important. In return, we listen to them and cooperate in working toward the goal.

AWARENESS

This sets the stage for commitment and is created by those ideas that build our interest and persuade us of the value of the subject.

Figure 16-23 The route to total commitment and the reinforcement cycle that maintains it. Germain, George L., *FEBCO, Safety and the Bottom Line*, Chapter 7, "Beyond Behaviorism to Holistic Motivation," pp. 224, Figure 7-4. Reproduced with permission.

The basic concept of behavior modification is the systematic use of positive reinforcement. The result of using positive reinforcement is improved performance in the area where the positive reinforcement is connected [1]. If a person does something and immediately following the act something pleasurable happens, he/she will be more likely to repeat that act. If a person does something and immediately following the act something painful occurs, he/she will be less likely to repeat that act again or will make sure he or she is not caught in that act next time [1].

According to Peterson, positive reinforcement is gaining wider and wider acceptance in industry. It has been used to increase productivity and quality, improve labor relations, meet Equal Opportunity Objectives (EEOs), and reduce absenteeism [1].

Safe behavior reinforcement is nothing more than recognizing employees when they do a good job at the time they are doing it. Recognition is the most underused management technique we have, and it is the most powerful. Some organizations provide a means of recognizing good safety performance.

However, at almost every company, perception surveys measure recognition as the lowest-rated category [1]. This is a reflection of the old, traditional management style. Many companies think of safety awards, contests, and trinkets as recognition. At best these might be giveaways or what you might call awards, but they have little to do with recognition. Recognition refers to whether workers feel like they are being told that they are doing a good job. It refers to the one-on-one interpersonal recognition that is so often missing in a traditional management system [1].

Forward-looking employers are unequivocal in their belief that safety responsibility must become fully integrated with other business processes.

REFERENCES

1. Peterson, Dan, *The Challenge of Change, Creating a New Safety Culture, Implementation Guide*, CoreMedia Development, Inc., 1993, Implementation Guide, "Recognition for Performance," pp. 57–58. Modified with permission.

2. Geller, Scott E., *The Psychology of Safety: How To Improve Behaviors and Attitudes on the Job*, Chilton Book Company, Radnor, Pennsylvania, 1996.

3. Topf, Michael D., *Making your Company SAFOR, Safety and Employee Involvement*, 1998, p. 56. Reprinted with permission.

4. Theune, Donald H., *Making your Company SAFOR, A Comprehensive Holistic Behavioral Safety Process, How It Works and Why (Safety: A Function of Responsibility)*, 1998, pp. 3–6. Reprinted with permission.

5. *FAA System Safety Handbook*, "Principles of System Safety," Chapter 3, p. 15, December 30, 2000, public domain.

6. Covey, Stephen, *The 7 Habits of Highly Effective People*, A Fireside Book, Simon and Schuster, 1990.

7. Deming, W. Edward, *Out of Crisis*, MIT Center for Advanced Engineering Study, 1986.

8. Cox, Sue, Tom Cox, *Safety Systems and People*, Butterworth–Heinemann, Linacre House, Jordan Hill, Oxford. Reprinted with permission.

9. Theune, Donald H. *The Safety Myth: "It Won't Happen to Me," Making Your Company SAFOR*[SM], written and edited by Michael D. Topf, 1998. Reproduced with permission.

Part 4
Measuring the Safety Culture

17

Safety and Health Program Evaluation: Assessing the Management System

INTRODUCTION

After you have developed and implemented your management system and related programs, it is time to test the system to see if it is working. We can do this by developing measurable goals and objectives—for example, work procedures; management, supervision, and employee activities; perception surveys; self-assessment; and independent reviews. To be successful these goals must include resources to achieve the results. This must include employees, managers, and supervisors [6].

A safety management evaluation takes a look at the management system to help you understand how it is working and how it can best be continually integrated into the management process. Therefore the ability to asses, analyze, and evaluate your management system is important to make sure that the highest success can be achieved.

As we have discussed, employee participation (refer to Chapter 7) is one of the most important elements in an effective management system. It will help to determine if the management system is working effectively and efficiently. However, you must realize that all elements are equally important. Therefore, all systems that support your safety program must be reviewed—for example, management leadership, hazard prevention and control programs, incident investigations (near misses and other loss-producing events), and training [1, 2].

There is another caution: measuring safety performance with only records of injuries not only limits evaluation to a reactive stance, it also sets up a negative motivational system that is apt to take a backseat to the positive systems that are used for productivity and quality [4].

We would like to set the record straight here: management system reviews must go beyond the typical plant inspections or audits. Inspections are valuable for reviewing a facilities process and/or individual jobs to identify and eliminate or control hazards. In addition, audits can be useful for identifying specific program activities that will help to determine if the objectives of the selected activity have been met [1, 2] but are limited when it comes to reviewing a management system. Basically, what we want to do is look at how management is functioning around these elements.

A management review system is concerned with a larger question: making sure that the quality of the management system and the safety programs is measurable. Therefore, the program review looks at the systems created to implement the safety program. It asks if these systems are working effectively and efficiently [6].

THE NATURE OF ALL SAFETY SYSTEMS

The key to remember is to fix the system and not the blame. Many of us lose sight of this because we are sometimes in the reactionary mode. We need to learn from the errors in the system and either stay the course or adjust the system to fix the need [3]. Figure 17-1 shows how systems combine to make up an organization.

Refer to Table 17-1 to understand the nature of all safety systems. This table provides you with a simple overview of how the inputs, processes,

within systems...
within systems...
within systems...

Figure 17-1 Organizations are composed of systems. OR-OSHA Web site, http://www.cbs.state.or.us/external/osha/educate/training/pages/materials.html, OROSHA 116, Safety and Health Program Evaluation, Rev. 1/00 dig., public domain. Modified with permission.

Table 17-1
Nature of All Safety Systems

Inputs: Act as antecedents or activators—clarify the process

- Standards = Vision, mission, objectives, strategies, policies, plans, programs, budgets, processes, procedures, appraisals, job descriptions, and rules
- People = Management and employees
- Resources = Tools, equipment, machinery, materials, facilities, and environment

Process: The steps in the transformation of inputs into outputs.

- Procedures = Steps in a process
- Consequences = Natural (incidents) and system (reward/discipline)

Outputs: Results of the process.

- Products and services = The intended system output
- Conditions = Objects/employees in the workplace. The "states of being" in the physical and psychosocial workplace environment
- Behaviors = Action of people taken at all levels of the organization. What they do or do not do in the workplace affects safety. Refer to Figures 17-2 and 17-3 for more on the input, outputs, and consequences.

OR-OSHA Web site, http://www.cbs.state.or.us/external/osha/educate/training/pages/materials.html, OR-OSHA 116, Safety and Health Program Evaluation, Rev. 1/00 sig., public domain. Modified with permission.

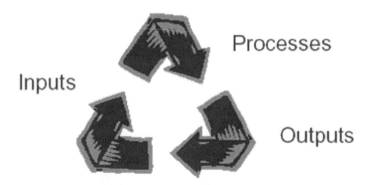

Processes

Inputs

Outputs

Figure 17-2 Safety and health program evaluation: organizations are composed of systems. OR-OSHA Web site, http://www.cbs.state.or.us/external/osha/educate/training/pages/materials.html, OR-OSHA, Safety and Health Program Evaluation, Rev. 1/00 dig., public domain. Modified with permission.

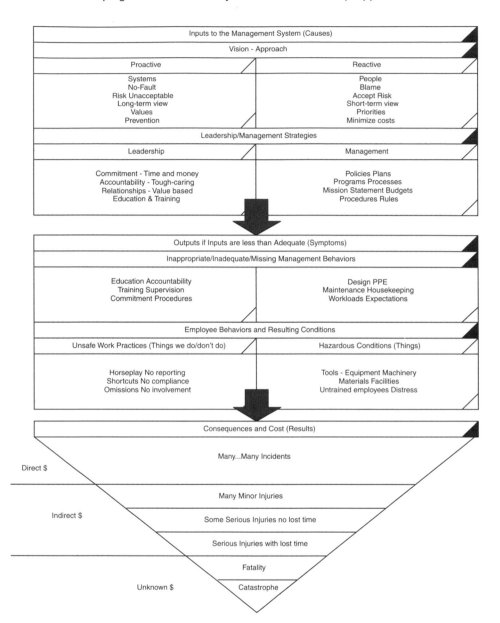

Figure 17-3 The big picture: inputs, outputs, consequences. OR OSHA 119, p. 29, http://www.cbs.state.or.us/external/osha/educate/training/pages/materials.html, public domain.

and outputs are the center of management and how it controls the system.

Refer to Table 17-2 and Figure 17-4 for a review of a typical management system review. You may have other additional elements that you

Table 17-2
Conducting a Safety and Health Review

You can assess your management safety system using the following five elements (Figure 17-4):

- Management leadership
- Employee involvement
- Worksite analysis
- Hazard prevention and control
- Safety and health training

OSHA Website,http://www.osha-slc.gov/SLTC/safetyhealth_ecat/mod3.htm,public domain.

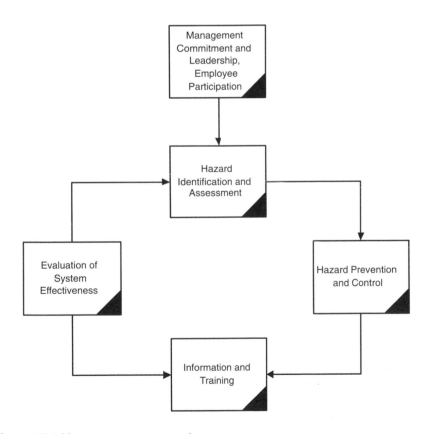

Figure 17-4 Management system review.

Table 17-3
Safety and Health Program Assessment

Program Elements

For the sake of simplicity, we will utilize the elements as outlined in OSHA's Safety Program Management Guidelines and performance evaluation procedure draft. Refer to Table 17-2.

Indicators and Measures

The indicators relate to the elements and describe how to tell if the subelement is in place and working. Refer to Appendix G.

Scoring

You have to decide if you want to develop a scoring system. If you choose to score your system, you can use the score as an indicator only. OSHA uses a scale of zero (0) to four (4). Pick the scoring system that best fits your operation. This will normally be an estimate based on your knowledge of the current operation and how it is working. Keep one point in mind. Each element is important. Some organizations use a total score for all the elements. As stated, it is important to look at each item, not at one huge element; otherwise you will not know what element needs to be modified.

Some Suggestions

Managers and employees may wish to conduct an assessment of their management system together and develop a joint rating. Sometimes, it may be important to have a selected number of your employees evaluate your system also. Use sampling techniques to verify what is important. If you talk with 20 employees and all 20 say the same thing, you have to make the assumption that all employees support that position. Whatever you do, be flexible and use your judgment.

OSHA Web site, http://www.osha-slc.gov/SLTC/safetyhealth_ecat/asmnt_worksheet_instructions.htm, public domain.

want to incorporate into your process—to measure, for example, communication, or behavior-based safety. These additional elements need to be evaluated before they are added to make sure that they add value to your process.

Table 17-3 shows an overview of a typical safety and health program assessment.

Reviewing and revising goals and objectives are critical to achieving your vision of a safe workplace. High-quality programs demand continuous improvement; therefore, organizations should never be complacent about safety. A periodic review of each program component will help you to achieve and maintain the vision reflected in your policy statement [6].

There are many methods to help assess the management system and safety program effectiveness. This chapter will provide an overview of each method. In addition, it will help to provide more detailed information on how to use these tools and how to evaluate each element and subelement of the management system.

We will focus on the three basic methods for assessing the management system effectiveness:

- Reviewing documentation of specific activities (safety committees, business contracts, activity-based safety system, behavior-based safety, etc.)
- Interviewing employees in the organization for their knowledge, awareness, and perceptions of what has happen to the management system and the safety program, using, for example, an employee safety perception survey, or a through an independent review
- Reviewing site conditions and, when hazards are found, finding the opportunities for improvement in the management systems that allowed the hazards to occur or to be uncontrolled [2]

Some elements of a management system are best reviewed using one of the suggested methods. Other elements lend themselves to be reviewed by other selected methods.

ASSESSMENT TECHNIQUES

As we have discussed, there are many tools that can be used to evaluate any workplace: independent review, employee safety perception survey, document (records) review, workplace evaluations, self-assessments, job hazard analysis, employee interviews, etc. These are some of the basic tools for an evaluation of the workplace.

The review should contain a list of the safety programs and/or management systems evaluated and a narrative discussion of the review of each element of the system. When you develop the report, always make sure that you start with all positive aspects of the review and then discuss

Table 17-4
Benefits of Performance Measurement

- Helps establish standards of performance through the organization
- Improves performance by identifying injury and illness potential
- Decreases operating costs by eliminating waste
- Allows decisions to be made and plans to be developed based on objective data without opinions and perceptions
- Improves problem identification
- Identifies system errors and allows problems to be corrected prior to system/equipment breakdowns
- Creates better understanding of the system in regard to data analysis and continuous improvement
- Allows system elements to be better understood
- Improves individual and team performance and constant feedback.

Manzella, James C., "Measuring Safety Performance to Achieve Long-Term Improvement," *Professional Safety*, September 1999, p. 36, Figure 3. Modified and reproduced with permission.

Table 17-5
Measuring Safety Performance

- The measurement must provide adequate data to evaluate the system
- Data gathered must be objective
- The measurement system must be credible with personnel providing, analyzing, and reviewing data
- The system must be cost justified
- Only well-documented systems can be objectively measured

Manzella, James C., "Measuring Safety Performance to Achieve Long-Term Improvement," *Professional Safety*, September 1999, p. 35, Figure 2. Modified and Reproduced with permission.

the opportunities for improvements to the system. It also should include a schedule of recommended changes, with target completion dates and responsible parties.

In addition, some evaluations may provide a grading system, so that each year's results can be compared to previous years. This report should be available to all employees and should be written in plain English and understandable by everyone. Avoid using jargon. Only use terms that all employees understand and use on a daily basis. Tables 17-4 and 17-5 present the benefits of safety performance measurement and a method to measure safety performance.

Table 17-6
Symptoms and Obvious and Underlying Clues to the Cause

Primary symptoms—Specific hazardous unsafe conditions or unsafe behaviors

- Conditions typically exist at the site of the potential incident
- Behaviors usually performed by the potential victim (employee)
- Usually the result of poor safety program (management system) design or implementation

Secondary symptoms—Underlying hazardous unsafe conditions and unsafe behaviors

- Inadequate implementation of policies, plans, processes, programs, and procedures
- Behaviors performed by employees, co-workers, supervisors, and middle management contribute to or produce the primary symptoms
- May exist and/or occur in any department

OR-OSHA Web site, http://www.cbs.state.or.us/external/osha/educate/training/pages/materials.html, OR-OSHA 116, Safety and Health Program Evaluation, Rev. 1/00 sig., public domain. Modified with permission.

HEALING A SICK SYSTEM

Safety management systems can suffer incidents for a number of reasons. It is important to implement an effective system wellness plan to make sure the prognosis for all systems in an organization remains positive. That is what an evaluation is all about: to determine the wellness of the system [3]. Table 17-6 lists symptoms of an ailing system, and Table 17-7 gives an overview of causes producing the symptoms.

When developing a successful safety management system, you must first see and understand the gap (opportunity for improvement). To make sure your system changes result in the effects you intend, you must jump over that gap [3]. To identify specific opportunities for improvement:

- Determine where you are now: analyze your management system
  Where are you now?
  What does our safety system look like now?
- Decide where your want to be in the future. Remember, "Begin with the end in mind."

Table 17-7
Obvious and Underlying Causes (Poor Corporate Nutrition)
Producing the Symptoms

System root causes—Management program defects

- Formulated by upper management
- Inadequate design of policies, plans, processes, programs, and procedures
- Produce secondary symptoms
- May exist in any program in any department

Causes—Poor corporate nutrition producing the symptoms
Deep root causes—Unsupportive vision, mission, strategies, objectives

- Formulated by top management and discussed with employees
- Reactive leadership, vision, mission, strategies, objectives, and budgets
- Produce program design defects

Factors external to the organization—materials, industry, community, society, etc.

OR-OSHA   Web   site,   http://www.cbs.state.or.us/external/osha/educate/training/pages/
materials.html, OR-OSHA 116, Safety and Health Program Evaluation, Rev. 1/00 sig., public
domain. Modified with permission.

> Have a vision where you want to be in the next 3–5 years or further
> What do you want your management system to look like?
> - Understand the opportunity for improvement: evaluate the management system
> Understand that there is always a better way
> What cultural values are missing?
> What system components are inadequate?
> - Jump the gap, plan/do/check/act. We will discuss this later in this chapter.
> Develop an action plan to get from here to there [3].

According to Dan Peterson, "To provide the difference we may need to consider some alternative ideas about what we know; to what we have always believed about safety; about what works and what does not. I cannot conceive of any organization getting a handle on its safety problems without turning in its old ideas and beliefs first" [5]. Remember when we discussed the definition of insanity?

Figure 17-5 The Deming Cycle. OR-OSHA Web site, http://www.cbs.state.or.us/external/osha/educate/training/pages/materials.html, OR-OSHA Safety and Health Program Evaluation, p. 116, Rev. 1/00 sig., public domain.

THE DEMING CYCLE

As we discussed the Deming 14 points and how they compared to safety, we saw a similar process. Let's take a different look at the Deming Cycle (Figure 17-5).

Step 1: Plan—Design the change or test
- Purpose: Take time to thoroughly plan the proposed change before it is implemented
- Pinpoint specific conditions, behaviors, and results you expect to see as a result of the change
- Plan to make sure that there is a successful transition of change

Step 2: Do—Carry out the change or test
- Purpose: Implement the change or test on a small scale to see how it works
- Educate, train, and communicate the change; help all employees transition to the new culture
- Keep the change small to better measure variable

Step 3: Study—Examine the effects or results of the change or test
- Purpose: To determine what was learned—what went right or wrong
- Statistical process analysis, surveys, questionnaires, interviews

Step 4: Act—Adopt, abandon, or repeat the cycle
- Purpose: Integrate what works into the system
- As we discussed before, ask not only if you are doing the right things, but if you are doing things right
- If the result was not as intended, abandon the change or begin the cycle again with the new knowledge gained [3].

WHAT SHOULD BE EVALUATED?

Everything that you know to be contributing to your management system should be evaluated. There are seven critical elements to a world-class management system as outlined in Table 17-8. These elements can help you with your evaluation.

WHO SHOULD CONDUCT THE REVIEW?

Although your employees can and do perform evaluations, in some cases other professionals who are knowledgeable about the workplace and who know how to manage safety systems need to provide an independent review. This type of review will help you to calibrate your own self-assessment. Refer to Table 17-9 for an overview of a proposed team that should be involved in conducting the review. In addition, the table provides the structure for the review.

Evaluations often cause anxiety for employees because of the uncertainty. You may be able to reduce that anxiety by discussing the evaluation with your employees. One thing to keep in mind is that the evaluation will be focusing on the management system and not on the employees.

Reviewers can be drawn from the safety department or the safety committee. The best reviewers will be individuals or employees who possess a new or fresh vision. These individuals should not be involved in the day-to-day operations of the facility. For example, some reviewers can be corporate safety professionals, another worksite, insurance companies (many loss-control sections of an insurance carrier offer their clients the services of a certified safety professional and a certified industrial hygienist), or OSHA's consultation to help review the system or for applying for VPP

Table 17-8
Seven Critical Elements of a World-Class Management System

Top Management Commitment and Accountability

Values	Leadership	Integrity	Character
Discipline	Service	Resources	Selfless
Opportunity Standards	Trust	Respect	Pride
Measurement	Communications	Resources Application	Consequences

Employee Involvement

Invitation	Suggestions	Reports	Consequences
Teams	Empowerment	Ownership	Everyone Engaged
Committees	Open Communications		

Hazard Identification and Control

Observation	Measurement	Assessment	Incidents
Accidents	Inspections	JHA	Maintenance
Surveys	Interviews	Records	Reports

Incident/Accident Investigation

Symptoms	Surface cause	Root cause Controls	Improvements

Education and Training

Continual	Skills	Knowledge	Impact
Everyone	Consequences	Attitude	

Periodic Evaluation

Proactive Behaviors	Prevention	Systems	Conditions
Engineering	Continual	Controls	Improvements
Communications	Design Consequences	Purchasing	Teams

Adapted from, OR OSHA 116, Seven Critical Elements, p. 10, http://www.cbs.state.or.us/external/osha/educate/training/pages/materials.html, public domain.

status. Another option is to have two managers switch roles, from different plant environments and locations, and have each assess the other's area [1]. Refer to Table 17-9.

The reviewers should have some knowledge of safety management concepts and have the technical expertise to review safety management

Table 17-9
Review Process Checklist and Structure; Suggested
Review Structure

- Suggested Team Members
  One safety/health professional (from different location)
  One manager, for example, production/plant manager
  An hourly employee on a selected basis, particularly from another
    environment
- Opening Meeting
  State purpose and objectives of review
  State scope of review
  Provide overview of the review process
  Discuss the logistics to facilitate meetings, interviews, small group meetings,
    one-on-ones, and facility tours
  Discuss any safety considerations for the Review Team
- Report of Findings
  The Review Team should prepare a preliminary report and present the
    findings and opportunities for improvement at the closing meeting
  A final report should be issued to the plant manager
- Action Plan Summary
  Once the final report is received an Action Plan Summary should be
    developed by the facility to outline how to close the gap in the process
  The Action Plan Summary should identify action to be taken, the person(s)
    responsible for the action, and the date the action is to be completed

Roughton, Mercurio, 2001.

systems and programs that support the system. Of these three areas, safety
management is the most important element.

EVALUATION TOOLS

We will now discuss how to evaluate the management system by using
various methods. There are some important points to keep in mind when
conducting a review. First, it is important to collect various forms of information for verification. The following is a proposed sequence of events:
opening meeting, plant tour, document review, employee interviews,
closing meeting, and a draft report for management review. Table 17-10

Table 17-10
Measuring Process Behaviors

Cultural Indicators

The cultural indicator measures the safety climate of the management system. Cultural indicators are more important than outcome statistics in developing long-term improvement plans.

Questions that need to be answered include:

- Is safety performance measured?
- Is safety perceived as a line function only?
- Are managers at all levels involved?
- Are all employees involved in the process?
- Is the safety process flexible to allow for changes?
- Is safety in general looked upon as positive?

The following are some suggested measurement tools:

- Independent reviews
- Perception surveys
- Interviews
- Observations

Refer to Figure 17-6 for an overview of the safety management review cycle, and to Figure 17-7 for a summary of the safety management review tools that can be used to assess your management system.

Adapted from OR OSHA 100w, "Measuring Process Behaviors," p. 37, http://www.cbs.state.or.us/external/osha/educate/training/pages/materials.html, public domain.

will help you to identify some process behaviors that will help you do a better review of the management system.

INDEPENDENT REVIEW

The objective of the independent review is to provide an assessment of the level of understanding and performance in each of the management system elements and to identify opportunities for improvement. If the facility has completed an internal self-assessment, the independent review will serve as verification (calibration) of management commitment to safety.

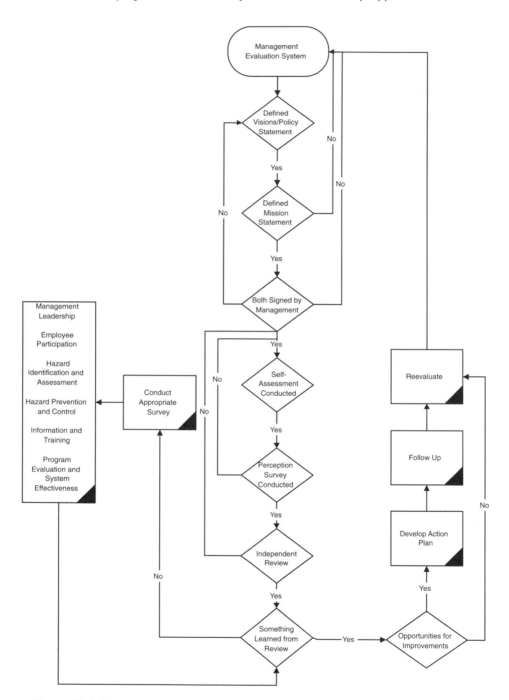

Figure 17-6 Management review cycle overview.

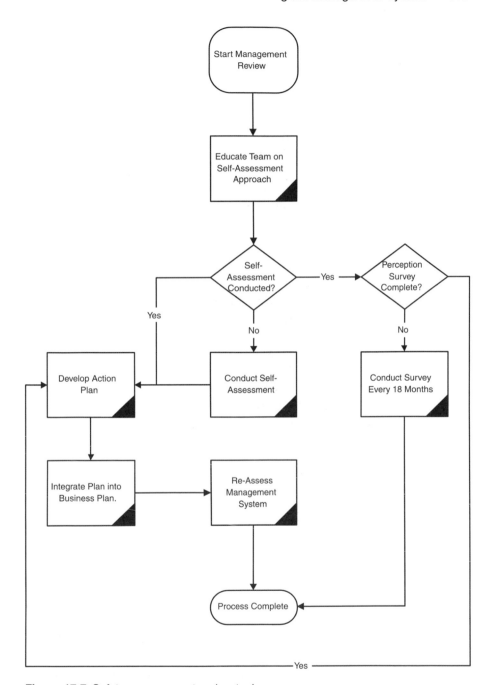

Figure 17-7 Safety management review tools.

To accomplish the stated objective, the independent review will:

- Provide for efficient and consistent measurement of the operations and performance relative to the management system
- Provide a system to track action plans and measurement of opportunities for improvement
- Provide for the development of a cross-functional team with information sharing and networking

This independent review should be administered by a third party who has some expertise in management system development and review.

Pre-review Activities

Once you have decided to conduct a management review, several pre-review activities should occur—for example, planning and preparation team selection and communications. The following activities should be completed leading up to a review:

- Set review dates and notify facilities of schedule
- Select review team leaders and members for scheduled reviews
- Send pre-review communication package
- Collect and review background information of facility:
  Results/action plan of safety perception survey
  Results/action plan of self-assessment
  Results of internal program reviews
  Trend analysis of OSHA recordable cases and workers' compensation claims

Review Method

A cross-functional team should conduct the review. The review team should collect information through formal and informal interviews and small group meetings with a representative sampling of employees across all levels using a variety of measurement tools. Based on lessons learned from the interviews and meetings, the team will evaluate performance in each of the selected elements to identify opportunities for improvement to achieve the next level of performance.

Opening Meeting

The review should begin with an opening meeting with the management team and selected employees. The purpose of this opening meeting is to:

- Discuss the objectives and scope of the review
- Discuss the review process and associated logistics (for example, scheduling of interviews and small employee group meetings)
- Introduce review team members
- Discuss safety considerations, such as personal protective equipment
- Establish lines of communication
- Develop schedule of the closing meeting

Facility Tour

Immediately following the opening meeting, the review team should complete a tour of the facility to familiarize themselves with the process. The reviewing process can help the team understand existing hazards and provide information about the breakdown of the management systems.

When evaluating each part of a safety program, use one or more of the above methods, as appropriate [2].

Employee Interviews

The primary means of collecting information during the review process will be meetings with selected departments. Talking to a number of randomly selected employees will provide an indicator of the quality of employee training and the perceptions of the management system [2]. Most of the time this is the best source of information on how the management system is working.

Meetings with small groups should be held with a representative number of employees to evaluate their understanding and application of the principles contained in your management system's key elements. A secondary approach could include the use of informal one-on-one interviews with employees in the production area where you can verify what you have discussed in the group interviews.

Employee interviews are very important in establishing what has occurred in the organization. There are several types of interviews, formal and informal. Formal interviews are conducted privately with randomly selected employees who are asked specific preselected questions that pertain to the organization. Informal interviews occur at employee workstations and generally follow a list of selected topics. For example, how well is the safety policy communicated and understood, and how is the disciplinary system working? You ask the employees to explain the policies and procedures.

To gauge the effectiveness of required training, you must interview employees and supervisors. For example, ask employees to describe what hazards they are exposed to, and how they are protected. In addition, ask employees to explain what they are supposed to do in several different types of emergency situations. Ask supervisors how they are trained, how they reinforce the training, enforce safety rules and work practices, and what their responsibilities are during emergency situations.

After you have conducted the employee and supervisor interviews, you now have a sense of the direction of the safety process. Interviews with the upper management should focus on their level of participation and commitment to the safety process. Ask them how the policy statement was created, and how the statement is communicated to all employees. For example, ask what information management receives about safety activities, and what action is taken based on the result of the information. Ask how management's commitment to safety is demonstrated to the employees. Ask what level of employee participation is used. Is employee participation voluntary or is it mandated? Do they participate in walk-arounds, inspections, follow-up committees, etc.?

If safety training is effective, employees will be able to discuss site-specific hazards they work with and know how they can protect themselves and others. They will understand the safety policy and what it means to the organization. Every employee should be able to say precisely what he or she is expected to do as part of the management system [2].

Employee perceptions can provide other useful information. An employee's opinion of how easy it is to report a hazard and get a response will tell you a lot about how well the hazard reporting system is working. If employees indicate that the system for enforcing safety rules and safe work practices is inconsistent or confusing, you will know that the system needs improvement [2]. We will discuss safety perception surveys later in this chapter.

Review of Site Conditions

Although physical inspections are not a typical function of a management system review, the conditions at the workplace does reveal much about the safety program effectiveness. Worksite conditions can be observed indirectly by examining documents such as inspection reports, hazards reported by employees, and incident investigations.

The site review also may reveal hazards. Be careful that this review does not become a routine inspection, with emphasis only on hazard correction. If a serious hazard is identified during the tour, take the necessary steps to make sure that it is corrected. In addition, ask what management systems should have prevented or controlled the hazard. Determine why the system failed, and either recommend changes or take other appropriate corrective measures. Refer to Chapter 12 for additional information on this technique.

Document Review

Reviewing documentation is a standard safety conformance appraisal (audit) technique. It is important to understand whether hazards are tracked to completion and if the tracking system is effective. In addition, it can be used to determine the quality of specific activities—for example, activity-based safety, behavior-based safety, self-inspections, or routine hazard analyses.

Inspection records allow the reviewer to understand if serious hazards are being identified, and if the same hazards are found repeatedly. If serious hazards are not being found and incidents keep occurring, there may be a need to train reviewers to look for other hazards. If the same hazards are being found repeatedly and incidents still occur, then the problem may be more complicated. In this case, hazards may not be corrected in a timely manner. The reviewer should evaluate the tracking system to see if it is capturing safety problems in regard to accountability for correcting reported hazards [7].

If certain hazards continue to occur after being corrected, the management systems are not performing properly. Someone is not taking responsibility for correcting the identified hazards. If this is the case, responsibilities may not be clearly defined or those who are responsible may not be held accountable [7].

Report of Findings

After the review, preliminary findings and opportunities for improvements should be presented at the closing meeting with the management team and selected employees.

The organization will be responsible for prioritizing and developing an action plan based on the review. The action plan should describe the specific actions that will be taken, the person(s) responsible, and the completion date. Once prepared, it is the facility's responsibility to update the action plan. Refer to Appendix B for a sample action plan.

Evaluation Judgments

The important work of gathering information about the management system is the most time consuming. However, it is also the easiest to understand and accomplish. The hardest part is making judgments about management system effectiveness.

You must decide which activities contribute to the safety goal, and which do not. Judgments and decisions made by reviewers should be driven by this quest for profitability and improvements in safety performance. Do not accept a report that only describes the program. The report should address any opportunities for improvement that have been identified in the management system.

Using the Evaluation

The evaluation is a valuable tool only if it leads to improved performance in meeting the safety goals and objectives of your organization. Some of the recommendations from the evaluation will be one-time corrections. However, recommendations may involve changing emphasis in the management system or developing new activities. Consider establishing, as a permanent objective, an audit of the procedures that your program sets for safety and health program activities. Refer to Chapter 5 for more information about establishing objectives. Figure 17-8 shows a typical fishbone diagram that you can use to determine how to set your goals for your action planning.

Refer to Appendix F for forms that can be used to conduct this survey.

Review Frequency

Frequency of reviews should be determined based on several factors—for example, the results from previous reviews, previous self-assessments, safety performance as measured by OSHA Incidence Rate (OIR), Lost Workday Incidence Rate (LWDIR), Workers' Compensation Claims

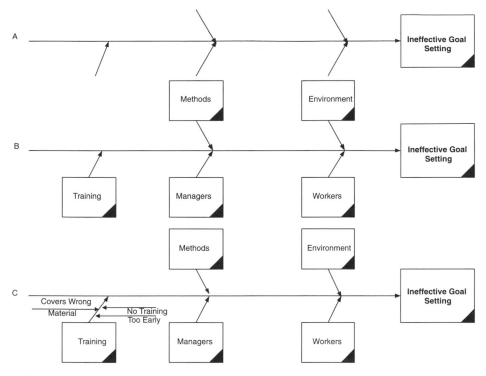

Figure 17-8 Fishbone diagram for goal setting. Peterson, Dan, *The Challenge of Change, Creating a New Safety Culture, Implementation Guide*, CoreMedia Development, Inc., 1993, Figure 6-2, p. 33. Reproduced with permission.

Rate (WCCR), and Incurred Workers' Compensation Costs per Hour Worked.

SAFETY PERCEPTION SURVEY

The Safety Perception Survey is a validated statistical tool that can be used to assess your safety culture using the perceptions of employees. This perception survey provides a detailed report on how a specific area is performing in 21 key management categories. These categories are critical in sustaining your safety performance. Once the data is collected, challenges in safety performance issues are identified between the perception (thinking) of the management group and what employees believe is the true culture. Refer to Appendix G for a sample survey form and the 100 questions that are used in the survey. Results from the survey can be used for problem solving and action planning to help improve safety performance [4].

What Is Perception?

Let us begin by trying to understand perception. Employees perceive their experiences in an organized framework that they have built out of their experiences (at home and at work) and values. Their own problems, own interests, and own backgrounds control their perception of each situation. Basically, each employee is saying, "I behave according to the facts as I see them, not as you see them. My needs and my wants are important, not yours, so I act on the basis of my perception of myself and the world where I live. I react not to an objective world, but to a world seen from my own beliefs and values" [5].

Since perceptions are strongly influenced by personal values, managers who motivate find it necessary to avoid excessive rationality. Employees insist on acting like human beings, rather than rational machines. We must accept them as the emotional beings they are and motivate them in their individual ways. We cannot easily change them to fit the motivational patterns we want them to have. Always we motivate employees in terms of their needs, not ours or what we want as management.

One of the difficulties with perception is perceptual set—that is, people tend to perceive what they previously have been led to believe they will perceive. If a new employee is told the supervisor is friendly, the employee will be more likely to see a friendly supervisor; if the perception is bad, then the employee sees a bad supervisor [5].

Administering the Safety Perception Survey

One thing that you must consider: if employees are not candid with their responses, there is a risk of biasing the outcome of the performance indicators. One problem is that the data collected may not represent the employees' real perceptions. If a sufficient number of employees are not included in the survey, the results may not be statistically valid for identifying trends. In addition, if employees are confused about how to properly complete the survey form, their responses may not be valid. In some cases, employees may view the perception as past history and not the new culture [5].

The key to the entire process is management commitment. Management must understand that they have to commit the necessary resources to implement of the survey. Both management and employees must be willing to promote the survey, encourage employee participation, and use the survey results as a tool to establish goals and objectives.

To get this buy-in from all individuals, a meeting with management and employee/labor can be used to inform employees of the intent of the survey and to answer any questions or concerns.

Management and employees must both agree to move ahead with the implementation, or the survey will not work. If there are still unresolved issues and either side is noncommittal, you must develop an action plan to address any issues. Appendix G contains a list of questions that are used for this survey. The questions will provide some insight into what to expect from the survey [5].

Selecting a Steering Team

To make sure that there is good employee participation (refer to Chapter 7), it is important that you select a cross-functional steering team that can run the process. The idea behind appointing this type of team is that it allows the distribution of work associated with the survey planning, administration, and analysis of results. In addition, the team can be used to promote a more efficient process and make sure that a variety of perspectives and opinions are considered along the way. The team members help to add credibility to the process.

When developing its purpose and project scope, the team should consider the following options:

- Are you trying to determine if safety performance meets company expectations?
- Are you trying to understand the safety culture created by the management systems?
- Do you want to train employees on the factors that affect safety performance?
- Do you want to help management and employees understand their specific safety roles and responsibilities?
- Do you want to measure how perceptions have changed since the last survey?
- Do you want to understand why there is a different perception between management and employees, department, shift, etc. [5]?

Planning the Survey

The comfort level of employees participating in the survey can be enhanced through a buy-in meeting. Other forms of communication should be considered to promote the survey:

- Newsletters
- Letter from management
- Announcements in weekly/monthly safety meetings
- Presenting safety topics during daily interactions with employees

Public relations/communication professionals can help to make sure that information is communicated in an easy and understandable manner, and that statements made are aligned with company values.

When planning to conduct the survey you may want to consider the following points:

- Are there large groups of employees at a single location, or are there small groups of employees over a broad area? This could be different department, shifts, locations, etc.
- What is the best method to survey top management?
- What location is suitable for conducting the survey? For example, is there a regularly scheduled safety meeting where the survey can be administered?
- Are there any language barriers?
- How much time will it take to conduct the survey? Will the survey require overtime?
- Who will be responsible for collecting the completed surveys?

The steering team should develop a schedule of dates, times, locations, meeting areas, and the number of employees to be surveyed.

Conducting the Survey

When employees are taking the survey, it is important to make sure that they are put at ease. Anything that might be perceived as a threat to anonymity should be considered. The following outlines various ways to give this sense of confidentiality to employees:

- Have one of the employees pass out forms
- Make it clear that forms should not be signed
- Do not pass around an attendance sheet

- Do not look over the shoulder of employees as they work on the questionnaires
- Answer any questions raised by employees as completely as possible
- Have an employee collect questionnaires
- Provide a box where completed surveys can be placed as employees leave the room
- Do not let the supervisor stay in the room

The important part is to provide an atmosphere free from distractions that will allow employees to feel comfortable.

Analyzing and Communicating Survey Results

Employee participation may raise many questions and expectations. The team should consider the following questions:

- What format will be used to analyze the data collected?
- What methods will be used, and who will be responsible for analysis and communication of results?
- How will you deal with the response you get from the survey?
- How will employees who participated in the survey be informed of the results?
- What is the procedure for action planning to close any limiting factors (gaps)?

It is important for the steering committee to review the results, action plans, and communication.

After the surveys are analyzed, the team should review the data, analyze results for strengths and challenges, communicate findings to the management staff, and develop an action plan to resolve issues. The survey results can be presented in several different ways:

- *Comparison graph.* A vertical bar graph that visually compares the perceptions of employees, supervisors, and managers in each category. This comparison graph is the most commonly used format for presenting (Figure 17-9).
- *Comparison table.* A table that contains percent favorable for each management category by variable selected. Numbers are presented in these tables, making them useful when conducting data analysis.
- *Survey responses by question.* This method provides responses for each question in the survey. This is useful when a category has been selected for improvement and more details are needed regarding the nature of concerns.

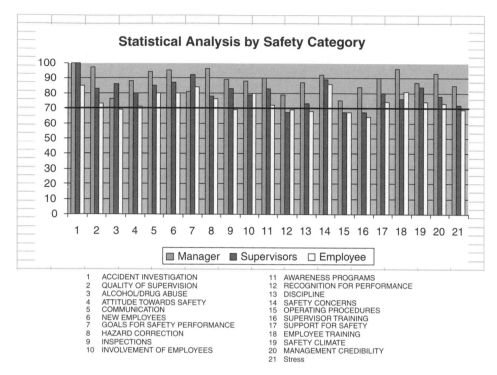

Statistical Analysis by Safety Category

1	ACCIDENT INVESTIGATION	11	AWARENESS PROGRAMS
2	QUALITY OF SUPERVISION	12	RECOGNITION FOR PERFORMANCE
3	ALCOHOL/DRUG ABUSE	13	DISCIPLINE
4	ATTITUDE TOWARDS SAFETY	14	SAFETY CONCERNS
5	COMMUNICATION	15	OPERATING PROCEDURES
6	NEW EMPLOYEES	16	SUPERVISOR TRAINING
7	GOALS FOR SAFETY PERFORMANCE	17	SUPPORT FOR SAFETY
8	HAZARD CORRECTION	18	EMPLOYEE TRAINING
9	INSPECTIONS	19	SAFETY CLIMATE
10	INVOLVEMENT OF EMPLOYEES	20	MANAGEMENT CREDIBILITY
		21	Stress

Figure 17-9 Sample safety perception survey results.

These results can be generated for an entire site, specific to any variable (for example, departmental comparison), and specific to any subset of a variable (for example, comparison of data among first, second, and third shifts).

The primary measurement used for the survey findings is "percent favorable." This is the ratio of positive responses to the total number of responses obtained.

Communicating Results

Once the action plan has been developed, the findings must be prepared and presented to all employees by:

- Providing the purpose and overview of the survey
- Providing an overview of who participated in the survey (for example, the number of managers, supervisors, and employees for a total number of participants)

- Listing the findings and opportunities for improvement
- Identifying the action plan that may have resulted from the survey

Communicating these findings in a timely manner fulfills the commitments made to employees and demonstrates management commitment. In addition, it should be posted in a prominent area so that all employees can review and comment.

DEVELOPING AND IMPLEMENTING THE ACTION PLAN

The key is to focus on only one challenge at a time. In preparation for action planning activities, you need to assemble and understand the following information:

- The scope of the problem—for example, management categories, department-specific, shift-specific
- Survey responses by question
- Favorable response numbers—for example, more than 70 percent positive and less than a 15 percent range indicates strengths; less than 70 percent positive and more than a 15 percent range indicates opportunities for improvement
- Action planning problem solving and steps to be followed

To develop an action plan, identify the specific action items that need to be implemented. Keep to the facts and do not get into too many details. Action items should pinpoint events and critical steps to consider during implementation. Once the plan is designed, responsibility for each step can be assigned, along with an estimated target date. Refer to Appendix B for a sample action plan document.

Discuss the action plan with management and all employees to get their buy-in. Once the action plan has been developed, you must make sure that individuals are assigned responsibilities for tracking action plans to completion. This will make sure that the implementations of all action items are managed properly.

SUMMARY

This chapter has defined what a successful management system evaluation should do. It has outlined what should be evaluated, who should do

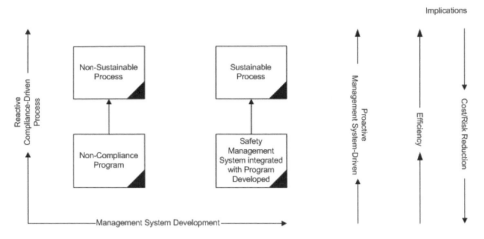

Figure 17-10 Journey to sustainable management system processes.

the evaluation, what tools are needed, how the evaluation should be conducted, and how to use the results.

By using this information and additional examples, guidelines, and instructions to perform annual evaluations, you will be able to compute your company's safety bottom line, just as you compute your financial bottom line. You will have the information needed to make knowledgeable and effective decisions promoting workplace safety [1].

The key to a successful and efficient assessment and program evaluation is to combine elements when using each technique. First, review the available documentation relating to each element. Then walk through the workplace and observe how effectively what is on paper is implemented. In other words, you are verifying compliance with the written documents. While walking around the facility, interview employees to verify that what you have read and what you saw reflects the state of the safety program [6].

An effective management system evaluation is a dynamic process. If you see or hear about aspects of the system not covered in your document review, ask to review the documents, if any, relating to these aspects. If the documents include program elements that were not visible during your walk around the site and/or not known to employees, research further. Utilizing this cross-checking technique should result in an effective, comprehensive evaluation of the workplace's safety program [6]. Figure 17-10 summarizes and highlights what this book has tried to detail: a management system that is sustainable with the process. The worst thing that we can do is to put a process in place and not allow it to be nurtured

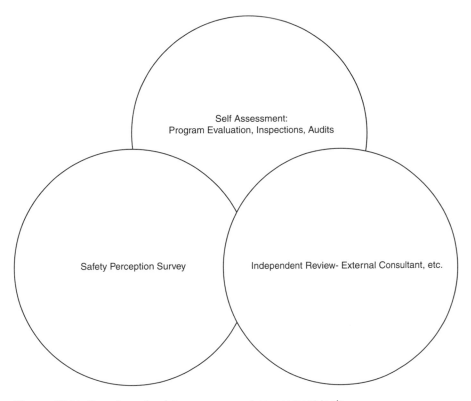

Figure 17-11 Overview of safety management assessment tools.

and mature into a sustainable process. The process needs to be seamless and integrated into other processes. To help you further understand the interaction of the management system, refer to Figure 17-11 for an overview of the safety management system review process.

REFERENCES

1. Oklahoma Department of Labor, Safety and Health Management: Safety Pays, 2000, http://www.state.ok.us/~okdol/, Chapter 2, pp. 12–15, public domain.

2. U.S. Department of Labor, Office of Cooperative Programs, Occupational Safety and Health Administration (OSHA), *Managing Worker Safety and Health*, November 1994, public domain.

3. OR-OSHA Web site, http://www.cbs.state.or.us/external/osha/educate/ training/pages/materials.html, OR-OSHA 116, Safety and Health Program Evaluation, Rev. 1/00 sig., public domain. Modified with permission.

4. Geller Scott E., *The Psychology of Safety: How To Improve Behaviors and Attitudes on the Job*, Chilton Book Company, Radnor, Pennsylvania, 1996.

5. Peterson, Dan, *The Challenge of Change, Creating a New Safety Culture, Implementation Guide*, CoreMedia Development, Inc., 1993, The Change Process Task List, pp. 41–50. Modified with permission.

6. Oklahoma Department of Labor, Safety and Health Management: Safety Pays, 2000, http://www.state.ok.us/~okdol/, Chapter 14, pp. 91–93, public domain.

7. OSHA Web site, http://www.osha-slc.gov/SLTC/safetyhealth_ecat/comp1_ review_program.htm#, public domain.

Final Words

DOES ANYONE REALLY WANT TO GET HURT?

Is there any manager or a supervisor who wants to see employees injured? Is there any employee who goes to work each day with the goal of getting hurt?

Not only do incidents cost money in the form of workers' compensation and medical costs, but there are also the indirect costs such as hiring and retraining new employees, repairing or replacing damaged equipment, or the loss of production. As we discussed in Chapter 1, however, even more important are the human costs of being unsafe. The pain and suffering of employees, and the toll it takes on their families and our society, can sometimes be the most damaging. Sometimes management can only see numbers (OIR). This is all that some managers understand. This is how they are trained to focus on the end result in many cases. As discussed, we need to get away from the numbers game and realize that people get hurt. Some managers now are trying to put all kinds of statistics on "injury numbers." Instead of looking at how the incident really occurred, there is a move in some managers to try and understand statistically how people get hurt.

CAN YOU REALLY DEVELOP A CULTURE THAT WILL SUSTAIN ITSELF?

With this book, we have provided you with basic information to establish and maintain a successful safety program and provided methods to integrate your programs into your management system (Chapters 4 and 5).

This book goes beyond what is required by OSHA. Once you have reviewed the individual chapters and sample information, you will be able to determine what your company's safety culture will be, and then begin the process of making it happen (Chapter 4).

This book is divided in four sections to provide you with a logical flow from where safety management begins and the cost of an incident, to methods for building and sustaining a safety culture, programs that support the safety culture, and finally methods that you can use to measure performance.

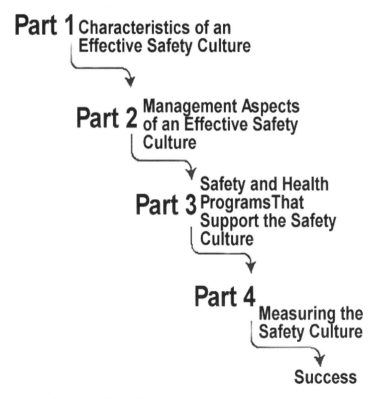

Developing an effective safety culture.

We hope that you have learned what it takes to develop a successful safety culture. We have defined a safety management system and how existing safety programs can be integrated into the safety process. The question is, has the recent interest in safety management systems created a new role for the safety professional, or is it a continuation of traditional past practices—"doing it the same way that we have always done it"?

BRIEF SAFETY HISTORY

We discussed the history of safety management with hopes that it will assist you in putting an additional spin on some of the present management approaches. Our objective here is to identify what is meant by a safety management system by noting its key elements (Chapter 3). In addition, we examined the concept of integration of safety management into broader management systems. Our final objective was to identify issues and themes to assist you in the categorization of the types of safety management systems.

The development of a safety management system is only part of the story. Over the years, the role of the safety professional may have broadened, but there appears to have been little change in the basic elements of developing a safety program. Safety programs early in the 1970s were due to the development of governmental regulations: companies were mandated to put in safety programs. This is where we discussed the difference between "I have got to do this" and "I want to do this" (Chapter 8).

We discussed Heinrich's theories and where they had an influence on the traditional safety practices. His theories have endured to the present day as one of the foundations of safety management. They were scientific and documented a specific approach to hazard recognition and control programs. His bottom-line theories saw employees, rather than their working conditions, as the primary cause of accidents.

Heinrich advocated a multidisciplinary approach to safety, focused on engineering, psychology, management, and salesmanship. The emphasis on psychology supported his theory that the majority of accidents were caused primarily by the unsafe acts or behavior of employees—the axiom on which his prevention philosophy was based. This axiom was central to Heinrich's domino model of accident causation, which depicted five dominos lined up in a sequence. As we discussed in Chapter 12 (Figures 12-7 and 12-8), unsafe acts/conditions were placed in the central position, preceded by inherited or acquired personal faults, and followed by an incident. The removal of the unsafe act/condition was expected to interrupt the sequence. The expected result was prevention of the incident. Control of the individual behavior of employees was the key. As we discussed in Chapter 12, Bird and Germain also used the domino theory for loss prevention, using a similar axiom.

The notion of integrating safety programs into the safety management system has become another catch phrase. The meaning of an integrated safety management system and the various strategies and techniques used to integrate safety have been detailed in this book. Some authors emphasize the integration of safety into broader organizational structures and functions. In this book, we have discussed several approaches to integration, and different approaches to safety management and employee participation (Chapter 7).

SAFETY AND QUALITY ARE SIMILAR

The renewed emphasis on the importance of management and the key role of top management in controlling the elimination of obvious and latent failures were discussed when we presented Deming's 14 points. Deming found that 90 percent of quality problems are caused by the system. If you translate this to safety, you will find that it is estimated that 94 percent of injuries are caused by unsafe acts of people.

The emphasis should be hazard prevention and imposed continuous improvement, as compared to trying to "inspect in" safety at the end of the process (when employees get hurt). Refer to Chapters 10 and 11. This can be accompanied by an emphasis on ongoing performance measurement, which requires an understanding of (positive and negative) variation, the use of statistical data and analytical techniques if you wish, and the use of measures reflecting system performance and not the end results. Involving employees gives them their best opportunity to contribute ideas, given their role as internal customers with a stake in the process, through to the extensive use of improvement teams and self-managed work teams. We discussed Deming's 14 point in Chapters 16 and 17.

Before you can develop your own safety culture, you first must know what your existing culture is. Failure to recognize the opportunities for improvement has been the downfall of many efforts to institute change [1]. To begin to change a culture, you need to seek to know the existing culture, not at the micro level, but down to the individual level (the employees) [1].

HOW LONG IS THE JOURNEY?

As we discussed, you need a road map to help chart your direction to your final destination. However, if you cannot identify where you are now, the map is useless [1] (Chapter 6).

Remember when we discussed "begin with the end in mind" [2]?

The question is, are you trying to build a safety culture in your organization when what you hear is "Safety is important," but you do not know how to begin [3]? Unfortunately, this is common. Most managers know the words "culture" and "excellence," but rarely understand what they mean and how to implement the needed changes.

According to Peterson, "It's a matter of accountability [3]. You must clarify what is expected of all employees in terms of safety and measure how well employees are achieving what is expected and the rewards there are achievement [3]. This is true in anything that you try to accomplish." Peterson continued to say "Define, Measure, and Reward," and "It is not Rocket Science" [3] (Chapter 9).

Some of the problems that you may encounter in your review of safety programs are that they are piecemeal without a solid, sustainable process [3]. Many organizations have many programs for all kinds of policies and procedures. However, they are usually there to make sure that supervisor and management performance are measured [3].

Once you have identified safety expectations, you have to measure the achievement of these expectations. As we have noted in this book, you need to put more emphasis on performance and measuring activities, rather than focusing on incident rates (OIR) [3]. Peterson suggests finding a way to put safety into the existing daily and weekly routines for the purpose of measurement [3]. Safety has got to be a part of your management system. You can integrate safety into your management system without creating additional paperwork [3] (Chapter 8).

LET'S LOOK AT THE OTHER SIDE OF SAFETY

On the other side of the coin, many safety professionals can develop great safety programs on paper and speak the safety language by citing OSHA regulations. The problem is that most do not know how to use the program to make a management system work. I like to call this the "Paint by numbers" concept. For example, give a child a red color for #1 and a blue color for #2, and they can paint a picture. On the other hand, give a safety professional without experience the project of implementing a management system, and he/she probably could not accomplish the task. However, a safety professional with this same level of knowledge could write a procedure based on a new requirement and do just fine. Safety professionals must understand the linking pins that join everything together—for example, management commitment and leadership and employee participation.

On a final note, we discussed the "traffic cop" mentality. The accompanying table will summarize some of the theories between the perceived Human Behavior and Safety. To understand the chart in more detail, review Chapters 16 and 17.

Self-Imposed Behavior versus Traffic Cop Mentality

Employee Attitude—Perceptions	Reason for Shortcut
Time	
An employee would rather be somewhere else and/or doing other things more important to them than working—even if they have to drive to get there.	Taking shortcuts is usually due to a "perceived" time gain to do other things: increase production, take more breaks, or talk to their co-workers.
Employee Perception	
Low risk; better employee than anyone else; me against them mentality; "incident won't happen to me"; "nothing has ever happened before. Therefore, I can take as many chances as I want."	Increases "free" time; increases production, which increases bonuses; medium risk; habits accepted/ignored by management and others, maybe even expected by management; "It can't happen to me."
Feelings/Emotions	
Excitement/thrill; pleasure; challenge (may be lessened without enforcers unless perceived as a physical ability challenge); control; freedom; choice/right.	Excitement/thrill; pleasure; challenge (man against machine); control; freedom; choice/right; (perceived) acceptance from management/co-workers; resistance to change.
Beliefs	
This is the way that it has always been; trained/taught; habit; proven safer technology (seatbelts, etc.); reactive.	"I have done this 100 times before and have never been hurt"; trained/taught; learned practice; habit; resistance to change; reactive.

Self-Imposed Behavior versus Traffic Cop Mentality

Hazard Recognition and Design Equipment runs faster; "If they couldn't, they wouldn't."	Unforeseen danger, hazard risk; "It has never happened to me," "I have done it thousands of times," "If they couldn't, they wouldn't."

I have heard many managers compare workplace safety to driving a car. This is used in the context of why employees take shortcuts. Many people believe that police (management) enforcement (discipline) is the main, and sometimes only, deterrent that keeps people (employees) from speeding (taking shortcuts), or at least driving (working) as fast as they would like to. To understand why individuals speed or take shortcuts, we need to understand and focus on the behavior of people and the reasons why they do what they do. The accompanying table compares two scenarios with some possible reasons why humans are the way they are. This table was constructed to just to make you think about your process. It is the author's way of trying to give you some additional thoughts. You should probably come up with your own concept that depicts your position. This is what the book is all about: a road map that you can design on your own that will get you where you want to be in the future.

Even with police enforcement, officers drive faster than posted "speed limits." Why? Because they can. This brings to mind a question: If police (management) were not a threat, some people (employees) would drive (take short cuts) to the extreme, and eventually some would reach a level of uneasiness due to increased level of risk and/or potential lack of control. Alternatively, the excess speed (unsafe acts, at-risk behaviors) simply exceeds their natural ability to remain in control.

Think about it: where police (management) are located, "pockets of compliance" (not at-risk behaviors) are seen. Once management leaves this pocket, speeding resumes (at-risk behaviors). Even with these efforts, speeding still occurs and unsafe shortcuts are still taken.

Remove the "traffic cop" (management), and what controls speed/shortcuts? Fear, time, engineering design, etc.? Thus, there is a need to address the factors that cause the behaviors. These are addressed through acceptance and an overall culture change through activity-based safety (behavior-based safety, employee participation, training, communication of past events, JHAs, etc.).

We have discussed many things in this book and hope that you can use some or all of the information presented. The question you may have in mind is: What happens if I follow this book and it does not work?

We can answer this question. The techniques detailed in this book are proven, are being used by many organizations, and work well. If these elements do not work in your environment, there may be several possible reasons:

1. Your organization may not be ready for a management system; therefore you have to use the old traditional way of doing things, being a "traffic cop." This is not the best method, but it can work for a while. When the "traffic cop" leaves, you are back to square one.
2. You may not have the management commitment and leadership that are needed. Management has to say more than the words. They have to demonstrate their commitment to safety, as we have described throughout the book.
3. You may not have a good employee base that will accept the new culture. You must understand how employees perceive things.

One caution as discussed is the emotional bank [2]. Think about it this way: employee perceptions change over time at work and at home. If you conduct an employee perception survey as we discussed, some of the old ways of thinking will still prevail. Sometimes no matter what you do, employees always look at the past. This is natural. This is where your challenge will begin. Help employees look at the positive side of the equation and not always focus on the negative side.

Bottom line: it is how the management of the organization supports the process. A culture is different in each organization. You have to find the system that works for you. Sometimes this can be difficult. The key is to keep trying until you find the right solution.

SUMMARY

This book has explored the meaning of the concept of a safety management system. A safety management system has been depicted as a combination of the planning and review, management organizational arrangements, the consultative arrangements, and the specific program elements that work together to improve safety performance. The historical overview of safety management systems suggests that the recent focus on safety management systems is not a new phenomenon but a renewed focus on the need for a managed approach to safety.

Whatever the case, we hope that you have good luck in your process. We know that with this road map and your concentrated effort, a management system will work for you.

I want to leave you with one final thought. This deals with something that I observed when I was in Australia a couple of years ago on business. One day I had the opportunity to walk around this little town outside of Sidney. As I begin to cross a street, I noticed a marking on the pavement that took me by surprise. It looked like an outline of a person, just like you see in police dramas (see the following image). I asked what was up with this. I was told that this is the way that the township creates safety awareness for people that cross the street. They have had a few people get hit by cars. The message here and the only reason I mention it is in the same context that I discussed the posters. Do not let your program come to this. This is not awareness.

I equate this to riding down a highway and you come upon a serious vehicle accident. At first, you feel bad and in your mind, it makes you think, "Am I glad that was not me." However, as you continue down the road, you soon forget the accident and you are back to your normal self, top speed. Some managers think that showing graphic pictures of fingers and hands being cut off is a way to prevent incidents. There are many vendors making good money producing these low budget movies. Do not get your self into this trap.

We need to get past this and learn how to develop a culture that will support your effort to sustain a safety process without resorting to tactics of these types of visual aids. In one of the author's opinion, this type of awareness only hurts the process. Now you have to make up your own mind as to what you want to do and how you are going to approach the safety culture process.

Thanks for purchasing this book. Good luck! We hope to see you in your new culture.

REFERENCES

1. Dunn, Richard, Editorial, *Plant Engineering*, p. 12, October 2000.

2. Covey, Stephen, *The 7 Habits of Highly Effective People*, A Fireside Book, Simon and Schuster, 1990.

3. Interview with Dan Peterson, *Industrial Safety and Hygiene News*, p. 42, October 1999.

Appendix A

Sample Policy Statement Worksheet: Sample Guidance in Writing a Complete Statement

INTRODUCTION

Generally, a written safety and health policy statement will run 6 to 12 sentences in length. It will include some or all of these five elements:

- An introductory statement
- A statement of the purpose or philosophy of the policy
- A summary of management responsibilities
- A summary of employee responsibilities
- A closing statement

INTRODUCTORY STATEMENT

The written policy statement generally starts with a clear, simple expression of your concern for and attitude about employee safety and health. Examples of introductions to policy statements include:

- This company considers no phase of its operation or administration more important than safety and health. We will provide and maintain safe and healthful working conditions, and we will establish and insist on safe work methods and practices at all times.
- Incident prevention is a primary job of management, and management is responsible for establishing safe and healthful working conditions.

- This company has always believed that our employees are our most important assets. We will always place the highest value on safe operations and on the safety and health of employees.
- The company will, at all times and at every level of management, attempt to provide and maintain a safe and healthful working environment for all employees. All safety and health protection programs are aimed at preventing incidents and exposures to harmful atmospheric contaminants.
- All members of management and all employees must make safety and health protection a part of their daily and hourly concern.

PURPOSE/PHILOSOPHY

An effective safety and health program will have a stated purpose or philosophy. This is included in the written policy statement so that both you and your employees are reminded of the purpose and value of the program. You may wish to incorporate into your policy such statements as:

- We have established our safety and health program to eliminate employee work-related injuries and illnesses. We expect it to improve operations and reduce personal and financial losses.
- Safety and health protection shall be an integral part of all operations, including planning, procurement, development, production, administration, sales, and transportation. Incidents and health hazard exposures have no place in our company.
- We want to make our safety and health protection efforts so successful that we make elimination of incidents, injuries, and illnesses a way of life.
- We aim to resolve safety and health problems through prevention.
- We will involve both management and employees in planning, developing, and implementing safety and health protection.

MANAGEMENT RESPONSIBILITIES

Your safety and health action plan will describe in detail who is to develop the program and make it work, as well as who is assigned specific responsibilities, duties, and authority. The policy statement may include a summary of these responsibilities. For example:

- Each level of management must reflect an interest in company safety and health and must set a good example by complying with company rules for safety and health protection. Management interest must be vocal, visible, and continuous from top management to departmental supervisors.
- The company management is responsible for developing an effective safety and health program.
- Plant superintendents are responsible for maintaining safe and healthful working conditions and practices in areas under their jurisdiction.
- Department heads and supervisors are responsible for preventing incidents and health hazard exposures in their departments.
- Foremen are responsible for preventing accidents and health hazard exposures on their lines.
- Supervisors will be accountable for the safety and health of all employees working under their supervision.
- The Safety Director has the authority and responsibility to provide guidance to supervisors and to help them prevent incidents and exposure to health hazards.
- Management representatives who have been assigned safety and health responsibilities will be held accountable for meeting those responsibilities.

EMPLOYEE RESPONSIBILITIES

Many companies acknowledge the vital role of their employees in the operation of a successful safety and health program by summarizing employee roles and contributions in the policy statement. Employees have the unique responsibility of assisting management with all incident prevention efforts through effective participation in all safety and health activities, especially providing direct feedback regarding the effectiveness of these efforts.

Here are some examples:

- All employees are expected to follow safe working practices, obey rules and regulations, and work in a way that maintains the high safety and health standards developed and sanctioned by the company.

- All employees are expected to give full support to safety and health protection activities.
- Every employee must observe established safety and health regulations and practices, including the use of personal protective equipment.
- All employees are expected to take an active interest in the safety and health program, participate in program activities, and abide by the rules and regulations of this company.
- All employees must recognize their responsibility to prevent injuries and illnesses and must take necessary actions to do so. Their performance in this regard will be measured along with their overall performance.

CLOSING STATEMENT

The closing statement is often a reaffirmation of your commitment to provide a safe and healthful workplace. It also may appeal for the cooperation of all company employees in support of the safety and health program.

- I urge all employees to make this safety and health program an integral part of their daily operations.
- By accepting mutual responsibility to operate safely, we all will contribute to the well-being of one another and consequently the company.
- We must be so successful in our efforts that total elimination of incidents, injuries, and illnesses becomes a way of life.

SUMMARY

Policy statements can vary in length and content. The briefest are typically basic statements of policy only. Longer statements may include company philosophy. Still others will address the safety and health responsibilities of management and other employees.

Some policy statements will cover in detail items such as specific assignment of safety and health duties, description of these duties, delegation of authority, safety and health rules and procedures, and encouragement of employee participation. While some companies may wish to include these additional items in the policy statement, OSHA believes it usually is best to leave these details for later discussion.

This worksheet is designed to help you develop your safety and health policy statement. It contains examples of specific statements often found in safety and health policies. These are examples only, but they may give you ideas for a policy statement that expresses your style, your attitudes, and your values.

One example of a safety and health policy statement is: This company considers no phase of its operation more important than safety and health protection. We will provide and maintain safe and healthful working conditions and establish and insist upon safe work methods and practices at all times. Safety and health shall be an integral part of all operations including planning, procurement, development, production, administration, sales, and transportation. Incidents have no place in our company.

We will work consistently to maintain safe and healthful working conditions, to adhere to proper operating practices and procedures designed to prevent injury and illness, and to comply with federal, state, local, and company safety and health regulations. Each level of management must reflect an interest in company safety and health objectives and is required to set a good example by always observing the rules as a part of the normal work routine. Management interest must be vocal, visible, and continuous, from top management to departmental supervisors. All employees are expected to follow safe working practices, obey rules and regulations, and work in a way that maintains the high safety and health standards developed and sanctioned by the company. We urge all employees to make our safety and health program an integral part of their daily operations. Then the total elimination of accidents and injuries will become not just an objective, but a way of life.

Appendix B

Action Planning: Sample Versions 1 and 2

Sample Action Plan #1

Activity	Responsibility	Target Date	Status
Management Leadership			
■			
■			
■			
■			
■			
Employee Participation			
■			
■			
■			
Job Hazard Analysis			
■			
■			
■			
■			
Hazard Reduction And Control Measures			
■			
■			
■			
Training			
■			
■			
■			
■			
■			

Sample Action Plan #2

Activity	Conduct monthly employee safety and health meetings	Establish procedures for management and employee participation in inspections and accident investigations	Provide hazard recognition & accident investigation training to management and employees
Time Commitment	Begin by June	Inspections and investigations begin by Sept. 30	Complete by December 31
Responsible Employees	Plant Manager	Plant/Safety Manager	Safety Manager
Resources Needed	1-hour safety meeting each month for all employees. audio-visual equipment	Time spent by employees on inspections and investigations	Time for developing training materials, possibly outside professional/trainer
Results Expected	Employee input on safety and health matters; volunteers for participation programs	Inspections & investigations being performed	Inspections that ID all hazards; investigations that uncover root causes; fewer incidents and accidents
Possible Roadblocks	Employee mistrust at first; limited knowledge about all potential hazards	Lack of employee interest	Cost of reduced production; unwillingness to spend funds for improvements
Status/Evaluation	Monthly reports to Plant Manager; Assessment Annually	Quarterly reports to CEO, look for patterns or trends	Keep training records; retrain periodically

Appendix C

Sample Forms for Employee Reporting of Hazards, Tracking Hazard Corrections, Follow-Up Documentation

Example #1
Form for Employees to Report Hazards

Part 1

Hazard or problem

Department where hazard observed	
Date	Hour
Suggested action	
Employee's Signature (Optional)	

Employee: Completes Part 1 and Gives to Supervisor

Part 2

Action taken
Department
Date
Supervisor's Signature

Supervisor: Completes and Gives to Manager

Part 3

Date
Review/Comments
Manager's signature

Example #2
Reporting Safety or Health Problem

Date

Description of problem (include exact location, if applicable)

Note any previous attempts to notify management of this problem and provide name of person notified:

Date	
Optional: Submitted by:	Name

Safety Department Findings:
Actions Taken
Safety Committee Review Comments:

All actions completed by:	Date

Sample #3
Employee Report of Hazard

NOTE: This form is provided for the assistance of the employee in reporting hazards. This form is not the only way to notify management of hazards.

I believe that a condition or practice at the following location is a safety or health hazard.

Is there an immediate threat of death or serious physical harm?	Yes	No
Provide information that will help locate the hazard, such as building or area of building or the supervisor's name.		

Describe briefly the hazard you believe exists and the approximate number of employees exposed to it.

If this hazard has been called to anyone's attention, as far as you know, please provide the name of the person or committee notified and the approximate date.
Signature (optional)
Type or print name (optional)
Date

Management evaluation of reported hazard	
Final action taken	
All actions completed by Initials	Date

Sample
Follow-Up Documentation
This form can be used as part of sample form 1, 2, or 3 or separately

Hazard			
Possible injury or illness			
Exposure	Frequency		
Duration			
Interim protection provided			
Corrective action taken			
Follow-up check made on	Date		
Any additional action taken?			
Signature of Manager or Supervisor	Date		
Is corrective action still in place?		Yes	No
Three month follow-up check made on	Date		

Sample
Tracking Hazard Corrective Actions Form

Instructions: Under the column headed "System," note how the hazard was found. Enter Insp. for inspection, ERH and name of employee reporting hazard, or inc. for incident investigation.

Under the column Hazard Description (column 2), use as many lines as needed to describe the hazard. In the third column, provide the name of the person who has been assigned responsibility for corrective actions (column 3). In column 4 list any interim corrective actions to correct the hazard and the date performed. In the last column, enter the completed corrective action and the date that final correction was made.

1	2	3	4	5
System	Hazard Description	Responsibility: Assigned	Interim Completed Action	Action With Date

Appendix D

Medical Providers

EVALUATING THE QUALIFICATIONS OF A MEDICAL PROVIDER

Whether you choose to employ a professional physician or contract with an outside medical provider, it is important to evaluate specific qualifications. Remember, medical providers are individuals selling a service just like any vendor. However, it is important to understand that you must contract a service that will best fit your needs. In some cases, that may not always be the cheapest method. You should use a bidding procedure just as you would for any service—for example, subcontractors or vendors.

The following are some questions to ask prospective medical providers:

- What type of training does the medical provider have?
- In what type of industries has the medical provider had experience?
- What type of information does the medical provider want to know about your business?
- What does this provider know about OSHA recordkeeping requirements?
- What could this medical provider do to contribute to the improvement of your safety program?
- Can this medical provider provide references?
- Has there ever been an OSHA inspection in a facility where the medical provider was associated? What was the outcome of the inspection?

On the other hand, a prospective medical provider should ask questions concerning the following:

- Your work processes
- Your known or potential hazards
- Your facilities, type, and location
- Number of employees

- Standards and/or regulations that apply in your business
- Medical surveillance programs, current or past
- Collective bargaining contracts
- Any previously issued OSHA citations
- Existence and specifics of a safety and health policy
- Current method of providing medical provider services
- Other health care providers involved in providing services

PROTOCOLS: ESTABLISHED STANDARDIZED PROCEDURES

Specific medical protocols are written and standardized (work) plans for providing medical treatment to employees. They are comparable to the standardized procedures that you may already use in some areas of your business—for example, your system for maintaining accounts or servicing company equipment. You must provide your medical provider a set of protocols for treatment of work-related injuries, for response to emergency situations, for collection of data from medical surveillance programs, and/or for all the other activities listed in your medical program.

These procedures are not designed to interfere with the medical provider's treatment of any work-related injuries. Instead the procedures are designed to help make sure that there is early detection of work-related issues through consistent and thorough evaluation of employee health complaints.

Standardized procedures also promote the use of the most up-to-date treatments. They are particularly important if you are using several contractors to provide your medical services, because they help to make sure that all of your employees receive the same type of care. Even if company employees provide their own medical services, there still should be standardized procedures written for the medical surveillance programs, health care, and first aid. These standardized procedures should be communicated to all medical providers who provide treatment for your employees. A Comprehensive Guide for Establishing an Occupational Health Service, published by the American Association of Occupational Health Nurses (AAOHN), includes information on developing protocols. To obtain a copy of this guide, contact AAOHN, 50 Lenox Point NE, Atlanta, Georgia 30324, telephone (404) 262-1162.

MEDICAL PROVIDER QUALIFICATIONS

Once you have decided which safety and health services are needed for your facility, you must decide who will provide these services. There are several factors to consider before contracting a medical provider:

- Your medical provider must be organized so that the individuals providing the services are not working alone. Most state laws require that a registered nurse or physician supervise these individuals.
- The occupational health professionals you use should have specialized, up-to-date training or experience in the methods of occupational health care.
- You must decide if you hire physicians as your own employees or if you want to contract to an outside medical provider.

Occupational Health Professionals

Occupational health professionals are medical doctors (M.D.), doctors of osteopathy (D.O.), and registered nurses (R.N.). They are certified and hold a license to practice their chosen professions.

Occupational Medicine Specialists

Occupational medicine specialists are medical doctors (M.D.) or doctors of osteopathy (D.O.) who have additional training or experience in the treatment of work-related illnesses and injuries.

Occupational Health Nurses

Occupational health nurses are registered nurses (R.N.) who have received specialized training in occupational health. Like physicians with specialties in occupational medicine, occupational health nurses can gain this training through formal college programs, continuing education, or experience working in the field. A registered nursing license allows nurses to perform many health evaluation and care functions independently. An occupational health nurse should be capable of performing or managing most of the medical activities.

Nurse Practitioners

Nurse practitioners are registered nurses who have completed formal advanced training in physical assessment and the management of minor injuries. In most states, they are licensed or certified for advanced practice by the state licensing boards. Nurse practitioners perform many health evaluation and care activities independently. They perform physical exams; diagnose health problems using laboratory tests, X-rays, or other tests; and treat employees. In most states, nurse practitioners can prescribe medications. Additionally, nurse practitioners perform many other activities using written protocols developed collaboratively with a physician. Nurse practitioners can take additional training and specialize in the treatment of occupational illnesses and injuries. When working with standard procedures, nurse practitioners should be capable of performing all the medical provider program activities.

Support Personnel

Support personnel provide more limited services. They have received specific training and usually are certified or licensed by the educational institution where they received the training.

Sometimes licensing or certification is granted by the state. The scope of practice for support personnel requires that they work under the supervision of licensed health professionals except when delivering first aid. Licensed vocational nurses (LVNs), licensed practical nurses (LPNs), emergency response personnel (sometimes called emergency medical technicians, or EMTs), and first aid personnel are in this category.

LVNs and LPNs are licensed by state agencies to perform specific health care activities, for example, taking blood pressures and applying dressings. These persons must practice under the supervision of a physician or a registered nurse.

"First aid personnel" perform the function of first response and provide temporary treatment until care of the employee can be transferred to someone with more advanced training—for example, performing treatments such as splinting and applying ice or pressure dressings, and also transporting the injured employee to a medical facility. A person does not need a background of formal health care education to be trained in first aid and cardiopulmonary resuscitation (CPR). A certification in first aid usually is granted by training providers such as the American Red Cross after the student completes a standard curriculum and demonstrates competence.

Emergency response personnel, sometimes called emergency medical technicians (EMTs), have received advanced training in first aid, CPR, and other life support techniques. With certain restrictions, they can perform sophisticated emergency procedures and transport injured employees to a medical facility.

The medical provider should retain records of the original training, licenses, update courses, and certification of all employees participating in the delivery of occupational health services, including first aid, CPR, and/or emergency response activities.

For all training related to the Bloodborne Pathogens standard, the standard requires that records be kept for 3 years, and that the records contain training dates, the content or a summary of the training, names and qualifications of trainers, and names and job titles of trainees.

MEDICAL PROVIDERS: CASE HISTORIES

The following provide success stories using medical providers. These cases, which are summarized from OSHA literature, illustrate how each organization's needs are site-specific and that their methods for providing services may differ in various ways. An important point demonstrated in these case histories: sometimes an individual whom you hire as your own employee provides the best services. In other cases, it may be better to contract a medical provider to get more consistent service.

CASE 1

Background Information

Of the employees employed by a larger manufacturing company, 1,000 are at one site where the components of their major product are manufactured. A second site 5 miles away has 500 employees in two buildings. One facility houses a production line where components are assembled, as well as the maintenance department and a garage where the company trucks and other vehicles are parked and serviced. The other building at the second site contains a small showroom and executive offices for all the administrative divisions, for example, accounting, human resources, and marketing.

The manufacturing site is 17 miles from the nearest medical facility. The light assembly and administrative site is 12 miles from the nearest medical facility. The company runs two 8-hour shifts consisting of a day shift and an evening shift. There are two security personnel working alone from midnight to 7 A.M.

The company employs a safety professional to head its safety department and a nurse practitioner to head its medical provider service. Both departments have other professional and non-professional staff to support various departments. The safety department and medical provider both have offices at the manufacturing site.

Hazard Analysis

The basic work of providing a comprehensive assessment (Chapter 10) was done by a committee composed of the safety director, the nurse practitioner, the director of manufacturing, the supervisor of maintenance, and two production employees, one from the day shift and one from the evening shift. In addition, consultation was requested from the loss control department of their insurance carrier. The loss control engineer was an industrial hygienist who confirmed the need to monitor for noise in the manufacturing area. She also helped the nurse practitioner and the human resources director write job descriptions for all major employee classifications. The descriptions emphasized important safety considerations—for example, the amount of weight lifted and the hazardous substances handled.

A second consultant was retained from the department of occupational medicine at the nearby university. The physician suggested a design for a health surveillance program to include cumulative trauma disorders (CTDs) for the shoulders, arms, and hands of the employees working in light assembly. The program was designed to utilize employees and other resources already available at the company.

Employee Training

The safety director and the nurse practitioner designed a training program to be delivered during employee orientation. The programs inform the employees about the company safety and health policies and procedures and program elements, and alerted them to specific hazards in their jobs and what they need to do to protect themselves. Safety department and medical provider staff members who were prepared by taking special train-the-trainer courses conduct the training. Employees are invited to contact the safety director, the nurse practitioner, or their supervisors if they have any further questions. The training includes a short quiz at the end to demonstrate whether the employees understood the specific risks associated with the job and how to protect themselves. There is a regular schedule for follow-up training.

Because the plants are more than 10 minutes away from the nearest medical facility, the decision was made to establish emergency response teams in both locations and on both shifts. The teams are set up on a volunteer basis and consist of five employees per shift per location plus all the security personnel. The company contracts with the American Red Cross to provide training and refresher classes in first aid and cardiopulmonary resuscitation (CPR) as applicable.

Because emergency response team members in the course of their assigned responsibilities could be exposed to infectious diseases such as hepatitis B and AIDS, they are covered by OSHA's Bloodborne Pathogens standard. Consequently, in addition to the training required by the standard, they also have been offered the hepatitis B vaccine, and personal protective equipment to protect them against exposure has been selected and distributed to them by the nurse practitioner.

Medical Surveillance

Machines in two departments were noisy. The safety director designed, and the maintenance department constructed, double-layered sheetrock walls with sound-reducing baffles between them around the two machines. Then the company contracted with an industrial hygienist to perform an environmental sound survey. This survey indicated that, even after construction of the sound baffles, the noise level in one of the departments was still too high. The company also contracted with a nearby audiologist to perform baseline pure tone hearing tests on all current employees. New employees for the department designated as too noisy are tested as part of their orientation. This same audiologist does the required annual audiometric testing. The nurse practitioner conducts the education program about hearing conservation. She also did the research necessary to purchase the best hearing protectors for the employees.

Analysis of Employee Use of the Medical Provider

A clerk in the medical provider's office is assigned to enter information about each employee's visit. This information is entered onto a spreadsheet that includes the date, the time of day, the employee's department, the employee's complaint, and the treatment rendered. Totals from the graph are examined by the nurse practitioner every month. The safety committee investigates any unusual clusters of complaints.

Establishing Standard Procedures

The company nurse developed procedures for the "first aiders" to use when administering first aid, CPR, and emergency transfer of injured employees. She also developed procedures that describe the standardized assessment and onsite treatment that she uses for employee illnesses or injuries and for all health surveillance programs.

Early Recognition and Treatment

The company nurse who heads this department has graduate-level training in assessment and management of illnesses and injuries. She is licensed to treat many of the employees' occupational injuries and illnesses using previously approved standard procedures. A referral relationship has been established by a contract with a local hospital that has an emergency room and a medical provider. Employees with illnesses or injuries that are assessed by the nurse practitioner as too severe to be treated onsite are transferred to the emergency room. Those employees who receive treatment by the nurse practitioner and do not respond to the treatment as expected are referred to the medical provider. In this way, a majority of the work-related injuries and illnesses are treated the medical surveillance system.

Medical Case Management

The nurse works with human resources to develop a case management system for all employees who are off work with illnesses or injuries lasting more than 5 days. The system consists of a method for prompt treatment authorization, a referral list for second opinions, assistance in filling out insurance forms, communication with the insurance carrier to make sure that benefit payments are timely, and ongoing contact with the employee and the family.

Coordination for Emergency Services

The safety director, the nurse practitioner, and the head of the security department worked together to develop a system where all employees in each department know their exact responsibilities in the event of an emergency. They also discussed their plan with the local fire department that will be responding to emergency calls.

Recordkeeping

The OSHA Log, the MSDSs, and the results from the noise surveillance are maintained in the nurse's office so employees' questions can be answered. All employee visits to the Medical Provider office are documented in the individual employee medical record.

CASE 2

Background Information

A meatpacking company with 500 employees has 460 employees who work 10 hours each in the slaughter department and the department where carcasses are dismantled. In addition, there are 20 employees who perform plant sanitation and maintenance functions on an overlapping evening shift. Another 20 employees are in supervisory positions and in administrative positions—for example, personnel, payroll, and safety and health.

The plant is 7 miles from the nearest health care facility. The company employs a full-time nurse (registered nurse) and a full-time safety director.

Hazard Analysis

The company's safety and health committee consisted of the safety director, the occupational health nurse, a supervisor, and four employees, one each from the slaughter, fabrication, sanitation, and maintenance departments. This group's hazard assessment included basic safety and industrial hygiene evaluations as well as a complete ergonomic review following the "OSHA Ergonomic Program Management

Guidelines for Meat Packing Plants," OSHA Publication 3123 (revised 1993). The assessment included a check for potential exposures to communicable diseases from the slaughtered animals. The committee developed a list of questions about safety and health conditions and potential hazards in the plant. To answer these questions, the committee performed a series of walk-throughs coupled with employee interviews.

In addition, the safety director and the nurse analyzed the actual jobs being performed for potential ergonomic problems. OSHA's ergonomic guidelines helped them identify those positions that involved specific activities associated with the development of CTDs. Furthermore, they reviewed all of the material safety data sheets (MSDSs) for all the chemicals used for cooling and sanitation at the plant.

Employee Training

Management decided to conduct the employee training program using their own company professional. The safety director designed the training to address hazards for example, fire, walking surfaces, cuts, elevations, and ammonia leaks. The nurse designed a training program that promotes hygienic practices to reduce the possibility of exposure to biologic hazards such as brucellosis, anthrax, and Q fever associated with animal handling. In many respects, the activities performed by the employees to protect the meat from contamination, as required by the U.S. Department of Agriculture, also protect them. Where this is not the case, the program is designed to emphasize what employees need to know and do to protect themselves. The nurse also developed material that informs the employees about the early signs and symptoms of CTDs and ways to help prevent them.

Together, the company health nurse and the safety director trained the employees about the hazardous substances and temperature hazards associated with the industry. Additional classes are held for the clerical, sanitation, and maintenance employees. In addition, supervisors are trained to recognize the early symptoms of ergonomic problems, so that they can encourage employees to report these problems as readily as other injuries.

Five volunteers from each plant and all supervisors from each shift make up the company's emergency response team. A commitment of 1 year is expected of members of the team. These employees are covered by OSHA's Bloodborne Pathogens standard. They all were given first aid training that included instruction and practice in how to protect themselves from exposure to bloodborne pathogens such as the hepatitis B and AIDS viruses. Retraining occurs for all team members at the anniversary date, when new members are added.

Medical Surveillance

A physician was consulted to help develop a cumulative trauma disorder (CTD) surveillance program. As one result, the physician made the portion of the pre-placement physical examination that dealt with the upper extremities and the back more detailed for the employees in the slaughter department and the division where carcasses are dismantled. At the 1-month and 1-year anniversary dates, randomly selected employees are invited back to be reexamined. Examination results that indicate early development of CTD are reported to the employees. In addition, management is informed about positions that need further evaluation. However, no personal information that identifies particular employees is released to management.

Analysis of Employee Use of the Medical Provider

The nurse maintains a spreadsheet of all employee visits to the health office. She combines this information with that received from the medical provider to form a report that is presented to the safety committee each month. This report and the incident reports become the basis for special safety emphasis programs in the company.

Establishing Standard Procedures

The company nurse and the medical provider worked together to develop procedures for all the treatment given in the medical provider's office, including the dispensing of over-the-counter medications. They also wrote procedures for hygienic practices for the employees exposed to biologic hazards. These procedures were intended to make sure that employees do not consume food, beverages, and tobacco products with contaminated hands, and that they do not accidentally contaminate their street clothes or shoes before leaving the plant. The safety director developed procedures for making sure that all contract employees performing pest eradication operations do not accidentally expose employees to pesticides. Finally, this same team developed procedures that included proper work techniques and frequent knife sharpening to prevent ergonomic problems.

Early Recognition and Treatment

A policy was established to encourage employees to promptly report symptoms of illness and injuries to the nurse. The nurse treats minor illnesses and injuries in the plant medical office using dressings, ice, and over-the-counter medications as applicable. She refers more severe problems to the medical provider. This nurse has taken a continuing education course in the recognition and conservative treatment of CTDs and is able to implement early treatment and referral. She is also able to review preventive measures with the employees at each visit.

Medical Case Management

The nurse and the medical provider maintain close communication about all employees with work-related injuries that are not responding to treatment as expected. The nurse makes sure that specialist referrals occur promptly. The nurse practitioner also works closely with the supervisors, proposes transitional duty positions, and clears these work proposals with the treating physician. This facilitates employees' returning to work as soon as possible.

Coordination for Emergency Services

The nurse and the safety director head the team and respond to each emergency. The plant receptionist is responsible for contacting outside emergency organizations, so this person is included in the emergency response team meetings.

Recordkeeping

The OSHA Log, the Material Safety Data Sheet (MSDS), and the results from the cumulative trauma disorder (CTD) surveillance are kept in the nurse's office so that employees' questions can be answered. All employee office visits are documented in the individual employee medical record.

CASE 3

Background Information

A small, independent janitorial service provides cleaning and light maintenance services for commercial buildings. There are 35 employees: the owner–manager, three clerical support personnel, an evening supervisor, and 30 service personnel (20 men and 10 women). The service personnel report to a central office where they are dispatched in teams of two or three in company vans. The service personnel all work an evening shift from 5 P.M. to 1:30 A.M. Two of the clerical employees work a day shift, 9 A.M. to 5:30 P.M. One clerical employee works a shift that spans the day and evening shifts. The supervisor works the same evening shift as the service personnel.

This company employs no safety, industrial hygiene, or occupational health professionals.

Hazard Analysis

The owner–manager of the janitorial service was aware, from reading the newspaper, that the Occupational Safety and Health Administration (OSHA) was enforcing protective measures for employees exposed to ergonomic hazards. She contacted the OSHA-funded, state-run consultation service and received information about the criteria that were being used for enforcement. This information guided her in organizing a management system that would meet the enforcement requirements. At the owner's request, consultation personnel examined this system and other aspects of the company's safety and health program, including the hazards to which employees were exposed. They helped develop protection against the hazards found and recommended a CTD training program with an emphasis on back injuries. They also recommended a driver safety program.

Employee Training

The State OSHA consultation office suggested contracting with a nearby medical provider, already providing work-related employee health services, to conduct training in the prevention of CTDs with an emphasis on back safety. In addition, the employees are informed about the contents of the cleaning solutions they are using along with proper mixing techniques and the use of gloves and protective eyeglasses. The MSDSs, which list toxic ingredients, are explained, and employees are told where these documents are kept.

The initial class was provided at the janitorial service company's central office. Since then, new employees have been instructed at the clinic as part of their pre-placement physical examination and orientation. Finally, the owner received booklets promoting safe defensive driving and the use of seat belts from the Automotive Occupant Restraint Council. These booklets were distributed to all current employees and are included in the orientation materials for all new employees. The employer plans to develop and distribute to all employees a brief self-test based on the booklet.

The employer encourages the service personnel to take a beginning first aid course offered through a local municipal adult education program by granting paid time for the class. The evening supervisor was required to take the beginning and advanced first aid course. Each van is supplied with a first aid kit, as is the office. The service employees are instructed to report all injuries to the supervisor at the end of their shift, or sooner by phone if they think that the problem needs more than minor first aid treatment. The supervisor then refers the employee to a nearby emergency room or a medical provider as necessary.

Medical Surveillance

In this case there was no need for health surveillance.

Analysis of Employee Use of the Medical Provider

The medical provider sends a monthly statement to the employer that summarizes all bills that have been submitted to the employees' compensation insurance carrier. This summary includes both diagnostic and treatment information. The employer analyzes this information for trends in injuries and illnesses as a way of determining if employees are being exposed to identified hazards, if hazards exist that have not been identified, or if employees need more training about hazards.

Establishing Standard Procedures

The State OSHA consultation staff worked with the owner to develop standardized procedures for first aid and emergency situations. They also developed specific procedures for mixing and using all the cleaning solutions, using the buddy system for lifting heavy objects, and rotating tasks involving lengthy repetitive motions, for example, vacuuming. Discussion of the procedures is included in the new employee orientation.

Early Recognition and Treatment

The medical provider has completed a mini-residency in occupational medicine and has ample knowledge of the risks to which company employees could be exposed.

Medical Case Management

The owner functions as the human resources director as well as the manager. As such, she is in close contact with any employee who experiences lost work time related to industrial injury. She does not have access to her employees' individual medical records that are maintained confidentially at the medical provider. She considers this adequate case management. Her employees' compensation insurance carrier assists her by providing information about helping injured employees return to work quickly. The suggestions have prompted her to increase the frequency with which she makes telephone contact with these employees.

Coordination for Emergency Services

Because employees move from one workplace to another instead of having a fixed worksite, no special arrangements were made with emergency organizations. The company did make its own emergency preparations.

Recordkeeping

The OSHA Log and the MSDSs are maintained by the owner and are available for the employees to see upon request. Individual employee medical records are kept at the contract medical provider and remain confidential.

SUMMARY

These examples demonstrate how three different employers provide medical services using a combination of in-house resources, subcontractors, and government agencies. Some of the subcontractors bring their services to the premises; in other situations the employees travel to the medical providers. In each case, the employer has selected services based on the special characteristics of the business process, the potential exposures within that process, the business location, and the employee population. Each medical program includes activities aimed at the prevention of exposures, the early recognition and treatment of work-related illnesses and injuries, and a reduction in the severity of and potential for disability from work-related illness and injury.

Appendix E

Sample Forms for Employee Training

SAMPLE LESSON PLAN

Organization:	Date:
Department:	Lesson Plan #
Title of Lesson Plan:	Time Allocation:
Instructor(s):	
Audience:	
Where:	
Training Objectives:	
Classroom Requirements:	
Training Aids and Equipment:	
Trainee Supplies:	
Trainee Handouts:	
References:	

Time	Content	Notes	A.V
8:00 am 8:15 am	Housekeeping	Rest rooms, emergency exits, breaks, etc	N/A
8:15 am 8:30 am	Introductions	Get to know students	N/A
8:30 am 9:15 am	Management Overview	Introduction– Discuss aspect of management system	Slides 1–10 Overhead Projector Flip Chart Slide Projector

Time	Content	Notes	A.V
9:15 am 9:30 am	Program Elements	Discuss the six aspects of the management system	Slides 11–15 Overhead Projector Flip Chart Slide Projector
9:30 am 9:45 am	Break		
9:45 am 10:00 am	Management leadership	Discuss management involvement	Slides 16–18 Overhead Projector Flip Chart Slide Projector
10:00 am 10:15 am	Employee participation	Discuss why employee participation is an important part of the process	Slides 18–19 Flip Chart Slide Projector
10:15 am 10:30 am	Hazard identification and assessment	Discuss methods of identification and how to perform an assessment. Provide handout #1	Slides 19–23 Overhead Projector Flip Chart Slide Projector
10:30 am 11:00 am	Break		N/A
11:00 am 12:00 pm	Hazard identification and assessment (continued)	Discuss Handout #2 Exercise #1	Video Overhead Projector Flip Chart
12:00 pm 1:00 pm	Hazard prevention and control	Discuss Handout #3	Slides 23–30 Overhead Projector Flip Chart Slide Projector
1:00 pm 2:00 pm	Break		N/A
2:00 pm 2:15pm		Discuss Handout #4 Exercise #2	Slides 31–35 Overhead Projector Flip Chart Slide Projector
2:15 pm 3:00 pm	Information and training		N/A
3:00 pm 4:30 pm	Evaluation of program effectiveness.	Discuss Handout #5 Exercise #3	Slides 36–41 Overhead Projector Flip Chart Slide Projector Video
4:30 pm 5:00 pm	Review and discussions		N/A

SAMPLE EMPLOYEE TRAINING ATTENDANCE RECORD

Meeting/training type: Safety Awareness, Incident Investigation, Monthly, Weekly, other. Attach Lesson Plan (refer to sample lesson plan, this appendix).

Subject	
Date	Length of Instruction
Department	Shift

Last Name	First Name	Middle Initial	Social Security Number	Signature

My signature indicates that I have received and understood the safety and health training/provided and acknowledge that I understand the content as presented.

SAFETY TRAINING PROGRAM EVALUATION: QUALITY IMPROVEMENT

Training Title:_____ Date: _____

Instructor #1: _____ Instructor #2:_____

It is our hope that this presentation has enhanced your job skills and knowledge and was able to further your professional development. We are striving to achieve a higher level of quality in training to meet the needs of operations and achieve the safety awareness to all employees. To meet our needs now and in the future, we appreciate your input and answers to a few questions about the subject material in this training program. Please be honest in your answers.

Please circle the number that best reflects your evaluation for the following topics. [1 = Poor, 5 = Excellent].

Topics	1	2	3	4	5
The objective of the training was met					
Safety Management					
Operations Information					
Motor Vehicle Operations					
Construction Safety					
Hazard Communication					
Hazard Review/Toxicology					
Air Monitoring					
Lockout/tagout					
Personal Protection					
Site Safety Plans					
What was the best feature of this training session?					
Do you have any suggestions for new sessions?					
Do you have any additional comments on how we can improve the program?					

Reference: Adapted from Roughton, James E., Nancy Whiting, *Safety Training Basics*

SAMPLE TRAINING SIGN-IN LOG

Location_____   Dates of Course_____   Trainer Signature_____

Course Name: _____

I certify that I have been trained in the above subject matter and have had the opportunity to ask questions, and that those questions have been answered to my satisfaction

Name (Please Print)	Employee Signature	Employee #	SS #	Total Hours

Appendix F

Evaluation and Review of Safety and Health Management Programs

SAFETY PROGRAM EVALUATION FORM

The goal of this book has been to present you with a comprehensive and easy-to-use tool and information about the management systems and safety programs that will help you evaluate your system's effectiveness. To help you make this evaluation, we have included important information.

This evaluation tool is designed as a convenient way to help your safety committee assist you in the evaluation of your management system, and look for opportunities for improvements.

Using This Evaluation Tool

Each of the elements contains a number of statements to evaluate the system. To help you understand what some of the questions mean, a set of indicators and measures have been developed to provide additional information about a particular element. The information should be used as a strategy for conducting your evaluation.

The elements are used to analyze indicators in each of the following categories to more accurately determine the performance rating.

- *Conditions*. Inspect the workplace for hazards. The absence of physical and environmental hazards indicates effectiveness.
- *Knowledge, attitudes*. Analyze what employees are thinking by conducting a survey. Full knowledge, positive attitudes, high trust, and low fear indicate effectiveness.

- *Behaviors, actions*. Observe both employee and manager behaviors. Consistent appropriate behavior and adherence to safety rules indicate effectiveness.
- *Standards*. Analyze system inputs–policies, plans, programs, budgets, processes, procedures, appraisals, job descriptions, and rules. Informative/directive, clear, concise, communicated inputs indicate probable effectiveness.

Continual/frequent monitoring and review is a best management practice. An effective program evaluation system does not wait until the end of the year before taking a look at how the safety program is working. That's like driving down the road using the rearview mirror to stay in your lane. Effective management (organizing, planning, controlling, leading) requires frequent program review.

Rating	Suggested Indicators
5– Fully Met	Analysis indicates the condition, behavior, or action described in this statement is fully met and effectively applied. There is room for continuous improvement, but workplace conditions and behaviors indicate effective application. Employees have full knowledge and express positive attitudes. Employees and managers not only comply, but exceed expectations. Effective leadership is emphasized and exercised. Safety policies and standards are clear, concise, fair, informative and directive, and communicate commitment to everyone. Results in this area reflect continual improvement is occurring. This area is fully integrated into line management. First line management reflect safe attitude and behavior. Safety is first priority (value).
3– Mostly Met	Analysis indicates the condition, behavior, or action described in this statement is adequate, but there is still room for improvement. Workplace conditions, if applicable, indicate compliance in this area. Employees have adequate knowledge, express generally positive attitudes. Some degree of trust between management and labor exists. Employees and managers comply with standards. Leadership is adequate in this area. Safety policies and standards are in place and are generally clear, concise, fair, informative, and directive. Results in this area are consistently positive, but may not reflect continual improvement.

Rating	Suggested Indicators
1– Partially Met	Analysis indicates the condition, behavior, or action described in this statement is partially met. Application is most likely too inadequate to be effective. Workplace conditions, if applicable, indicate improvement is needed in this area. Employees lack adequate knowledge, express generally negative attitudes. Mistrust may exist between management and labor. Employees and managers fail to adequately comply or fulfill their accountabilities. Lack of adequate management and leadership in this area. Safety policies and standards are in place and are generally clear, concise, fair, informative, and directive. Results in this area are inconsistent, negative, and do not reflect continual improvement.
0–Not Present	Analysis indicates the condition, behavior, or action described in this statement does not occur.

Score	Management Commitment and Leadership
	A written policy that sets a high priority for safety exists.
	A written safety goal and supporting objectives exist.
	The workplace safety policy is supported by management.
	Management supports safety goals and objectives.
	Management supports safety rules.
	Managers personally follow safety rules.
	Managers personally intervene in the safety behavior of others.
	Managers set a visible example of safety leadership.
	Managers participate in the safety training of employees.
	Management enforces safety rules.
	Safety program tasks are each specifically assigned to a person or position for performance or coordination.
	Each assignment of safety responsibility is clearly communicated.
	Individuals with assigned safety responsibilities have the necessary knowledge, skills, and timely information to perform their duties.
	Individuals with assigned safety responsibilities have the authority to perform their duties.

Score	Management Commitment and Leadership
	Individuals with assigned safety responsibilities have the resources to perform their duties.
	An accountability mechanism is included with each assignment of safety responsibility.
	Individuals are recognized and rewarded for meeting safety responsibilities.
	Individuals are disciplined for not meeting safety responsibilities.
	Supervisors know whether employees are meeting their safety responsibilities.

Score	Employee Participation
	There is a process designed to involve employees in safety issues.
	Employees are aware of the safety involvement process at the workplace.
	Employees believe the process that involves them in safety issues is effective.
	The workplace safety policy is effectively communicated to employees.
	Employees support the workplace safety policy.
	Safety goals and supporting objectives are effectively communicated to employees.
	Employees support safety goals and objectives.
	Employees use the hazard reporting system.
	Injury/illness data analyses are reported to employees.
	Hazard control procedures are communicated to potentially affected employees.
	Employees are aware of how to obtain competent emergency medical care.

Score	Hazard Identification and Control
	A comprehensive baseline hazard survey has been conducted within the past 5 years.
	Effective job hazard analysis (JHA) is performed, as needed.
	Effective safety inspections are performed regularly.

Score	Hazard Identification and Control
	Effective surveillance of established hazard controls is conducted.
	An effective hazard reporting system exists.
	Change analysis is performed whenever a change in facilities, equipment, materials, or processes occurs.
	Expert hazard analysis is performed, as needed.
	Hazards are eliminated or controlled promptly.
	Hazard control procedures demonstrate a preference for engineering methods.
	Effective engineering controls are in place, as needed.
	Effective administrative controls are in place, as needed.
	Safety rules are written.
	Safe work practices are written.
	Personal protective equipment is effectively used as needed.
	Effective preventive and corrective maintenance is performed.
	Emergency equipment is well maintained.
	Engineered hazard controls are well maintained.
	Housekeeping is properly maintained.
	The organization is prepared for emergency situations.
	The organization has an effective plan for providing competent emergency medical care to employees and others present on the site.
	An early-return-to-work program is in place at the facility.

Score	Incident Investigation
	Incidents/accidents are investigated for root causes.
	Investigations are conducted to improve systems.
	Investigators are trained in procedures.
	Teams conduct serious incident/fatality investigations.
	Analysis involves all interested parties.
	Disciplinary actions are not automatically tied to incidents.

Score	Education and Training
	An organized safety training program exists.
	Employees receive safety training.
	Employee training covers hazards of the workplace.
	Employee safety training covers all OSHA-required subjects.
	Employee training covers the facility safety system.
	Appropriate safety training is provided to every employee.
	New employee orientation includes applicable safety information.
	Workplace safety policy is understood by employees.
	Employees understand safety goals and objectives.
	Employees periodically practice implementation of emergency plans.
	Employees are trained in the use of emergency equipment.
	Supervisors receive safety training.
	Supervisors receive all training required by OSHA standards.
	Supervisors are effectively trained on all applicable hazards.
	Supervisors are trained on all site-specific preventive measures and controls relevant to their needs and supervisory responsibilities.
	Supervisor training covers the supervisory aspects of their safety responsibilities.
	Safety training is provided to managers, as appropriate.
	Managers are aware of all relevant safety training mandated by OSHA.
	Managers understand the organization's safety system.
	Relevant safety aspects are integrated into all management training.

Score	Periodic Program Review
	Workplace injury/illness data are effectively analyzed.
	Safety training is regularly evaluated.
	Post-training knowledge and skills for safety are tested or evaluated.

Score	**Periodic Program Review**
	Hazard incidence data are effectively analyzed.
	Hazard controls are monitored to assure continued effectiveness.
	A review of in-place OSHA-mandated programs is conducted at least annually.
	A review of the overall safety management system is conducted at least annually.

OR OSHA 116, Safety & Health Program Evaluation Form, pp. 20–23, http://www.cbs.state.or.us/external/osha/educate/training/pages/materials.html, public domain.

INDICATORS AND MEASURES

These are indicators and measures that you can use to supplement your assessment of your management system.

Management Commitment

1. A written policy that sets a high priority for safety exists

An effective policy will be both informative and directive. It will express a commitment to safety and the intent of the company to carry out the policy. A written policy will clarify expectations. The policy will be accessible to all employees. "Safety First" will be meaningful only if management doesn't reprioritize safety down when the going gets tough. Prioritizing safety may also have the affect of communicating the message that either safety or production is the top priority. Safety is considered a corporate value and is not prioritized over production.

2. A written safety goal and supporting objectives exist

Each of the management system and program elements contain certain goals and supporting objectives.

A goal describes an end-state. Objectives should be (1) measurable, (2) observable, and (3) completed in a specific time. For example, a safety goal might be "Increase employee participation in safety." A supporting safety objective would be "Complete joint supervisor/ employee job hazard analyses on all jobs in Plant A by July 30th (current year)."

Written plans containing goals and objectives are better able to communicate purpose and function clearly. Consequently, duties, responsibilities, criteria, specifications, and expectations will be more effectively met in a consistent manner throughout the organization.

3. The workplace safety policy is support by management

What does that support look like? Support needed for an effective safety and health program must include a substantial commitment to safety. Top management commitment must go beyond "lip service." It's more than just moral support, just talking up safety. Management needs to walk the talk by investing serious time, their own and staff, and money into proactive safety initiatives.

4. Safety goals and objectives are supported by management

Goals and objectives are created by management and workers. If your primary safety goal is "zero incidents," management will establish objectives to achieve that goal. One objective to reach this goal might be, "Train all employees on proper lifting techniques by December 30." Management will then support this objective by developing an effective training system.

5. Management supports safety rules

Management best supports safety rules by educating all employees on those rules and insisting that everyone, at all organizational levels, follow the rules. Management must also provide everyone with the resources necessary for compliance.

6. Management personally follows safety rules

Management cannot expect employees to follow safety rules if they, themselves, do not. If supervisors and managers ignore safety rules, they actually "rewrite" those rules.

7. Managers personally intervene in the safety behavior of others

Supervisors and managers are responsible to personally intervene when they see an employee performing unsafe behaviors. They must stop the unsafe behavior, find out why it occurred, and do whatever it takes to make sure it doesn't happen again.

8. Management sets a visible example of safety leadership

Management and leadership are not the same. Management is an organizational skill. Leadership is a human relations skill. Leaders establish positive relationships with their followers. Opportunities for safety leadership exist at all levels. The greater the responsibility, the more opportunity exists to demonstrate safety leadership. Effective safety programs maximize opportunities for safety leadership. Safety leaders send powerful and important messages to others when they themselves set a good example. Setting a good example is accomplished through action and demonstrates commitment.

The tough–caring leaders are tough because they care. They set high standards and insist that their employees meet those standards because they care about each employee's safety and overall success on the job.

9. Managers participate in the safety training of employees

It's smart business involving supervisors in the safety training process. A policy of employee participation supports the principle, and communicates to each supervisor that it's their job to manage safety as well as production. Employees will more likely perceive safety as an important area of accountability when supervisors are directly involved in safety training.

10. Management enforces safety rules

Safety is too important to "encourage." Management must insist on it. Enforcing safety rules requires that management must first (1) establish clear standards, (2) provide the resources to achieve those standards, (3) devise a fair system of measurement against those standards, (4) administer appropriate consequences, and (5) consistently apply accountability throughout the company, top down and laterally.

11. Safety program tasks are each specifically assigned to a person or position for performance and coordination

Line managers should clearly understand that it is their obligation to perform the daily responsibilities for safety as well as production. The safety committee and the safety officer coordinate in a consultative role.

12. Each assignment of safety responsibility is clearly communicated

Formal written responsibilities will most clearly assign safety responsibilities. All employees should be educated and trained on their responsibilities.

13. Individuals with assigned safety responsibilities have the necessary knowledge, skills, and timely information to perform their duties.

Without the proper education, employees will not be competent or qualified to perform their safety responsibilities. To determine qualifications, systems should be in place to measure individual knowledge and skills.

14. Employees with assigned safety responsibilities have the authority to perform their duties

Accountability follows control or authority. Supervisors and managers generally have control/authority over the physical conditions in the workplace and should have the authority to successfully carry out their assigned responsibilities to provide resources, educate and train, recognize, discipline, etc. Employees also have authority/control over their own personal behavior.

15. Employees with assigned safety responsibilities have the resources to perform their duties

Managers and employees depend on others to provide the resources necessary to carry out their responsibilities. The employer should develop systems to make sure that management and labor at all levels receive the necessary resources so that everyone can achieve established safety standards of behavior and performance.

16. An accountability mechanism is included with each assignment of safety responsibilities

Safety responsibilities assigned to each manager, supervisor, and employee should be tied to appropriate consequences. If the employee fails to meet established standards, consistent, fair corrective actions should be administered. If the employee meets or exceeds those standards, positive recognition should occur. Remember, if you regularly recognize, you'll rarely have to reprimand.

17. Employees are recognized and rewarded for meeting safety responsibilities

Recognition—the expression of appreciation for a job well done—is always appropriate. It provides an opportunity for leadership. Recognition should (1) occur soon after the behavior, (2) be certain—employees know they will be recognized, and (3) be significant—a sincere, genuine expression. The most effective recognition is accomplished in private. It is most effectively thought of as a function of leadership, not management. It's personal. Recognition does not always include reward.

Reward, on the other hand, is a token or symbol of appreciation. It is usually tangible. A reward system must be developed thoughtfully and carried out carefully and fairly or it will fail. Rewards can be thought of as "entitlements."

18. Employees are held accountable for not meeting safety responsibilities

As with recognition, accountability is more a function of leadership than management. If corrective actions are effectively administered, the end result will be a positive, character-building experience for the employee. Success depends on the approach of the leader. Accountability should never be punitive in nature. Its intended purpose is to increase discipline—doing the right thing all the time. It should also help to improve character—making the right decision when no one is watching. If not appropriately carried out, corrective actions will function to create a more negative relationship between the manager and employee.

19. Supervisors know if employees are meeting their safety responsibilities

To accomplish this goal, management must be adequately supervising or overseeing work being accomplished. "Supervision" is required. Supervision is adequate when hazardous conditions and unsafe behaviors are being effectively detected. Supervision is effective when injuries and illnesses are prevented.

Employee Participation

20. There is a process designed to involve employees in safety issues

Enough can't be said about the importance of employee participation in safety. The more participation, the more ownership. When employees believe they own something, they will value it. Management must develop a system that invests time and money in employee participation. The system should include and encourage participation in safety education, membership in the safety committee, making safety suggestions, and communication.

21. Employees are aware of the safety participation process at the workplace

Employees have been educated about the various ways they may be involved in safety. When asked, employees know who their safety committee representative is. They understand safety suggestion program procedures. They participate in job hazard analyses.

22. Employees believe the process that involves them in safety issues is effective

Safety must be perceived as a positive system or it will not succeed. Employees will believe the process is effective when the safety and health system operates in a culture of safety leadership. Managers, supervisors, and employees all express leadership by not only doing the right things, but doing the right things right! Managers will provide resources, recognize, correct, and set a proper example. The safety committee will respond, provide useful information, and make effective recommendations. Bottom line, employees will perceive the system as effective when they experience the benefits.

23. The workplace safety and health policy is effectively communicated to employees

When asked, employees are able to answer questions about the workplace safety policy. Policy is communicated formally through written statements and presentations. Policy is communicated and reinforced informally through what managers say and do daily.

24. The workplace safety and health policy is supported by employees

Employees at all levels of the company are involved in some or all of the elements of the safety and health program. They are supporting policy by carrying out their safety responsibilities.

25. Safety and health goals and supporting objectives are effectively communicated to employees

When asked, employees are able to correctly answer questions about the workplace safety and health goals and objectives. Goals and objectives are communicated formally in writing and during educational classes. They are communicated informally through what managers say and do daily.

26. Safety and health goals and objectives are supported by employees

Employees are involved in planning and carrying out goals and objectives. They are supporting goals and objectives by carrying out their related safety responsibilities.

27. Employees use the hazard reporting system

Employees understand and feel comfortable using the company's hazard reporting system. Employees report hazards to their supervisors and/or safety committee representatives. Reporting hazards is encouraged and considered professional behavior. Hazard reports are not considered "complaints."

28. Injury/illness data analyses are reported to employees

The OSHA Log and Summary Form is posted where employees have easy access. The safety committee reviews the OSHA Log and other statistical data.

29. Hazard control procedures are communicated to potentially affected employees

A system to educate and train affected employees on safety programs such as lockout/tagout, hazard communication, and bloodborne pathogens is in place. When asked, employees are able to correctly answer questions about hazard control procedures that affect them.

30. Employees are aware of how to obtain competent emergency medical care

When asked, employees are able to correctly answer questions about how to respond to a workplace emergency. A system is in place to educate and train all employees on emergency procedures and how to obtain emergency medical care.

Hazard Identification and Control

31. A comprehensive baseline hazard survey has been conducted within the past five (5) years

Comprehensive surveys that evaluate the entire facility should be periodically conducted. Baseline surveys are important workplace activities that identify, analyze, and evaluate the current status of:

- *Safe/hazardous conditions*. This may include such things as chemical inventories, machine guarding, housekeeping, exposure to noise and hazardous atmospheres (levels). Hazardous conditions represent the surface causes for accidents.
- *Safe/unsafe work practices*. Through observation and interviews, the degree of safe employee and management-level behaviors is

assessed and evaluated. Unsafe behaviors also represent the surface causes for incidents. Hazardous conditions and unsafe practices are really the symptoms of deeper root (system) weaknesses.

• *Safety program elements.* Analysis and evaluation of the safety system elements determine the effectiveness of the current safety policies, programs, plans, processes, and procedures. These represent the root causes that have allowed surface causes to develop and exist. Always evaluate systems in your comprehensive baseline survey.

32. Effective job hazard analysis (JHA) is performed, as needed

The job hazard analysis is a very important and effective process to determine hazardous conditions, unsafe practices, and system weaknesses. In this process the supervisor and an employee work together to analyze the specific task. The employee is observed performing the task. Next, they break the task down into steps. At this point, they jointly assess each step to identify any conditions and practices that might cause an injury/illness. Next, they work together on means and methods to eliminate those hazards. They revise procedures to make the task less hazardous. The supervisor, safety director, or safety committee should then take the process to its completion by uncovering the systems weaknesses related to the hazards found during the JHA.

It's very important to include the employee in the JHA process. Employees know the job and may have many excellent ideas for improvement. Including employees also increases ownership. We value what we own. Consequently, employees will be more likely to use the revised procedures when not being directly supervised.

33. Effective safety and health inspections are performed regularly

How do you know the inspection process is effective? Hazards, practices, and systems are uncovered, corrected, and improved so that they do not recur.

Regular safety inspections should be conducted. Some employers rely solely on the safety committee. If the safety committee is not thoroughly educated and trained, this practice may not be effective.

In an effective safety system, everyone is involved in the inspection process. Employees inspect their workstations daily. Supervisors inspect their departments as often as necessary, depending on the nature of the hazards. Managers conduct inspections of their various departments with supervisors on a scheduled basis. They jointly analyze and evaluate the safety systems in place that affect those conditions and practices.

34. Effective surveillance of established hazard controls is conducted

Once workplace hazards are corrected, a system must be in place to make sure they stay corrected. The safety committee quarterly inspection process can be a great way to monitor the workplace to make sure that corrective actions are permanently established. Employee participation in this process should be instituted.

35. An effective hazard reporting system exists

Hazard reporting should be an ongoing activity. However, beyond that, it's a professional behavior that every employer should desire and expect from their employees. Some questions to ask in evaluating this item include:

- Are employees reporting hazards?
- Who are they reporting hazards to? The safety committee representative? The immediate supervisor? Both? Most effectively, the supervisor should get the report so that he or she can take immediate corrective action.
- Do employees feel comfortable reporting hazards? The more comfortable they are, the more trust they display in management.
- Are reports called "complaints"? Hopefully they are not. Reporting hazards is a best practice that saves lives and money.
- Are employees being recognized and possibly rewarded when they report hazards? Remember, employee behavior reflects the consequences expected.
- Are reporting procedures simple? The less complicated they are, the more effective the response.

36. Change analysis is performed whenever a change in facilities, equipment, materials, or processes occurs

It's important to realize that any change in the workplace may introduce hazardous conditions and/or unsafe work practices. A system must be in place to make sure that safety is considered in all phases of the change process. Process Safety Management (PSM) guidelines can serve as an excellent guide in developing such a system.

37. Expert hazard analysis is performed, as needed

Companies should utilize third-party professionals to help identify and control hazards. It's important to know that the workers' compensation premium assessed to your company each year also "pays" for a very important resource: consultations conducted by your workers' compensation insurance carrier. Using this resource or the resources provided by private consultants to perform hazard analysis is smart business and may result in greatly reduced accident costs.

38. Hazards are eliminated or controlled promptly

When hazards are reported, a system to promptly correct them must be in place. The longer it takes to identify and correct hazards, the greater the probability of an incident. A successful system is usually established when supervisors are effectively held accountable for making sure hazards are corrected promptly.

The safety committee and management may develop a prioritized schedule detailing a time frame for correcting hazards. The more serious the hazard, the more quickly it would be corrected.

39. Hazard control procedures demonstrate a preference for engineering methods

One effective strategy to eliminate or reduce hazards is called the "Hierarchy of Controls." The strategy includes three general methods:

- *Engineering controls*. The best method is to eliminate the hazard. Two conditions must be present for an incident to occur: (1) the hazard, and (2) exposure to the hazard. Engineering controls address the first condition. Hazards are "engineered out" through initial design redesign, substitution, enclosure, etc. You are changing the "thing" (tool, equipment, machinery, facility) to eliminate the hazard. If the hazard is successfully eliminated, the exposure to that hazard is also eliminated.
- *Administrative controls*. Also called "work practice" or "procedural" controls, this method attempts to "administrate out" the exposure to the hazard by designing safe work procedures. You are changing the "things we do or don't do" (process, procedure). You're not address-

ing the hazard itself, but only the "exposure" to the hazard. Administrative controls are effective only as long as employees follow established safe procedures.

- *Personal protective equipment.* PPE sets up a barrier between the employees and the hazard. It does not eliminate or reduce the hazard. This method is usually applied in conjunction with administrative controls. Again, it's important to understand that, as with administrative controls, this method will be successful only to the degree that employees follow established procedures.

40. Effective engineering controls are in place, as needed

Conduct a workplace inspection to make sure tools, equipment, and machinery are properly guarded and that wherever possible those hazards are controlled by engineering methods. Some ways to tell if your engineering controls are effective are to determine if they (1) do not result in less efficient operation of the equipment, (2) are able to prevent an injury or illness at all times, even when the worker is distracted, and (3) protect employees from environmental hazards.

41. Effective administrative controls are in place, as needed

How do you know administrative controls are effective? Managers and employees are displaying appropriate behaviors. Administrative controls change the way we do things in the workplace. They include employee job rotations, exercise programs, procedural changes, and breaks.

42. Safety and health rules are written

The purpose of written safety and health rules is to clarify required safety behaviors to everyone in the company. However, just having written safety rules does not in any way ensure that employees will follow those rules. Written safety rules are most effective when they include the reasons why the rule exists . . . why it's important to follow it. Rules must be tied to accountability.

43. Safe work practices are written

Written safety policies, plans, and procedures address safe work practices. The job hazard analysis (JHA) is another document that details safe work practices.

44. Personal protective equipment is effectively used as needed

In most companies, this item will require some work. Effective use, here, means 100 percent use. Effective PPE use will not be obtained unless many safety systems are working properly. Employees will need effective education and training, and they must be working within a culture of safety accountability. If you find that the company is "encouraging" PPE use when it's required, it's a red flag that the program is not going to be truly effective. Safety is too important to encourage . . . it must be required.

45. Effective preventive and corrective maintenance is performed

A preventive maintenance program makes sure that tools, equipment, and machinery operate properly so that unexpected starts/stops or breakdowns do not occur. Effective preventive maintenance programs include adequate scheduling and reporting. If tools, equipment, or machinery fail or become defective, causing a safety hazard, it's important that they be corrected as soon as possible to prevent an injury. When failed equipment causes a safety hazard, an effective procedure that processes safety work orders is an important part of the corrective maintenance program.

An effective preventive maintenance system can ensure that equipment and machinery operates properly, which reduces the chance of an accident. A quick-response corrective maintenance program is also very important. Faulty equipment and machinery that could cause a serious injury or fatality should be taken immediately out of service. Maintenance work orders that correct hazards should be identified as unique from other work orders. You don't want safety work orders at the bottom of a stack of work orders.

46. Emergency equipment is well maintained

It's important that emergency equipment be adequate for the specific purpose intended. Fire extinguishers, personal protective equipment, chemical spill containment equipment and materials, and other emergency equipment should be inspected regularly and properly positioned.

47. Engineered hazard controls are well maintained

Through adequate inspection and maintenance programs and existing hazard control systems are properly maintained so that they eliminate or help reduce exposure to hazards that might cause injury or illness. Ventilation systems, machine guards, enclosures, and guardrails are all examples of engineered hazard controls.

48. Housekeeping is properly maintained

Poor housekeeping is a common cause of injuries. Effective housekeeping must be a daily effort by everyone to make sure clutter is eliminated. Proper training and continual inspection will help make sure that the workplace is clean and organized.

49. The organization is prepared for emergency situations

Depending on the nature of the work and location of the workplace, many different types of emergencies might be possible. A comprehensive plan(s) must be developed to address all emergency scenarios. The company must make sure their emergency action plan or emergency response plan is in place and tested regularly. If the workplace contains confined spaces, an emergency rescue plan may be required.

50. The organization has an effective plan for providing competent emergency medical care to employees and others present on the site

Effective plans should always be written and convey information about the importance of the program and who is responsible for carrying out program responsibilities. Drills are essential.

51. An early-return-to-work program is in place at the facility

Early-return-to-work is an important concept. The key is that the employee returns to work as soon as possible and participates in *productive* work, not make-work. An effective program assigns health-provider-approved transitional duties to employees. Some companies enlist the help of injured employees in evaluating safety programs, conducting inspections, and problem solving. Whatever the employee does, the nature of the work should not be perceived as punitive.

Incident Investigation

52. Incidents are investigated for root causes

When incidents are investigated (analyzed for cause, not blame) it's important that the investigator uncover the root causes that represent the underlying system weaknesses. Hazardous conditions and unsafe behaviors are called the surface causes for the incident.

How do you know you're addressing root causes? If you're describing faulty tools, equipment, machinery, facilities, materials (things) in the workplace, or employee/manager behaviors (things done or not done), you're describing surface causes. If you're describing inadequate or missing policies, programs, plans, processes, or procedures, you're addressing the system weaknesses—the root causes of failure. Only by addressing and correcting system weaknesses will permanent improvement in the safety system exist.

Education and Training

53. An organized safety and health training program exists

An organized management system will include a written plan that informs and directs employees' responsibilities. There are many training program strategies. However, the most effective programs integrate safety into operations. Because safety is an important component of quality in the production process, it is considered too important to be taught separately from other operations. Line management is directly involved in training safety procedures.

Supervisors/trusted mentors train involved employees. It is not delegated to a training staff. They know the company has high standards and insist that all employees meet those standards. All this sends the message that safety is an important line function.

54. Employees receive safety and health training

Employees should be trained before being exposed to any new hazard. Examples of appropriate times to train include new employee orientation, change of job assignment, and new equipment, materials, or procedures. Strong training documentation is necessary that certifies employee understanding, abilities, and skills.

55. Employee training covers hazards of the workplace

New employee training needs to include the general hazards that may be encountered in the workplace. Once the new employee is assigned to a particular department, it's important that he or she receive education and training about the specific hazards of the work station and area. Effective training will provide the employee the necessary knowledge so that they are able to detect, report, and/or correct associated hazards. The JHAs developed are important tools and must be used in the specific training.

56. Employee safety and health training covers all OSHA-required subjects

57. Employee training covers the facility safety system

All employees should be introduced to and have a good understanding of the management system, the safety program, and the various elements that make up the system. They should be encouraged to make suggestions for management system improvements, and participate in safety committee activities.

58. Appropriate safety and health training is provided to every employee

Education of employees is just as important as training. The purpose of safety education is to create a positive safety attitude and change behavior. Effective training makes sure the employees understand *why* the safety rule, procedure, or requirement is important. Safety training is concerned with helping the employee develop the needed abilities, knowledge, and skills to use safety equipment or perform procedures safely.

59. New employee orientation includes applicable safety and health information

It's important that top management communicate their commitment to safety as a core value to new employees. Job security depends more on working safely than on working fast. Safety policies, programs, and rules are discussed. It is very important to explain why each safety rule is necessary. New employees should be introduced to their safety committee representative during orientation.

60. Workplace safety and health policy is understood by employees

The only real way to know if your employees understand the safety policy is to question them and observe them. Most/all employees should be able to answer questions to your satisfaction and be observed complying with policy.

61. Safety and health goals and objectives are understood by employees

Do employees participate in goal setting? Do they know what the goals are? Are employees involved in carrying out stated objectives?

62. A review of the overall safety and health management system is conducted at least annually

Continual/frequent monitoring and review is a best management practice. An effective program evaluation system does not wait until the end of the year before taking a look at how the safety program is working. That's like driving down the road using the rearview mirror to stay in your lane. Effective management (organizing, planning, controlling, leading) requires frequent program review.

The following emergency plans are examples of those that should be practiced as often as required or necessary:

- Emergency response plans (medical emergencies, fire, earthquake)
- Chemical release and evacuation
- Confined space rescue
- Fall rescue

63. Employees are trained in the use of emergency equipment

Training should include demonstrations to make sure all employees have the skills required to use the equipment. Drills should be performed.

64. Supervisors receive safety and health training

Management at all levels requires safety education and training. In the most effective safety systems, supervisors are expert in their knowledge of the hazards and safe procedures required in their departments. Supervisors and managers need to be knowledgeable and skilled in managing safety programs, including recognition and corrective actions.

Supervisors must receive all training mandated by regulatory requirements. In the most effective safety systems, supervisors are expert in their knowledge of the hazards and safe procedures required in their departments.

65. Supervisors receive all training required by OSHA Standards

Several OSHA Standards have mandatory training components. A partial list includes:

- Hazard communication
- Forklift
- Confined space rescue

66. Supervisors are trained on all site-specific preventive measures and controls relevant to their needs and supervisory responsibilities

They have the knowledge of preventive measures using effective control strategies. They also have the skills to effectively eliminate or reduce hazards and unsafe behaviors.

67. Supervisor training covers the supervisory aspects of their safety and health responsibilities

Supervisors need training on specific regulations and site-specific hazards. Supervisors need the knowledge and organizational skills necessary to demonstrate effective safety management, including conducting training, safety meetings, inspecting, enforcing policy and rules, and correcting hazards and unsafe actions. They demonstrate a tough–caring leadership approach by setting high standards and insisting everyone meet those standards, including themselves.

68. Safety and health training is provided to managers, as appropriate

Managers require appropriate education in safety. Managers should understand the means and methods to provide necessary resources.

To evaluate this item, you may want to administer a survey or interview a sample population of managers.

69. Managers are aware of all relevant safety and health training mandated by OSHA

To evaluate this item, you may want to administer a survey or interview a sample population of managers.

70. Managers understand the organization's safety and health system

To evaluate this item, you may want to administer a survey or interview a sample population of managers.

71. Relevant safety and health aspects are integrated into all management training

Managers need to get the "big picture." They need to understand safety as a "profit center" activity, and how to effectively support the safety system. They should receive education about the legal, fiscal, and moral obligations they have, as managers and leaders, to the law, stakeholders, and society at large.

Periodic Program Review

72. Workplace injury/illness data are effectively analyzed

The OSHA Log, safety committee minutes, and baseline surveys all contain potentially valuable statistical data. Each of these may be assessed to determine the presence/absence of hazards. They may be analyzed to discover trends. They may be evaluated to determine effectiveness of related safety systems.

73. Safety and health training is regularly evaluated

At the conclusion of training, employee knowledge of the subject should be measured through oral or written exams. Ability and skill are most effectively measured by requiring the employee to demonstrate a task or procedure.

74. Post-training knowledge and skills for safety and health are tested or evaluated

Evaluation of safety training usually occurs through interviews and observation. Interviews and surveys may be used to determine knowledge and attitudes. Observation is required to evaluate employee skills and safety behaviors. The information gathered from these evaluation methods may be used to determine how effective safety training is and where improvements may need to be made.

75. Hazard incidence data are effectively analyzed

Incident reports, hazard reports, and inspection results provide a wealth of data about the types and locations of hazardous conditions in the workplace. This data should be gathered and analyzed to better reach conclusions about how to correct underlying system weaknesses that may be allowing them to exist. This is an excellent activity for a well-trained safety committee to conduct as a continuous effort.

76. Hazard controls are monitored to ensure continued effectiveness

In the most effective safety cultures, everyone is involved in monitoring hazard controls in the workplace. It's not usually adequate to simply rely on the safety director or safety committee to monitor hazard controls. Employees and supervisors, especially, need to be directly involved in this activity.

77. A review of in-place OSHA-mandated programs is conducted at least annually

It's going to be difficult to conduct an effective review of the safety culture/system if it's conducted annually. You may want to consider using the services of expert third-party resources in conducting this compliance review.

78. A review of the overall safety and health management system is conducted at least annually

Continual/frequent monitoring and review is a best management practice. An effective program evaluation system does not wait until the end of the year before taking a look at how the safety program is working. That's like driving down the road using the rearview mirror to stay in your lane. Effective management (organizing, planning, controlling, leading) requires frequent program review.

Source: OR OSHA Web site, Safety and Health Program Evaluation Form, http://www.cbs.state.or.us/osha/consult/evalform.html, public domain.

TWENTY-SEVEN ATTRIBUTES OF EXCELLENCE OF A SAFETY AND HEALTH MANAGEMENT PROGRAM

1. Written Safety and Health Policy

- There is a policy that promotes safety and health.
- The policy is available in writing.
- The policy is straightforward and clearly written.
- Top management supports the policy.
- The policy can be easily explained or paraphrased by others in the organization.
- The policy is expressed in the context of other organizational values.
- The policy statement goes beyond compliance to address the safety behavior of all members of the organization.
- The policy guides all employees in making a decision in favor of safety when apparent conflicts arise with other values and priorities.

2. Clear Safety Goals and Objectives Are Established and Communicated

- A set of safety goals exists in writing.
- The goals relate directly to the safety policy or vision.
- The goals incorporate the essence of "a positive and supportive management safety system integrated into the workplace culture."
- The goals are supported by top management and can be easily explained or paraphrased by others in the organization.

- Objectives exist that are designed to achieve the goals.
- The objectives relate to opportunities for improvement identified in a management review or when using other comparable assessment tools.
- The objectives are clearly assigned to responsible individual(s).
- A measurement system exists that indicates progress on objectives toward the goal.
- The measurement system is consistently used to manage work on objectives.
- Others can easily explain the objectives in the organization.
- All employees know measures used to track objective progress.
- Members of the workforce are active participants in the objective process.

3. Management Commitment and Leadership

- The positive influence of management is evident in all elements of the safety program.
- Employees of the organization perceive management to be exercising positive leadership and can give examples of management's positive leadership.

4. Authority and Resources for Safety and Health

- Authority to meet assigned responsibilities exists for all employees.
- Authority is granted in writing.
- Authority is exclusively in the control of the individual holding the responsibility.
- Personnel believe in the authority granted to them.
- Personnel understand how to exercise the authority granted to them.
- Personnel have the will to exercise the authority granted to them.
- Responsibilities are being met appropriately and on time.

5. Resources

- Adequate resources (personnel, methods, equipment, and funds) to meet responsibilities are available to all personnel.
- Necessary resources are exclusively in the control of the individual holding the responsibility.

- All personnel are effectively applying resources to meet responsibilities.

6. Accountability

- All personnel are held accountable for meeting their safety responsibilities.
- Methods exist for monitoring performance of responsibilities.
- Failure to meet assigned responsibilities is addressed and results in appropriate coaching and/or negative consequences.
- Personnel meeting or exceeding responsibilities are appropriately reinforced for their behavior with positive consequences.
- Data related to key elements of safety performance are accumulated and displayed in the workplace to inform all employees of progress being made.
- Individuals and teams to revise goals and objectives so as to facilitate continuous improvement in safety use accountability data.

7. Management Example

- All managers know and understand the safety rules of the organization and the safe behaviors they expect from others.
- Managers throughout the organization consistently follow the rules and behavioral expectations set for others as a matter of personal practice.
- Members of the workforce perceive management to be consistently setting positive examples and can illustrate why they hold these positive perceptions.
- Members of management at all levels consistently address the safety behavior of others by coaching and correcting poor behavior and positively reinforcing good behavior.
- Members of the workforce credit management with establishing and maintaining positive safety values in the organization through their personal example and attention to the behavior of others.

8. Company-Specific Work Rules

- The rules are clearly written and relate to the safety policy.
- The rules address potential hazards.

- Safe work rules are understood and followed as a result of training and accountability.
- Top management supports work rules as a condition of employment.
- Methods exist for monitoring performance.
- All personnel, including managers, are held accountable to follow the rules.
- Employees have significant input to the rules.
- Employees have authority to refuse unsafe work.
- Employees are allowed access to information needed to make informed decisions.
- Documented observations demonstrate that employees at all levels are adhering to safe work rules.

9. Employee Participation

- Employees accept personal responsibility to make sure that there is a safe workplace.
- The employer provides opportunities and mechanism(s) for employees to influence the safety program design and operation.
- There is evidence of management support of employee safety interventions.
- Employees have a substantial impact on the design and operation of the safety program.
- There are multiple avenues for employee participation. These avenues are well known, understood, and utilized by employees.
- The avenues and mechanisms for participation are effective in reducing incidents and enhancing safe behaviors.

10. Structured Safety and Health Forum That Encourages Employee Participation

- A written charter or standard operating procedure (SOP) outlines the safety committee structure and other forums.
- There is a structured safety forum in the goals. All employees throughout the company are aware of the forums.
- Meetings are planned, using an agenda, and remain focused on safety.
- Safety committees and/or crew hold regularly scheduled meetings.
- Employees on the committee are actively participating and contributing to discussion.

- Minutes are kept and made available to all employees.
- Top management actively participates in committee and crew meetings.
- A method exists for systematic tracking of recommendations, progress reports, resolutions, and outcomes.
- Employees are involved in selecting topics.
- Participation in the committee is respected and valued in the organization.
- The safety committee is supplemented with other forums such as crew and toolbox meetings as needed.
- Clear roles and responsibility are established for the committee and officers.
- There are open lines of communication between employees and forum meetings.
- The Safety Committee analyzes safety hazards to identify deficiencies in the injury and illness prevention program.
- The safety committee makes an annual review of the injury and illness prevention program.
- Reviewed results are used to make positive changes in policy, procedures, and plans. This review includes all facets of the facility.

11. Hazard Reporting System

- A system for employees to report hazards is in place and is known to all employees.
- The system allows for the reporting of physical and behavioral hazards.
- Supervisors and managers actively encourage use of the system and employees feel comfortable using the system in all situations.
- The system provides for self-correction through empowerment.
- The system involves employees in correction planning, as appropriate.
- The system provides for rapid and regular feedback to employees on the status of evaluation and correction.
- Employees are consistently reinforced for using the system.
- Appropriate corrective action is taken promptly on all confirmed hazards.
- Interim corrective action is taken immediately on all confirmed hazards where delay in final correction will put employees or others at risk.

- The system provides for data collection and display as a means to measure the success of the system in resolving identified hazards.

12. Hazard Identification (Expert Survey)

- Surveys are completed at appropriate intervals, with consideration to more frequent surveys in more hazardous, complex, and highly changing environments.
- These surveys are performed by individuals competent in hazard identification and control, especially with hazards that are present at the worksite.
- The survey drives immediate corrective action on items found.
- The survey results in optimum controls for hazards found.
- The survey results in updated hazard inventories.

13. Hazard Controls

- Hazard controls are in place at the facility.
- Hazard controls are selected in appropriate priority order, giving preference to engineering controls, safe work procedures, administrative controls, and personal protective equipment.
- Once identified, hazards are promptly eliminated or controlled.
- Employees participate in developing and implementing methods for the elimination or control of hazards in their work areas.
- Employees are fully trained in the use of controls and ways to protect themselves in their work area, and utilize those controls.

14. Hazard Identification (Change Analysis)

- Operational changes in space, processes, materials, or equipment at the facility are planned.
- Planned operational changes are known to responsible management and affected workers during the planning process.
- A comprehensive hazard review process exists and is used for all operational changes.
- The comprehensive hazard review process involves competent, qualified specialists appropriate to the hazards anticipated and the operational changes being planned.

- Members of the affected workforce actively participate in the comprehensive hazard review process.
- The comprehensive hazard review process results in recommendation for enhancement or improvement in safety elements of the planned operational change that are accepted and implemented prior to operational startup.

15. Hazard Identification (Job Hazard Analysis)

- Members of management and of the workforce are aware that hazards can develop in existing job task, processes, and/or phases of activity.
- One or more hazard analysis systems designed to address routine job, process, or phase hazards are in place at the facility.
- All jobs, processes, or phases of activity are analyzed using the appropriate hazards analysis system.
- All jobs, processes, or phases of activity are analyzed when there is a change, when a loss incident occurs, or on a schedule of no more than 3 years.
- All hazard analyses identify corrective or preventive action to be taken to reduce or eliminate the risk of injury or loss, where applicable.
- All corrective or preventive actions identified by the hazard analysis process have been implemented.
- Upon implementation of the corrective or preventive actions identified by the hazard analysis process, the written hazard analysis is revised to reflect those actions.
- All employees of the workforce have been trained on the use of appropriate hazard analysis systems.
- A representative sample of employees is involved in the analysis of the job, process, or phase of activity that applies to their assigned work.
- All employees of the workforce have ready access to, and can explain the key elements of, the hazard analysis that applies to their work.

16. Hazard Identification (Routine Inspection)

- Inspections of the workplace are conducted in all work areas to identify new, reoccurring, or previously missed safety hazards and/or failures in hazard control systems.

- Inspections are conducted routinely at an interval determined necessary based on previous findings or industry experience (at least quarterly at fixed worksites, weekly at rapidly changing sites such as construction, as frequently as daily or at each use where necessary).
- Employees at all levels of the organization are routinely involved in safety inspections.
- All employees involved in inspections have been trained in the inspection process and in hazard identification.
- Standards exist that outline minimum acceptable levels of safety and are consistent with federal or state requirements as applicable.
- Standards cover all work and workplaces at the facility and are readily available to all employees.
- All employees involved in inspections have been trained on the workplace safety standards and demonstrate competence and their application to the workplace.
- All inspections result in a written report of hazard findings, where applicable.
- All written reports of inspections are retained for a period required by law or sufficient to show a clear pattern of inspections.
- All hazard findings are corrected as soon as practically possible and are not repeated on subsequent inspections.
- Statistical summaries of all routine inspections are prepared, charted, and distributed to management and employees to show status and progress of hazard elimination.

17. Emergency Preparation

- All potential emergency situations that may affect the facility are identified.
- A facility plan to deal with all potential emergencies has been prepared in writing.
- The plan incorporates all elements required by law, regulation, and local code.
- The plan is written to complement and support the emergency response plans of the community and adjacent facilities.
- The plan is current.
- All employees at the facility can explain their role under the plan and can respond correctly under exercise or drill situations.
- Community emergency response commanders know the plan.
- The plan is tested regularly with drills and exercises.

- Community emergency responders are involved, where appropriate, in the facility drills and exercises.
- The plan is implemented immediately when an emergency at or affecting the facility is known.
- The plan is effective at limiting the impact of the emergency on the facility and the workforce.

18. Emergency Communication

- Emergency communications systems are installed at the facility.
- The communication systems are redundant (such as alarm boxes, emergency telephones, PA systems, portable radios).
- The communication systems are operational.
- The communication systems are tested at regular intervals (at least monthly).
- All employees are trained in the use of the communication systems and can demonstrate their proper use.
- Exit signs, evacuation maps, and other emergency directions are installed at the facility.
- Emergency directions are available, correct and accurate in all spaces, corridors, and points of potential confusion.
- Employees are aware of the emergency directions and can accurately describe the action they are to take in an emergency based on the directions available to them in their work area.
- Emergency equipment appropriate to the facility (including sprinkler systems, fire extinguishers, first aid kits, fire blankets, safety showers and eye washes, emergency respirators, protective clothing, spill control and cleanup material, chemical release computer modeling, etc.) is installed or available.
- Emergency equipment is distributed in sufficient quantity to cover anticipated hazards and risks, is operational, and is tested at regular intervals (at least monthly).
- All employees are trained in the use of emergency equipment available to them and can demonstrate the proper use of the equipment.

19. Emergency Medical Assistance

- The facility has a plan for providing emergency medical care to employees and others present in the workplace.

- The plan provides for competent emergency medical care that is available on all work shifts.
- Competent emergency medical care, when needed, is provided in accordance with the plan.
- All emergency medical delivery is done in accordance with standardized protocols.
- Competent emergency medical care, if provided on-site, is certified to at least the basic first aid and CPR levels.
- Off-site medical providers of emergency medical care, if utilized, are medical doctors, registered nurses, paramedics, emergency medical technicians, or certified first responders.
- All employees of the workforce are aware of how to obtain competent emergency medical care.

20. Facility/Equipment Maintenance

- A preventive maintenance program is in place at the facility.
- Manufacturers' or builders' routine maintenance recommendations have been obtained and are utilized for all applicable facilities, equipment, machinery, tools, and/or materials.
- The preventive maintenance system ensures that maintenance for all operations in all areas is actually conducted according to schedule.
- Operators are trained to recognize maintenance needs and perform or order maintenance on schedule.

21. Incident Investigation and Control

- Workplace policy requires the reporting of all actual and "near miss" incidents.
- All employees are familiar with the policy on incident reporting.
- All incidents are reported as required by policy.
- Workplace policy requires a thorough investigation of all incidents.
- All incidents are investigated as required by policy.
- Personnel trained in incident investigation techniques conduct all investigations.
- All investigations include input from affected parties and witnesses, where possible.
- All investigations determine "root causes."

- Recommendations designed to adequately address root causes are made as a result of all investigations and result in prompt corrective action.
- Completed investigative reports are routed to appropriate levels of management and knowledgeable staff for review and are available to government officials, as applicable.

22. Injury/Illness Analysis

- A system exists that tracks safety trends at the facility.
- The system addresses trailing indicators, including incidents, injuries and illnesses, hazards identified, and complaints from employees and others.
- The system addresses leading indicators of safety effectiveness, including employee attitudes and employee behaviors.
- All employees are aware of the need to provide incident and activity information to the system, and do so systematically, accurately, and consistently.
- An individual, or group, is assigned responsibility for compiling and analyzing records for safety trends.
- Trend data is consistently provided to all employees.
- All employees are aware of safety trends, causes, and means of prevention.
- Trend data is utilized to drive improvement and prevention activities.
- Employees are active participants in the determination of collection methods, collection, analysis, and intervention selection.

23. Employees Learn Hazards, How to Protect Themselves and Others

- An employee safety-training program exists at the facility.
- The training is provided to all employees, unless proficiency in the knowledge and skills being taught has been effectively demonstrated.
- The training covers all legally required subjects.
- The training covers hazards (awareness, location, identification, and protection or elimination).
- Training covers the facility safety system (policy, goals and objectives, operations, tools and techniques, responsibilities, and system measurement).

- Training is regularly evaluated for effectiveness and revised accordingly.
- Post-training knowledge and skills are tested or evaluated to make sure of employee proficiency in the subject matter.
- The training system makes sure that employees consistently and correctly apply knowledge and skills taught.

24. Understanding Assigned Safety and Health Responsibilities

- All elements of the company's safety program are specifically assigned to a job or position for coordination.
- Assignments are in writing.
- Each assignment covers broad performance expectations.
- All personnel with program assignments are familiar with their responsibilities.

25. Supervisors Know Safety and Health Responsibilities and Underlying Reasons

- A supervisory safety-training program exists.
- The training is provided to all supervisors, unless proficiency in the knowledge and skills being taught has been effectively demonstrated.
- The training covers all subject matter delivered to employees to the extent necessary for supervisors to evaluate employee knowledge and skills and to reinforce or coach desired employee safety and health behaviors.
- The training covers the facility safety system (policy, goals and objectives, operations, tools and techniques, responsibilities, and system measurement).
- The training covers supervisory safety and health responsibilities.
- Training is regularly evaluated for effectiveness and revised accordingly.
- Post-training knowledge and skills are tested or evaluated to ensure supervisory proficiency in the subject matter.
- The training system ensures that supervisors consistently and correctly apply knowledge and skills taught.

26. Managers/Supervisors Learn Safety and Health Program Management

- A management safety-training program exists at the facility.
- The training is provided to all managers, unless proficiency in the knowledge and skills being taught has been effectively demonstrated.
- The training covers all subject matter delivered to employees and supervisors to the extent necessary for managers to evaluate employee and supervisory knowledge and skills and to reinforce or coach desired safety behaviors.
- The training covers the facility safety system (management concepts and philosophies, policy, goals and objectives, operations, tools and techniques, and system measurement).
- The training covers management safety responsibilities.
- Training is regularly evaluated for effectiveness and revised accordingly.
- Post-training knowledge and skills are tested or evaluated to ensure management proficiency in the subject matter.
- The training system makes sure that knowledge and skills taught are consistently and correctly applied by managers.

27. Safety and Health Program Review

- The safety and health program is reviewed at least annually.
- The criteria for the review are against established guidelines or other recognized consensus criteria in addition to the facility goals and objectives and any other facility-specific criteria.
- The review samples evidence over the entire organization.
- The review examines written materials, the status of goals and objectives, records of incidents, records of training and inspections, employee and management opinion, observable behavior, and physical conditions.
- Review is conducted by an individual (or team) determined competent in all applicable areas by virtue of education, experience, and/or examination.
- The results of the review are documented and drive appropriate changes or adjustments in the program.
- Identified deficiencies do not appear on subsequent reviews as deficiencies.

- A process exists that allows deficiencies in the program to become immediately apparent and corrected in addition to a periodic comprehensive review.
- Evidence exists which demonstrates that program components actually result in the reduction or elimination of incidents.

Source: Adapted and modified from 25 Attributes of an Effective Safety & Health Program 24, OR OSHA 116, p. 24, http://www.cbs.state.or.us/external/osha/educate/training/pages/materials.html, public domain.

SAMPLE QUESTIONS FOR EVALUATING A MANAGEMENT SYSTEM

The objective is to ask questions that will help to uncover employee knowledge, skills, attitudes, and workplace conditions.

Tell Me about Your Safety Policy.

Listen for: how much is known about the policy; whether it reflects concern for employees' health and safety; whether the role(s) and broad responsibilities are defined for management, supervision, employees, and the safety committee; whether there is a positive and enthusiastic attitude when talking about it; who was it communicated to on all levels; how it was communicated (action, example, explanation, writing); whether it is used as a guide to make decisions; whether the policy is accessible to all employees; who was involved in developing it; whether it is signed by the highest decision maker, i.e., employer, CEO, plant manager, or superintendent. Note: Responses should be fairly consistent between management, supervision, and employees.

What Are Your Company Goals and Objectives?

Listen for: how much is known about them; whether they are clear and concise; whether they are talked about with a positive attitude and enthusiasm; whether they are written; who they were communicated to; how they were communicated; who provided input (should include employees); whether and when they are reviewed or reevaluated; who is involved in the evaluation. Note: Responses should be fairly consistent between management, supervision, and employees.

What Are the Primary Roles of Upper, Middle, and Lower Management, and Supervision?

Listen for a clear description of what the roles are; whether it refers to the safety/health policy; whether each level knows the role of the others above and below them. Note: Responses should be fairly consistent between all levels of management and supervision.

Describe Your Safety/Health Responsibilities and Level of Authority

Listen for a clear description/understanding of what they are; whether it refers to their job description; whether they give examples confirming appropriate authority; whether they refer to some form of accountability; their level of participation.

Tell Me about the General Safety/Health Responsibilities of Management, Supervision, and Employees

Listen for a clear description/understanding of what they are; whether they distinguish between upper, middle, and lower management; whether they refer to their job description and safety/health policy; whether each level knows the general safety/health responsibilities of the others above and below them; how they were communicated; how often safety/health responsibilities are updated; whether individuals are held accountable. Note: Responses should be fairly consistent between all levels of management and supervision.

In What Ways Do You Participate in Safety/Health Activities?

Listen for types of activities (i.e., meetings, training, inspections, and investigations); the level of participation; resources and support provided (i.e., time, effort, money); whether these are talked about with a positive attitude and enthusiasm.

How Do You Get Your Employees Involved in Safety/Health?

Listen for employee input, providing timely feedback on input, methods of involvement, level of participation, level of communication.

Describe Your Process for Evaluating and Updating Your Safety/Health System

Listen for a good understanding of the process; reference to policy and procedure; who is involved; how they are involved; what triggers an update (i.e., periodic, incident, inspection); who reviews it; writing and submission of recommendations; what happens to recommendations; how long it takes for management to respond; the feeling of support by all levels of personnel; how long it takes to implement recommendations.

Tell Me about Your System for Identifying and Correcting Safety/Health Hazards

Listen for a good understanding of the system; reference to a policy or procedure; who is involved; how they are involved; writing and submission of recommendations; what happens to the recommendations; how long it takes management to respond; the feeling of support by all levels of personnel.

Suggested Questions for One-on-Ones

The following questions were generated by discussing safety in a group. These questions can be used to verify what you hear in the group when doing one-on-ones.

- What is your job and exactly what do you do?
- What type of work goes on in a typical workday?
- What processes are important to your job?
- How long have you been on this job?
- What kind of training did you get?
- Who taught you to do that?
- When did you last receive training on the process?
- Can you explain this process to me?
- How long has this process been used?

- How often do you do this?
- How has the process changed in the last (year, month)?
- How many people do this job?
- Who checks on your work?
- Who is your supervisor? How often?
- How is this organization to work for?
- Why is this a safe place to work?
- Have you been through a safety/health inspection before?
- Have you ever been hurt on this job?
- What types of new chemicals are you working with?
- Have you ever been hurt doing this?
- Who is trained in first aid here?
- What do you do if someone gets hurt here?
- What's the emergency evacuation plan?
- What is your company's position/policy on safety and health?
- How do you view the company's safety program?
- Has anyone on the job been injured in the past year?

Source: Sample Questions for Evaluating a Safety System, OR OSHA 116, p. 34–35, http://www.cbs.state.or.us/external/osha/educate/training/pages/materials.html, public domain.

SAMPLE REPORT OF FINDINGS

Management Leadership
Findings
Opportunities for Improvements (Next Step)
Rating
Employee Participation
Findings

Opportunities for Improvements (Next Step)
Rating
Hazard identification and assessment
Findings
Opportunities for Improvements (Next Step)
Rating
Hazard prevention and control
Findings
Opportunities for Improvements (Next Step)
Rating
Information and training
Findings
Opportunities for Improvements (Next Step)
Rating
Evaluation of program effectiveness
Findings
Opportunities for Improvements (Next Step)
Rating

Appendix G

Sample Safety Perception Survey Form and Questions

Sample
Safety Perception Survey Form and Questions

Safety Perception Survey	Location	Category

Instructions
Background Information
Complete the boxes to the
  right that apply to you.
 	
Survey Questions
Circle either "yes" or "no" to
  each question
		Example Categories
		
		Employee
		
		Supervisors
		
		Managers

	Department	
Questions		
1	Do you feel you received adequate job training?	Y N
2	Do supervisors discuss accidents and injuries with employees involved?	Y N
3	Is discipline usually assessed when operating rules are violated?	Y N
4	Would a safety incentive program cause you to work more safely?	Y N
5	Do you perceive the major cause of accidents to be unsafe conditions?	Y N
6	Does your company actively encourage employees to work safely?	Y N
7	Is safety considered important by management?	Y N

8	Are supervisors more concerned about their safety record than about accident prevention?	Y	N
9	Do you think penalties should be assessed for safety and health violations?	Y	N
10	Have you used the safety committee to get action on a complaint or hazard which concerned you?	Y	N
11	Is high hazard equipment inspected more thoroughly than other equipment?	Y	N
12	Is amount of safety training given supervisors adequate?	Y	N
13	Have you been asked to perform any operations which you felt were unsafe?	Y	N
14	Are records kept of potential hazards found during inspections?	Y	N
15	Are employees influenced by your company's efforts to promote safety?	Y	N
16	Are employees provided information on such things as cost, frequency, type, and cause of accidents?	Y	N
17	Does your company deal effectively with problems caused by alcohol or drug abuse?	Y	N
18	Are unscheduled inspections of operations made?	Y	N
19	Is off-the-job safety a part of your company's safety program?	Y	N
20	Does management insist upon proper medical attention for injured employees?	Y	N
21	Are safety rules regularly reviewed with employees?	Y	N
22	Are you interested in how your company's safety record compares with other companies in your industry?	Y	N
23	Does your company hire employees who do not have the physical ability to safely perform assigned duties?	Y	N
24	Do your co-workers support the company's safety program?	Y	N
25	Do supervisors pay adequate attention to safety matters?	Y	N
26	Is safe work behavior recognized by supervisors?	Y	N
27	Do employees participate in the development of safe work practices?	Y	N
28	Are supervisors supported by management in their decisions affecting safety?	Y	N
29	Do the people in your department understand the relationship between what they do and the company's safety program?	Y	N

30	Is your family more concerned about off-the-job safety as a result of the company's safety program?	Y	N
31	Did you receive adequate safety training related to your job?	Y	N
32	Do you think your company has too many rules and regulations governing operations and safety?	Y	N
33	Are regular contacts made to all employees by supervisors on safety?	Y	N
34	Do employees understand the hazards of the operations they perform?	Y	N
35	Has the Employee Assistance program helped to eliminate alcohol and drug abuse in this company?	Y	N
36	Do employees participate in setting goals for safety?	Y	N
37	Do you think your company seeks prompt correction of problems found during inspections?	Y	N
38	Are you interested in how your division's safety record compares with other divisions?	Y	N
39	Can first line supervisors reward employees for good safety performance?	Y	N
40	Does alcohol or drug use increase accident risks?	Y	N
41	Do employees caution other employees about unsafe practices?	Y	N
42	Do you initiate action to correct hazards?	Y	N
43	Is safety stressed in interviews with prospective employees?	Y	N
44	Are accidents and injuries thoroughly investigated?	Y	N
45	Is discipline usually assessed when safety rules are broken?	Y	N
46	Do supervisors provide a safety orientation for newly assigned employees?	Y	N
47	Is safe work behavior recognized by your company?	Y	N
48	Do employees have a regular opportunity to attend safety meetings?	Y	N
49	Is the Safety Department adequately staffed and funded?	Y	N
50	Do you feel that safety meetings have a favorable effect on safety performance?	Y	N
51	Are employees with personal problems effectively handled by supervisors?	Y	N
52	Do you have problems obtaining support for the correction of hazardous conditions?	Y	N

53	Are checks made to be sure required protective equipment is being used?	Y	N
54	Does your company have established goals for safety performance?	Y	N
55	Are employees who are using alcohol or drugs on the job able to work without detection?	Y	N
56	Are risks involved sometimes overlooked in order to get the job done?	Y	N
57	Do employees participate in inspections for potential hazards?	Y	N
58	Does compliance with safety rules and regulations slow down the operation?	Y	N
59	Are safe workers picked to train new employees?	Y	N
60	Do supervisors discuss safety goals and performance with employees regularly?	Y	N
61	Do the company's safety rules and regulations protect the employee?	Y	N
62	Have your company's efforts encouraged you to work more safely?	Y	N
63	Is information that is needed to operate safely made available to employees?	Y	N
64	Do supervisors relate well to employees of a different generation?	Y	N
65	Are new employees assigned to work with experienced employees for job instruction?	Y	N
66	Are employees checked on a routine basis to see whether they are doing their jobs safely?	Y	N
67	Do unions support safety programs (if applicable)	Y	N
68	Do employees feel free to discuss causes of accidents with investigating officers?	Y	N
69	Does the safety committee or involvement team have the ability to correct unsafe conditions?	Y	N
70	Is promotion to higher level jobs dependent upon good safety performance?	Y	N
71	Do supervisors show a personal interest in having a safe operation?	Y	N
72	Are maintenance programs at a level that help prevent accidents?	Y	N

73	Do most supervisors have a good knowledge of the safety aspects of their jobs?	Y	N
74	Does the company have a uniform procedure for dealing with employees who violate rules?	Y	N
75	Do you feel overloaded at work?	Y	N
76	Are you often bored on the job?	Y	N
77	Is there a feeling of security that your job will be there in the future?	Y	N
78	Do you feel satisfied with your job?	Y	N
79	Do you feel you are pushed on the job?	Y	N
80	Are you forced to work overtime?	Y	N
81	Is it clear what you are responsible for?	Y	N
82	Do you feel like you can achieve on your job?	Y	N
83	Are you recognized when you do a good job?	Y	N
84	Are you given enough responsibility?	Y	N
85	Are you given too much responsibility?	Y	N
86	Do you have fun at work?	Y	N
87	Does your boss ask for your input?	Y	N
88	Does your boss use your input?	Y	N
89	Can you set your own pace?	Y	N
90	Do you feel a loyalty to your company?	Y	N
91	Is your company loyal to you?	Y	N
92	Is there a feeling of family at work?	Y	N
93	Have you had enough training?	Y	N
94	Are you judged by things beyond your control?	Y	N
95	Do you have enough authority?	Y	N
96	Is your boss fair?	Y	N
97	Would you like to be more involved on the job?	Y	N
98	Can you see a good future for yourself?	Y	N
99	Does it bother you to be absent from work?	Y	N
100	Do you like working for this company?	Y	N

Peterson, Dan, *The Challenge of Change, Creating a New Safety Culture, Implementation Guide*, CoreMedia Development, Inc., 1993, Table 4-1, pp. 21–22, public domain. Reprinted with permission.

Index